NUMERICAL METHODS
FUNDAMENTALS

NUMERICAL METHODS
FUNDAMENTALS

R. V. DUKKIPATI, PHD

MERCURY LEARNING AND INFORMATION
Dulles, Virginia
Boston, Massachusetts
New Delhi

Publisher: David Pallai
MERCURY LEARNING AND INFORMATION
22841 Quicksilver Drive
Dulles, VA 20166
info@merclearning.com
www.merclearning.com
(800) 232-0223

R. V. Dukkipati. *Numerical Methods Fundamentals.*
ISBN: 978-1-68392-871-3

The publisher recognizes and respects all marks used by companies, manufacturers, and developers as a means to distinguish their products. All brand names and product names mentioned in this book are trademarks or service marks of their respective companies. Any omission or misuse (of any kind) of service marks or trademarks, etc. is not an attempt to infringe on the property of others.

Library of Congress Control Number: 2023933078

232425321   Printed on acid-free paper in the United States of America.

Our titles are available for adoption, license, or bulk purchase by institutions, corporations, etc. For additional information, please contact the Customer Service Dept. at (800) 232-0223(toll free). Digital versions of our titles are available at numerous electronic vendors.

CONTENTS

PREFACE

INTRODUCTION

The objective of *numerical analysis* is to solve complex numerical problems using only the simple operations of arithmetic, to develop and evaluate methods for computing numerical results from given data. The methods of computation are called *algorithms*. An *algorithm* is a finite sequence of rules for performing computations on a computer such that at each instant the rules determine exactly what the computer has to do next. *Numerical methods* tend to emphasize the implementation of the algorithms. Thus, numerical *methods* are methods for solving problems on computers by numerical calculations, often giving a table of numbers and/or graphical representations or figures. The purpose of this book is to impart a basic understanding, both physical and mathematical, of the basic theory of numerical analysis/methods and their applications. In this book, an attempt is made to present in a simple and systematic manner the techniques that can be applied to the study of numerical methods. Special emphasis is placed on analytical developments, algorithms and computational solutions.

The objective of this text book is to introduce students from a variety of disciplines and backgrounds to the vast array of problems that are amenable to numerical solutions. The emphasis is placed on application rather than pure theory, which, although kept to a minimum and presented in a mostly heuristic and intuitive manner. This is deemed sufficient for the student to fully understand the workings, efficiency and shortcomings or failings of each technique. Since I intended this book as a first course on the numerical methods, the concepts have been presented in simple terms and the solution procedures have been explained in detail.

AUDIENCE

This book is a comprehensive text on numerical methods. It is self-contained and the subject matter is presented in an organized and systematic manner. No previous knowledge of numerical analysis and numerical methods is assumed. This book is quite appropriate for several groups of audience including:

– Undergraduate and graduate students in mathematics, science, and engineering taking the introductory course on numerical methods.

– Practicing engineers and managers who want to learn about the basic principles and concepts involved in numerical methods and how they can be applied at their own work place concerns.

– The book can be adapted for a short professional course on numerical methods.

– Design and research engineers will be able to draw upon the book in selecting and developing numerical methods for analytical and design purposes.

Because the book is aimed at a wider audience, the level of mathematics is kept intentionally low. All the principles presented in the book are illustrated by numerous worked examples. The book draws a balance between theory and practice.

CONTENTS

Books differ in content and organization. I have striven hard in the organization and presentation of the material in order to introduce the student gradually to the concepts and in their use to solve problems in numerical methods. The subject of numerical methods deals with the methods and means of formulation of mathematical models of physical systems and discusses the methods of solution. In this book, I have concentrated on both of these aspects: the tools for formulating the mathematical equations and also the methods of solving them.

The study of numerical methods is a formidable task. Each chapter in this book consists of a concise but thorough fundamental statement of the theory; principles and methods, followed by a selected number of illustrative worked examples. There are ample unsolved exercise problems for student's practice, to amplify and extend the theory, and methods are also included. The bibliography provided at the end of the book serves as helpful source for further study and research by interested readers.

In Chapter 1, Taylor's theorem, a few basic ideas and concepts regarding numerical computations, number representation, including binary, decimal, and hexadecimal numbers, errors considerations, absolute and relative errors, inherent errors, round-off errors and truncation errors, machine epsilon, error propagation, error estimation, general error formulae including approximation of a function, stability and condition, uncertainty in data or noise, sequences: linear convergence, quadratic convergence, and Aitken's acceleration formulae are described.

Chapter 2 deals with the solution of linear system of equations. The topics covered are the methods of solution, the inverse of a matrix, matrix inversion method, augmented matrix, Gauss elimination method, Gauss Jordan method, Cholesky's triangularization method, Crout's method, the Thomas algorithm for triangular system, Jacobi's iteration method, and Gauss-Seidal iteration method.

Chapter 3 deals with the solution of algebraic and transcendental equations. Here, we cover the topics such as the bisection method, method of false position, Newtonian-Raphson method, successive approximation method, secant method, Muller's method, Chebyshev method, Aitken's method, and comparison of iterative methods.

In Chapter 4, we cover the topics on numerical differentiation. The topics covered include the derivatives based on Newton's forward interpolation formula, the derivatives based on Newton's backward interpolation formula, the derivatives based on Stirling's interpolation formula, maxima and minima of a tabulated function, and cubic spline method.

Chapter 5 deals with finite differences and interpolation. It includes topics on finite differences, forward differences, backward differences, central differences, error propagation in a difference table, properties of operator delta, difference operators, relations among the operators, representation of a polynomial using factorial notation, interpolation with equal intervals, missing values, Newton's binomial expansion formula, Newton's forward interpolation formula, Newton's backward interpolation formula, error in interpolation formula, interpolation with unequal intervals, Lagrange's formula for unequal intervals, inverse interpolation, Lagrange's formula for inverse interpolation, central difference interpolation formulae, Gauss's forward interpolation formula, Gauss's backward interpolation formula, Bessel's formula, Stirling's formula, Laplace- Everett's formula, divided differences. Newton's divided differences, interpolation formula, selection of an interpolation formula, and cubic spline interpolation.

In Chapter 6, we present curve fitting, regression, and correlation. Here, we discuss the topics on linear equation, curve fitting with a linear equation, criteria for a "best" fit, linear least-squares regression, linear regression analysis, interpretation of a and b, standard deviation of random errors, coefficient of determination, linearization of nonlinear relationship, polynomial regression, quantification of error of linear regression, multiple linear regression, weighted least squares method, orthogonal polynomials and least squares approximation, least squares method for continuous data, approximation using orthogonal polynomials, and Gram-Schmidt orthogonalization process.

Chapter 7 presents numerical integration. Here, we cover the topics on the Newton-Cotes closed quadrature formula, trapezoidal rule, error estimate in trapezoidal rule, Simpson's 1/3 rule, error estimate in Simpson's 1/3 rule, Simpson's 3/8 rule, Boole's and Weddle's rules, Romberg's integration, Richardson's extrapolation, and Romberg's integration formula.

In Chapter 8, we discuss the numerical solution of ordinary differential equations. The methods covered include one-step methods or single-step methods, Picard's method of successive approximations, Taylor's series method, step-by-step methods or marching methods, Euler's method, modified Euler's method, Runge-Kutta methods, Runge-Kutta method of order two and four, predictor-corrector methods, Adam-Moulton predictor-corrector method, and Milne's predictor-corrector method.

An important requirement for effective use and application of numerical methods is the ease and proficiency in partial fraction expansion, engineering mathematics, and Cramer's rule. A basic review of partial fraction expansions, basic engineering mathematics, and Cramer's rule are outlined in Appendices A, B, and C, respectively.

In this edition four model question papers are appended at the end.

The bibliography provided at the end of the book serves as helpful sources for further study and research by interested readers. Answers to all end-of-chapter problems are given in the book.

I sincerely hope that the final outcome of this book will help the students in developing an appreciation for the topic of numerical methods.

Rao V. Dukkipati

ACKNOWLEDGMENTS

I am grateful to all those who have had a direct impact on this work. I am greatly indebted to my colleagues and to numerous authors who have made valuable contributions to the literature on the subject of numerical methods and more so to the authors of the articles listed in the bibliography of this book. My sincere thanks to Mr. P.R. Naidu of Andhra University and Andhra Pradesh for the excellent preparation of the complete manuscript. Finally, I express my heartfelt thanks to my family members: Sudha, Ravi, Madhavi, Anand, Ashwin, Raghav, and Vishwa, not only for their encouragement and support but also for sharing all the pain, frustration, and fun of producing a textbook manuscript.

I would appreciate being informed of errors, or receiving other comments and helpful suggestions about the book.

NUMERICAL COMPUTATIONS

Numerical methods are methods for solving problems on computers by numerical calculations, often giving a table of numbers and/or graphical representations or figures. Numerical methods tend to emphasize the implementation of algorithms. The aim of numerical methods is therefore to provide systematic methods for solving problems in a numerical form. The process of solving problems generally involves starting from initial data, using high-precision digital computers, following the steps in the algorithms, and finally obtaining the results. Often the numerical data and the methods used are approximate. Hence, the error in a computed result may be caused by the errors in the data, the errors in the method, or both.

In this chapter, we will describe Taylor's theorem, a few basic ideas and concepts regarding numerical computations, number representation, (including binary, decimal, and hexadecimal numbers), errors considerations, absolute and relative errors, inherent errors, round-off errors and truncation errors, error estimation, general error formulae (including approximation of a function), stability and condition, uncertainty in data, linear convergence, quadratic convergence, and Aitken's acceleration formulae.

1.1 TAYLOR'S THEOREM

Taylor's theorem allows us to represent, exactly, and fairly the general functions in terms of polynomials with a known, specified, and boundable error. Taylor's theorem is stated as follows:

Let $f(x)$ have $n + 1$ continuous derivatives on $[a, b]$ for some $n \geq 0$, and let $x, x_0 \in [a, b]$. Then

$$f(x) = p_n(x) + R_n(x) \tag{1.1}$$

for
$$p_n(x) = \sum_{k=0}^{n} \frac{(x-x_0)^k}{k!} f^{(k)}(x_0) \tag{1.2}$$

and
$$R_n(x) = \frac{1}{n!} \int_{x_0}^{x} (x-t)^n f^{n+1}(t)\,dt \tag{1.3}$$

Also, there exists a point ξ_x, between x and x_0 such that

$$R_n(x) = \frac{(x-x_0)^{n+1}}{(n+1)!} f^{(n+1)}(\xi_x) \tag{1.4}$$

where $R_n(x)$ is the *remainder*. Taylor's series is an associated formula of Taylor's theorem.

Taylor's series gives us a means to predict a function value at one point in terms of the function value and its derivatives at another point.

Taylor's series expansion is defined by

$$f(x_{i+1}) = f(x_i) + f'(x_i)(x_{i+1} - x_i) + \frac{f''(x_i)}{2!}(x_{i+1} - x_i)^2$$

$$+ \frac{f'''(x_i)}{3!}(x_{i+1} - x_i)^3 + \cdots + \frac{f^n(x_i)}{n!}(x_{i+1} - x_i)^n + R_n \tag{1.5}$$

We note that Equation (1.5) represents an infinite series. The remainder term R_n is included to account for all terms from $(n+1)$ to infinity:

$$R_n = \frac{f^{(n+1)}(\xi)}{(n+1)!}(x_{i+1} - x_i)^{n+1} \tag{1.6}$$

where the subscript n connotes that this is the remainder for the nth order approximation and ξ is a value of x that lies somewhere between x_i and x_{i+1}.

We can rewrite the Taylor's series in Equation (1.6) by defining a step size $h = x_{i+1} - x_i$ as

$$f(x_{i+1}) = f(x_i) + f'(x_i)h + \frac{f''(x_i)}{2!}h^2 + \frac{f'''(x_i)}{3!}h^3 + \cdots$$

$$+ \frac{f^{(n)}(x_i)}{n!}h^n + R_n \tag{1.7}$$

where the remainder term R_n is given by

$$R_n = \frac{f^{(n+1)}(\xi)}{(n+1)!}h^{n+1} \tag{1.8}$$

The estimation of function at a point b which is fairly close to a is desired, then the Taylor's series is written as an infinite series:

$$f(b) = f(a) + (b - a)f'(a) + \frac{(b-a)^2}{2!} f''(a) + \dots$$

$$+ \frac{(b-a)^n}{n!} f^{(n)}(a) + \dots \tag{1.9}$$

If b is very close to a, then only a few terms can give good estimation. The Taylor's series expansion for e^x, $\sin x$ and $\cos x$ are given below:

$$e^x = 1 + x + \frac{1}{2!}x^2 + \frac{1}{3!}x^3 + \dots + \frac{1}{n!}x^n + \frac{1}{(n+1)!}x^{n+1}e^{\xi x}$$

$$= \sum_{k=0}^{n} \frac{1}{k!}x^k + R_n(x) \tag{1.10}$$

$$\sin x = x - \frac{1}{3!}x^3 + \frac{1}{5!}x^5 + \dots + \frac{(-1)^n}{(2n+1)!}x^{2n+1} + \frac{(-1)^{n+1}}{(2n+3)!}x^{2n+3}\cos\xi x$$

$$= \sum_{k=0}^{n} \frac{(-1)^k}{(2k+1)!}x^{2k+1} + R_n(x) \tag{1.11}$$

$$\cos x = 1 - \frac{1}{2!}x^2 + \frac{1}{4!}x^4 + \dots + \frac{(-1)^n}{(2n)!}x^{2n} + \frac{(-1)^{n+1}}{(2n+2)!}x^{2n+2}\cos\xi x$$

$$= \sum_{k=0}^{n} \frac{(-1)^k}{(2k)!}x^{2k} + R_n(x) \tag{1.12}$$

The error in Taylor's series when the series is terminated after the term containing $(x - a)^n$ will not exceed

$$\left| f^{(n-1)} \right|_{max} \frac{[|x - a|]^{n-1}}{(n+1)!} \tag{1.13}$$

where max corresponds to the maximum magnitude of the derivative in the interval a to x. When the Taylor's series is truncated after n terms, then $f(x)$ will be accurate to $O(x - a)^n$.

EXAMPLE 1.1

Use the Taylor series expansion to approximate $f(x) = \cos x$ at $x_{i+1} = \pi/3$ with $n = 0$ to 6 on the basis of the value of $f(x)$ and its derivatives at $x_i = \pi/4$ which implies that $h = \dfrac{\pi}{3} - \dfrac{\pi}{4} = \pi/12$.

Solution:

The zero approximation is given by

$$f\left(x_{i+1}\right) \triangleq f\left(x_i\right) + f'\left(x_i\right)\left(x_{i+1} - x_i\right)$$

$$f(\pi/3) \triangleq \cos\frac{\pi}{4} = 0.70710678; \quad \cos\left(\frac{\pi}{3}\right) = 0.5$$

The % relative error is

$$\in_t = \frac{0.5 - 0.70710678}{0.5}(100) = -41.4\%$$

$$f'(x) = -\sin x$$

$$f\left(\frac{\pi}{3}\right) \cong \cos\left(\frac{\pi}{4}\right) - \sin\left(\frac{\pi}{4}\right)\left(\frac{\pi}{12}\right) = 0.52198666$$

$$\in_t = -4.4\%$$

$$f''(x) = -\cos x$$

$$f\left(\frac{\pi}{3}\right) \cong \cos\left(\frac{\pi}{4}\right) - \sin\left(\frac{\pi}{4}\right)\left(\frac{\pi}{12}\right) - \frac{\cos(\pi/4)}{2}\left(\frac{\pi}{12}\right)^2 = 0.49775449$$

with $\in_t = 0.449\%$.

Table 1.1 shows the Taylor series approximation for $n = 0$ to 6.

TABLE 1.1

Order n	$f^{(n)}(x)$	F($\pi/3$)	ε_t
0	cos x	0.70710678	−41.4
1	−sin x	0.52198666	−4.4
2	−cos x	0.49775449	0.449
3	sin x	0.49986915	2.62×10^{-2}
4	cos x	0.50000755	-1.51×10^{-3}
5	−sin x	0.50000030	-6.08×10^{-5}
6	−cos x	0.49999999	2.40×10^{-6}

1.2 NUMBER REPRESENTATION

A *base-b number* is made up of individual *digits*. In the *positional numbering system*, the position of a digit in the number determines that digit's contribution to the total value of the number.

For *decimal numbers*, the base (radix) is 10. Hence $(a_n \, a_{n-1} \, \ldots \, a_2 \, a_1 \, a_0)_b =$ $a_n \, b^n + a_{n-1} \, b^{n-1} + \ldots + a_2 \, b^2 + a_1 b + a_0$. a_n contributes to the number's magnitude and is called the *most significant digit* (MSD). Similarly, the right-most digit, a_0, contributes the least and is known as the *least significant digit* (LSD). Conversion of base-*b* fractions to base-10 is done by $(0.a_1 \, a_2 \, \ldots \, a_m)_b = a_1 b^{-1} +$ $a_2 b^{-2} + \ldots + a_m b^{-m}$. This is known as the *expansion method.*

There are two binary digits (bits) in the *binary number system*: zero and *one*. The left-most bit is called the *most significant bit* (MSB), and the right-most bit is the *least significant bit* (LSB). The rules of bit additions are: $0 + 0 = 0$; $0 + 1 = 1$; $1 + 0 = 1$; $1 + 1 = 0$ carry 1. The first ten digits 1, 2, 3, . . . 10 in base 10 and their representation in base-2 are shown in Figure 1.1.

Base 10	Base 2			
	2^3	2^2	2^1	2^0
1	0	0	0	1
2	0	0	1	0
3	0	0	1	1
4	0	1	0	0
5	0	1	0	1
6	0	1	1	0
7	0	1	1	1
8	1	0	0	0
9	1	0	0	1
10	1	0	1	0

FIGURE 1.1 Representation of numbers in decimal and binary forms.

Most computer languages use *floating-point arithmetic*. Every number is represented using a (fixed, finite) number of binary digits, called *bits*. Each binary digit is referred to as a bit. In this method, the computer representation a number in the following form:

$$\text{Number} = \sigma \, mb^{t-p} \qquad (1.14)$$

where σ = sign of the number ($\pm$), denoted by a single bit.

m = mantissa or a fraction (a value which lies between 0.1 and 1).

b = the base of the internal number system ($b = 2$ for binary, $b = 10$ for decimal or $b = 16$ for hexadecimal computers).

t = shifted exponent (the value that is actually stored).

$p =$ shift required to recover the actual exponent. Shifting in the exponent is normally done to avoid the need for a sign bit in the exponent itself.

The number is then stored by storing only the values of σ, m, and t. The normal way to represent and store numbers is to use a binary or base 2 number system which contains the following two digits.

$$\text{binary digits} = \{0 \ 1\} \tag{1.15}$$

For positive integers the binary form is

$$d_n 2^n + d_{n-1} 2^{n-1} + \ldots + d_1 2^1 + d_0 2^0 \tag{1.16}$$

while for positive numbers less than one it is

$$d_{-1} 2^{-1} + d_{-2} 2^{-2} + d_{-3} 2^{-3} + \ldots \tag{1.17}$$

with all binary digits d_i either 0 or 1. Such representations are unique.

Conversion between base 10 and base 2 is performed automatically by programming languages. Thus, conversion of an n-bit binary integer $b = b_{n-1} \ldots b_0$ to its decimal equivalent x is done as a sum of n powers of 2:

$$x = \sum_{k=0}^{n-1} b_k 2^k \tag{1.18}$$

A positive decimal integer x, in the range 0 to $2^n - 1$ is converted to its n-bit binary equivalent $b = b_{n-1} \ldots b_0$ by conducting a sequence of n divisions by decreasing powers of 2. In other words, the digits of the binary numbers are computed starting with the most significant bit, b_{n-1}, and ending with the least significant, b_0.

Noting that the hexadecimal numbers have a larger base or radix than decimal numbers, the first six letters of the alphabet are used to augment the decimal digits as follows:

$$\text{Hexadecimal digits} = \{0, 1, 2, 3, 4, 5, 6, 7, 8, 9, A, B, C, D, E, F\} \tag{1.19}$$

The conversion between binary, decimal, and hexadecimal numbers can be accomplished using Table 1.2.

TABLE 1.2 Binary, decimal, and hexadecimal numbers.

Binary	Decimal	Hexadecimal	Binary	Decimal	Hexadecimal
0000	00	0	1000	08	8
0001	01	1	1001	09	9
0010	02	2	1010	10	A
0011	03	3	1011	11	B
0100	04	4	1100	12	C
0101	05	5	1101	13	D
0110	06	6	1110	14	E
0111	07	7	1111	15	F

EXAMPLE 1.2

Determine the decimal values of the following numbers:

a) $x = (10010110)_2$

b) $x = (777)_8$

Solution:

a) $x = \sum_{k=0}^{7} b_k 2^k$ using Equation (1.18)

$\quad = 2^1 + 2^2 + 2^4 + 2^7$

$\quad = 2 + 4 + 16 + 128 = 150$

b) $x = (777)_8$

$\quad x = \sum_{k=0}^{2} b_k 8^k$

$\quad = 7(8)^0 + 7(8)^1 + 7(8)^2$

$\quad = 7 + 56 + 448 = 511$

EXAMPLE 1.3

Convert $(1011)_2$ to base-10.

Solution:

$(1)(2)^3 + (0)(2)^2 + (1)(2)^1 + 1 = 11$

The remainder method is used to convert base-10 numbers to base-b numbers. Converting a base-10 fraction to base-b requires multiplication of the base-10 fraction and subsequent fractional parts by the base. The base-b fraction is formed from the integer parts of the products taken into same order in which they were determined.

The *octal (base-8) numbering system* is one of the alternatives to working with long binary numbers. Only the digits 0 to 7 are employed. For instance,

$$7 + 1 = 6 + 2 = 5 + 3 = (10)_8$$

$$7 + 2 = 6 + 3 = 5 + 4 = (11)_8$$

$$7 + 3 = 6 + 4 = 5 + 5 = (12)_8$$

EXAMPLE 1.4

Perform the following operations:

a) $(7)_8 + (6)_8$

b) Convert $(0.14)_{10}$ to base-8

c) Convert $(27.52)_8$ to base-10.

Solution:

a) The sum of 7 and 6 in base-10 is 13. This is greater than 8. Using the remainder method, we have

13/8 = 1   remainder 5

1/8 = 0   remainder 1

The answer is $(15)_8$.

b) $0.14 \times 8 = 1.12$

$0.12 \times 8 = 0.96$

$0.96 \times 8 = 7.68$

$0.68 \times 8 = 5.44$

$0.44 \times 8 = $ etc.

The answer is $(0.1075....)_8$ which is obtained from the integer parts of the products above.

c) $(2)(8)^1 + (7)(8)^0 + (5)(8)^{-1} + (2)(8)^{-2}$

$$= 16 + 7 + \frac{5}{8} + \frac{2}{64} = (23.656)_{10}$$

The *hexadecimal (base-16) system* is a shorthand way of representing the value of four binary digits at a time.

EXAMPLE 1.5

a) Convert $(1475)_{10}$ to base-16.

b) Convert $(0.8)_{10}$ to base-16.

Solution:

a) Using the remainder method

$$\frac{1475}{16} = 92 \text{ remainder } 3$$

$$\frac{92}{16} = 5 \text{ remainder } 12$$

$$\frac{5}{16} = 0 \text{ remainder } 5$$

Now, $(12)_{10}$ is $(C)_{16}$ or (hex C).

Hence, the answer is $(5C_3)_{16}$.

b) $0.8 \times 16 = 12.8$

$0.8 \times 16 = 12.8$

$0.8 \times 16 = $ etc.

Since $(12)_{10} = (C)_{16}$, we have the answer as $(0.CCCCC\ldots)_{16}$.

EXAMPLE 1.6

a) Convert $(5431)_8$ to base-2.

b) Convert $(1011111101111001)_2$ to base-16.

Solution:

a) First convert each octal digit to binary digits.

$(5)_8 = (101)_2$

$(4)_8 = (100)_2$

$(3)_8 = (011)_2$

$(1)_8 = (001)_2$

Hence, the answer is $(101100011001)_2$.

b) Grouping the bits into fours starting at the right-hand bit, we have 1011 1111 0111 and 1001. Converting these groups into their hexadecimal equivalents, we get

$(1011)_2 = (B)_{16}$

$(1111)_2 = (F)_{16}$

$(0111)_2 = (7)_{16}$

$(1001)_2 = (9)_{16}$

Therefore, the answer is $(BF79)_{16}$.

EXAMPLE 1.7

a) Convert the following base-2 numbers to base-10: 1011001 and 110.00101

b) Convert the following base-8 numbers to base 10: 71563 and 3.14.

Solution:

a) $(1011001)_2 = (1 \times 2^6) + (0 \times 2^5) + (1 \times 2^4) + (1 \times 2^3) + (0 \times 2^2) + (0 \times 2^1) + (1 \times 2^0)$

$= 1(64) + 0(32) + 1(16) + 1(8) + 0(4) + 0(2) + 1(1) = 89$

$(110.00101)_2 = (1 \times 2^2) + (1 \times 2^1) + (0 \times 2^0) + (0 \times 2^{-1}) + (0 \times 2^{-2}) + (1 \times 2^{-3}) + (0 \times 2^{-4}) + (1 \times 2^{-5})$

$= 1(4) + 1(2) + 0(1) + 0(0.5) + 0(0.25) + 1(0.125) + 0(0.0625) + .03125) = 6.15625$

b) $(71563)_8 = (7 \times 8^4) + (1 \times 8^3) + (5 \times 8^2) + (6 \times 8^1) + (3 \times 2^0) = 7(4096) + 1(512) + 5(64) + 6(8) + 3(1) = 29{,}555$

$(3.14)_8 = (3 \times 8^0) + (3 \times 8^{-1}) + (4 \times 8^{-2}) = 3(1) + 1(0.125) + 4(0.015625)$

$= 3.1875$

1.3   ERROR CONSIDERATIONS

Sources of Errors: When a computational procedure is involved in solving a scientific-mathematical problem, errors often will be involved in the process. A rough classification of the kinds of original errors that might occur is as follows:

Modeling errors: *Mathematical modeling* is a process when mathematical equations are used to represent a physical system. This modeling introduces errors called *modeling errors*.

Blunders and mistakes: Blunders occur at any stage of the mathematical modeling process and consist of all other components of error. Blunders can be avoided by sound knowledge of fundamental principles, and also in taking proper care in approach and design to a solution. Mistakes are due to programming errors.

Machine representation and arithmetic errors: These errors are inevitable when using floating-point arithmetic while using computers or calculators. Examples are rounding and chopping errors.

Mathematical approximation errors: These errors are also known as a *truncation errors* or *discretization errors*. These errors arise when an approximate formulation is made to a problem that otherwise cannot be solved exactly.

Accuracy and Precision: Accuracy refers to how closely a computed or measured value agrees with the true value. *Precision* refers to how closely individual computed or measured values agree with each other. *Inaccuracy* (also known as *bias*) is the systematic deviation from the truth. *Imprecision* (uncertainty) refers to the magnitude of a scatter. These concepts are illustrated using an analogy from target practice as shown in Figure 1.2.

Figure 1.2 Illustrating the concepts of accuracy and precision from marksmanship example (a) inaccurate and imprecise, (b) accurate and imprecise, (c) inaccurate and precise, and (d) accurate and precise.

Errors are introduced by the computational process itself. Computers perform mathematical operations with only a finite number of digits. If the number x_a is an approximation to the exact result x_e, then the difference $x_e - x_a$ is called *error*. Hence

Exact value = approximate value + error

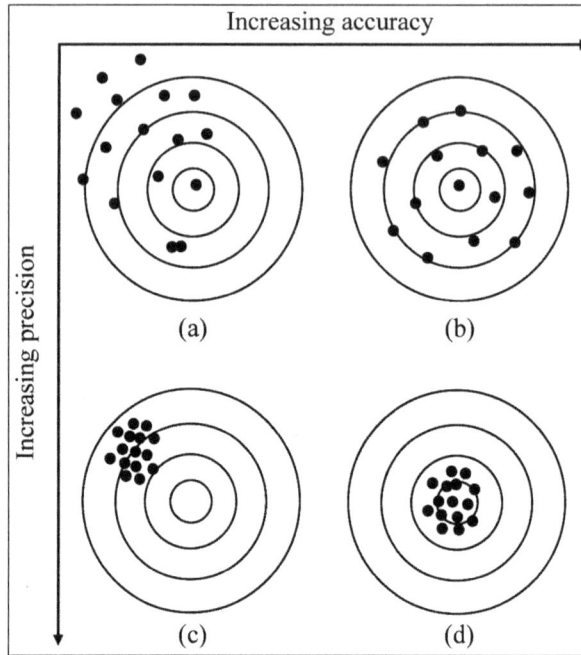

FIGURE 1.2 Concepts of accuracy and precision.

In numerical computations, we come across the following types of errors:

a) absolute and relative errors

b) inherent errors

c) round-off errors

d) truncation errors

1.3.1 Absolute and Relative Errors

If X_E is the exact or true value of a quantity and X_A is its approximate value, then $|X_E - X_A|$ is called the *absolute error E_a*. Therefore absolute error

$$E_a = |X_E - X_A| \tag{1.20}$$

and relative error is defined by

$$E_r = \left|\frac{X_E - X_A}{X_E}\right| \tag{1.21}$$

provided $X_E \neq 0$ or X_E is not too close to zero. The percentage relative error is

$$E_p = 100E_r = 100\left|\frac{X_E - X_A}{X_E}\right| \tag{1.22}$$

Significant digits

The concept of a significant figure, or digit, has been developed to formally define the reliability of a numerical value. The *significant digits* of a number are those that can be used with confidence.

If X_E is the exact or true value, and X_A is an approximation to X_E, then X_A is said to approximate X_E to t significant digits if t is the largest nonnegative integer for which

$$\frac{X_E - X_A}{|X_E|} < 5 \times 10^{-t} \tag{1.23}$$

EXAMPLE 1.8

If $X_E = e$ (base of the natural algorithm = 2.7182818) is approximated by $X_A = 2.71828$, what is the significant number of digits to which X_A approximates X_E?

Solution:

$$\frac{X_E - X_A}{|X_E|} = \frac{e - 2.71828}{e} \text{ which is } < 5 \times 10^{-6}$$

Hence X_A approximates X_E to six significant digits.

EXAMPLE 1.9

Let the exact or true value = 20/3 and the approximate value = 6.666.

The absolute error is 0.000666... = 2/3000.

The relative error is (2/3000)/ (20/3) = 1/10000.

The number of significant digits is four.

1.3.2 Inherent Errors

Inherent errors are the errors that preexist in the problem statement itself before its solution is obtained. Inherent errors exist because of the data being approximate or due to the limitations of the calculations using digital computers. Inherent errors cannot be completely eliminated but can be minimized if we select better data or employing high-precision computer computations.

1.3.3 Round-off Errors

Round-off errors are happens when inaccuracies arise due to a finite number of digits of precision used to represent numbers. All computers represent numbers, except for integers and some fractions, with imprecision. Digital computers use floating-point numbers of fixed word length. This type of representation will not express the exact or true values correctly. Error introduced by the omission of significant figures due to computer imperfection is called the *round-off error*.

Round-off errors are avoidable in most of the computations. When n digits are used to represent a real number, then one method is keep the first n digits and *chop off* all remaining digits. Another method is to *round* to the nth digit by examining the values of the remaining digits. The two steps involved in rounding to n digits are as follows:

1. Add $\mathrm{sgn}(x)\,\dfrac{b}{2}$ to digit $n+1$ of x.

2. Chop x to n digits.

where x is the nonzero real number, b is the base and $\mathrm{sgn}(x) = x/|x|$ denotes the *sign* of x with $\mathrm{sgn}\,(0) \approx 0$. Thus, the effect of the add and chop method of rounding is to round digit n up (away from zero) if the first digit to be chopped, digit $n+1$, is greater than equal to $b/2$, otherwise digit n is left as is. Errors which result from this process of chopping or rounding method are known as *round-off errors*.

Rounding to *k* decimal places

To round x, a positive decimal number, to k decimal places, we chop $x + 0.5 \times 10^{-5}$ after k^{th} decimal digit. Similarly, to a round a negative number, we round its absolute value and then restore the sign. Table 1.3 illustrates the rounding the numbers 234.0065792 and −234.00654983 to k decimal digits.

TABLE 1.3 Rounding numbers to k decimal digits.

k	234.0065792	−234.00654983
0	234	−234
1	234.0	−234.0
2	234.01	−234.01
3	234.007	−234.007
4	234.0065	−234.0065
5	234.00658	−234.00655
6	234.006579	−234.006550
7	234.0065792	−234.0065498
8	234.0065792	−234.00654983

Accurate to k decimal places

When we state that Y approximates y to k decimal places provided $|y - Y| < \frac{1}{2} \times 10^{-k}$ and if both y and Y are rounded to k decimal places, then the kth decimals in the rounded versions differ by no more than one unit. Consider for instance, the two numbers $y = 57.34$ and $Y = 57.387$ differ by $|y - Y| = 0.047 < 0.5 \times 10^{-1} = 0.05$ hence Y approximates y to 1 decimal place. Rounding y to Y to the $k = 1$ decimal place, we find $y_r = 57.3$ and $Y_r = 57.4$, respectively. Therefore, y_r and Y_r differ in the first decimal place by no more than one unit. Also, when Y approximate y to k decimal places, then these two numbers are said to *agree* to k decimal places. It should be noted here that these two numbers are not necessarily the *same* when rounded to k decimal places.

The most significant figure in a decimal number is the leftmost nonzero digit and the least significant figure is the rightmost digit. Significant figures are all the digits that lie in between the most significant and least significant figures. However, it should be noted here that zeros on the left of the first significant figure are not treated as significant digits. For instance, in the number $Y = 0.0078560$, the first significant digit is 7 and the rightmost zero is the fifth significant digit. Table 1.4 shows the results of rounding Y to k significant figure.

TABLE 1.4 Significant figures.

k significant digits	Y = 0.0078560
1	0.008
2	0.0078
3	0.00786
4	0.007856
5	0.0078560

Accurate to k significant figures

If
$$|x - X| < \frac{1}{2} \times 10^{-k} |x|$$

or
$$x - \frac{1}{2} \times 10^{-k} |x| < X < x + \frac{1}{2} \times 10^{-k} |x|$$

then we say that the floating-point number X approximates x to k significant figures. Table 1.5 shows k, the intervals $[x - d(k), x + d(k)]$, where $d(k) = \frac{1}{2} \times 10^{-k} |x|$, interval rounded, x rounded for $x = \pi = 3.141592654$. The last column in Table 1.4 shows the value of π rounded to k significant digits.

TABLE 1.5 Approximation of π to k significant figures.

k	Interval $[x - d(k), x + d(k)]$	Interval rounded	x rounded
1	[2.984513021, 3.298672287]	[3.0, 3.3]	3.0
2	[3.125884691, 3.157300617]	[3.1, 3.2]	3.1
3	[3.140021858, 3.143163450]	[3.14, 3.14]	3.14
4	[3.141435574, 3.141749734]	[3.141, 3.142]	3.142
5	[3.141576946, 3.141608362]	[3.1416, 3.1416]	3.1416
6	[3.141591083, 3.141594225]	[3.14159, 3.14459]	3.14159
7	[3.141592497, 3.141592811]	[3.141592, 3.141593]	3.141593
8	[3.141592638, 3.141592670]	[3.1415926, 3.1415927]	3.1415927

EXAMPLE 1.10

Given the number π is approximated using $n = 5$ decimal digits.

a) Determine the relative error due to chopping and express it as a percent.

b) Determine the relative error due to rounding and express it as a percent.

Solution:

a) The relative error due to chopping is given by
$$E_r(\text{chopping}) = \frac{3.1415 - \pi}{\pi} = 2.949 \times 10^{-5} \text{ or } 0.002949\%$$

b) The relative error due to rounding is given by
$$E_r(\text{rounding}) = \frac{3.1416 - \pi}{\pi} = 2.338 \times 10^{-6} \text{ or } 0.0002338\%.$$

EXAMPLE 1.11

If the number $\pi = 4 \tan^{-1}(1)$ is approximated using 5 decimal digits, find the percentage relative error due to

a) chopping   b)   rounding

Solution:

a) Percentage relative error due to chopping

$$= \left(\frac{3.1415 - \pi}{\pi} \right) 100 = \left(-2.949 \times 10^{-5} \right) 100 \text{ or } -0.002949\%.$$

b) Percentage relative error due to rounding

$$= \left(\frac{3.1416 - \pi}{\pi} \right) 100 = \left(2.338 \times 10^{-6} \right) 100 = 0.00023389\%.$$

EXAMPLE 1.12

Use Taylor series expansions (zero through fourth order) to predict $f(2)$ for $f(x) = \ln(x)$ with a base point at $x = 1$. Determine the true percentage relative error for each approximation.

Solution:

The true value of $\ln(2) = 0.693147\ldots$

Zero order:

$$f(2) \cong f(1) = 0$$

$$E_t = \left| \frac{0.693147 - 0}{0.693147} \right| 100\% = 100\%$$

First order:

$$f'(x) = \frac{1}{x} \qquad\qquad f'(1) = 1$$

$$f(2) \cong 0 + 1(1) = 1$$

$$E_t = \left| \frac{0.693147 - 1}{0.693147} \right| 100\% = 44.27\%$$

Second order:

$$f''(x) = -\frac{1}{x^2} \quad f''(1) = -1$$

$$f(2) = 1 - 1\frac{1^2}{2} = 0.5$$

$$E_t = \left|\frac{0.693147 - 0.5}{0.693147}\right| 100\% = 27.87\%$$

Third order:

$$f'''(x) = \frac{2}{x^3} \quad f''(1) = 2$$

$$f(2) \cong 0.5 + 2\frac{1^3}{6} = 0.833333$$

$$E_t = \left|\frac{0.693147 - 0.833333}{0.693147}\right| 100\% = 20.22\%$$

Fourth order:

$$f''''(x) = -\frac{6}{x^4} \qquad f'''(1) = -6$$

$$f(2) \cong 0.833333 - 6\frac{1^4}{24} = 0.583333$$

$$E_t = \left|\frac{0.693147 - 0.583333}{0.693147}\right| 100\% = 15.84\%$$

The above results show that the series is converging at a slow rate. A smaller step size would be required to obtain more rapid convergence.

EXAMPLE 1.13

Given two numbers $a = 8.8909 \times 10^3$ and $b = 8.887 \times 10^3$. Calculate the difference between the two numbers $(a - b)$ using the decimal floating point approximation (scientific notation) with three significant digits in the mantissa by (a) chopping and (b) rounding.

Solution:

a) In chopping, when three significant digits are in the mantissa, then

$$a = 8.890 \times 10^3 \text{ and } b = 8.887 \times 10^3$$

and $\qquad a - b = 8.890 \times 10^3 - 8.887 \times 10^3 = 0.003 \times 10^3 = 3.$

b) In rounding, we have

$$a = 8.890 \times 10^3 \text{ and } b = 8.887 \times 10^3$$

and $\qquad a - b = 8.890 \times 10^3 - 8.887 \times 10^3 = 0.004 \times 10^3 = 4.$

The exact (true) difference between the numbers is 3.8, which shows that the rounding gives a value much closer to the real answer.

1.3.4 Truncation Errors

Truncation errors are defined as those errors that result from using an approximation in place of an exact mathematical procedure. Truncation error results from terminating after a finite number of terms known as *formula truncation error* or *simply truncation error*.

Let a function $f(x)$ is infinitely differentiable in an interval which includes the point $x = a$. Then the Taylor series expansion of $f(x)$ about $x = a$ is given by

$$f(x) = \sum_{k=0}^{\infty} \frac{f^{(k)}(a)(x-a)^k}{k!} \tag{1.24}$$

where $f^{(k)}(a)$ denotes the kth derivative of $f(x)$ evaluated at $x = a$

or $\qquad f^{(k)}(a) \underline{\underline{\Delta}} \left. \dfrac{d^k f(x)}{dx^k} \right|_{x=0}$ $\tag{1.25}$

If the series is truncated after n terms, then it is equivalent to approximating $f(x)$ with a polynomial of degree $n - 1$.

$$f_n(x) \underline{\underline{\Delta}} \sum_{k=0}^{n-1} \frac{f^{(k)}(a)(x-a)^k}{k!} \tag{1.26}$$

The error in approximating $E_n(x)$ is equal to the sum of the neglected higher-order terms and is often called the *tail* of the series. The tail is given by

$$E_n(x) \underline{\underline{\Delta}} f(x) - f_n(x) = \frac{f^{(x)}(\xi)(x-a)^n}{n!} \tag{1.27}$$

It is possible sometimes to place an upper bound on x of $E_n(x)$ depending on the nature of function $f(x)$.

If the maximum value of $|f_n(x)|$ over the interval $[a, x]$ is known or can be estimated, then

$$M_n(x) \underline{\underline{\Delta}} \max_{a \le \xi \le x} \left[\left| f^{(n)}(\xi) \right| \right] \tag{1.28}$$

From Equations (1.27) and (1.28), the worst bound on the size of the trunca-tion error can be written as

$$| E_n(x) | \leq \frac{M_n(x) | x - a |^n}{n!} \tag{1.29}$$

If $h = x - a$, then the truncation error $E_n(x)$ is said to be of order $O\ (h^n)$. In other words, as $h \to 0$, $E_n(x) \to 0$ at the same rate as h^n.

Hence $\quad O\ (h^n) \approx ch^n \quad |h| \ll 1 \tag{1.30}$

where c is a nonzero constant.

The *total numerical error* is the summation of the truncation and round-off errors. The best way to minimize round-off errors is to increase the number of significant figures of the computer. It should be noted here that round-off error increases due to subtractive cancellation or due to an increase in the number of computations in an analysis. The truncation error can be reduced by decreasing the step size. In general, the truncation errors are *decreased* as the round-off errors are *increased* in numerical differentiation.

There exists no systematic and general approaches in evaluating numeri-cal errors for all problems. In most cases, error estimates are based on experi-ence and judgment of the engineer or scientist.

Model errors relate to bias that can be ascribed to incomplete mathe-matical models. Errors also enter into the analysis due to uncertainty in the physical data on which a model is based.

EXAMPLE 1.14

Given the trigonometric function $f(x) = \sin x$,

   a) expand $f(x)$ about $x = 0$ using the Taylor series

   b) truncate the series to $n = 6$ terms

   c) find the relative error at $x = \pi/4$ due to truncation in (b)

   d) determine the upper bound on the magnitude of the relative error at $x = \pi/4$ and express it as a percent

Solution:

   a) Using Equation (1.23), the Taylor series expansion is given by

$$f(x) = \sum_{k=0}^{\infty} \frac{f^{(k)}(a)(x-a)^k}{k!} = x - \frac{x^3}{3!} + \frac{x^5}{5!} - \frac{x^7}{7!} + \cdots$$

b) Truncation of the Taylor series to $n = 6$ terms.

$$f_6(x) = x - \frac{x^3}{3!} + \frac{x^5}{5!}$$

c) The relative error at $x = \pi/4$ due to truncation in (b) is given by

$$E_{r6} = \frac{f_6(\pi/4) - \sin(\pi/4)}{\sin(\pi/4)}$$

$$= \frac{\frac{\pi}{4} - \left(\frac{\pi}{4}\right)^3 \Big/ 3! + \left(\pi/4\right)^5 / 5! - \sin(\pi/4)}{\sin(\pi/4)}$$

$$= 5.129 \times 10^{-5} \quad \text{or} \quad 0.005129\%$$

d) Here $f^6(x) = -\sin x$. From Equation (1.27) using $a = 0$ and $x = \pi/4$, we obtain

$$M_6(x) \leq \sin(\pi/4)$$

Now from Equation (1.28), we have the upper bound on the truncation error given by

$$|E_6| \leq \frac{\sin(\pi/4)(\pi/4)^6}{6!} = 2.305 \times 10^{-4} \quad \text{or} \quad 0.02305\%$$

1.3.5 Machine Epsilon

Digital computers are *fixed-precision* devices and the number of digits the device can manipulate depends on its hardware configuration. *Machine epsilon*, ϵ_M is the smallest positive number that the device can add to 1 while recognizing the sum as different than 1.

ϵ_M is determined computationally by finding the smallest positive ϵ is for which $1 + \epsilon \neq 1$. For instance, if a particular computing device computes 1.000000001 for $1 + 10^{-9}$ but 1 for $1 + 10^{-10}$, then we conclude that $10^{-10} < \epsilon_M < 10^{-9}$ and the device in this case would be known as a 10 *significant-digit device*.

1.3.6 Error Propagation

Table 1.6 summaries the errors attributed to the *round-off errors* due to the limited number of digits using fixed-precision devices. In order to illustrate these errors, we consider the following numbers: $a = 237.6581$, $b = 238.2389$,

$c = 0.014789$, $d = 137469$; and $A = 238.0$, $B = 238.2$, $C = 0.01480$, and $D = 1.375 \times 10^5$.

TABLE 1.6 Possible types of round-off errors on a finite-precision computing device.

S.No.	Error	Comments
1	Negligible addition	When two numbers of notable different magnitudes are added or subtracted, then the result rounds to the largest number.
2	Creeping round-off	Repeated rounding to k significant digits will result in accumulation of errors.
3	Error magnification	Occurs when an erroneous number is multiplied/divided by a number of large/small magnitude.
4	Subtractive cancellation	Due to the subtraction of two nearly equal numbers where the difference lies in significant digits well beyond the devices capacity to record it.

Tables 1.7 and 1.8 show the five exact arithmetic calculations (answers rounded-off to four significant digits) and the same calculations performed on a device with four significant digits respectively.

1.4 ERROR ESTIMATION

Few computer methods are available to provide error estimates. These methods are briefly mentioned here.

1. *Double precision method*: In this method, the problem is solved twice, once in single precision and then in double precision. The estimate on the round-off error is then simply given by the difference between the two results obtained.

2. *Interval arithmetic method*: Each number in this method is represented by two machine numbers corresponding to the estimated maximum and minimum values. Two solutions are obtained at every step corresponding to the maximum and minimum values. The true solution is assumed to lie in about the centre of the range. The range here is the difference between the solutions corresponding to the maximum and minimum values.

3. *Significant digit arithmetic method*: In this method, the digits lost due to the subtraction of two nearly equal machine numbers are tracked. Only the significant digits in a number are kept and the rest are rejected or ignored. In this way, all digits retained or kept are assumed to be significant. The results obtained with this method are considered to be very conservative.

4. *Statistical approach*: This method starts with the assumption that the round-off error is independent. A stochastic model for the propagation of round-off errors is then adapted in which the local errors are considered as random variables. The local round-off errors are assumed to be either uniformly or normally distributed between their extreme values. Using standard statistical analysis methods, the standard deviation, the variance and the accumulated round-off errors are estimated.

5. *Backward error analysis*: In this method, based on the result of a computation the possible range of input data that could have produced it is determined. If the results found with this approach are consistent with the input data within the range of observational or round-off error, then there is some confidence placed on the result. If this does not happen, then a major source of error is assumed to exist somewhere else, presumably within the algorithm itself.

6. *Forward error analysis*: The method can be illustrated by means of an example.

   Suppose the value of $A\,(B + C)$ is to be computed when a, b and c are the approximations to A, B and C respectively, and the respective error amounts are e_1, e_2, and e_3.

   The true value is

   $$A\,(B + C) = (a + e_1)\,(b + e_2 + c + e_3) = ab + ac + \text{error}$$

   where $\text{error} = a\,(e_2 + e_3) + be_1 + ce_1 + e_1e_2 + e_1e_3$

   Now assuming the uniform bound $|e_i| \leq e$ and that error products can be ignored, we get

   $$|error| \leq [2|a| + |b| + |c|]\,e$$

   This procedure can be carried out for any algorithm. It is a tedious analysis. The resulting bounds are generally very conservative.

1.5 GENERAL ERROR FORMULA

1.5.1 Function Approximation

Consider the function

$$F = f(x_1, x_2, x_3, ..., x_n) \tag{1.31}$$

where $x_1, x_2, x_3, ..., x_n$ are variables.

Suppose Δx_i represents error in each x_i, so that the error in F is

$$F + \Delta F = f(x_1 + \Delta x_1, x_2 + \Delta x_2, ..., x_n + \Delta x_n) \qquad (1.32)$$

Taylor's series expansion of the right-hand side of Equation (1.31) gives

$$F + \Delta F = f(x_1, x_2, ..., x_n) + \sum_{i=1}^{n} \frac{\partial f}{\partial x_i} \Delta x_i + O(\Delta x_i^2) \qquad (1.33)$$

If we assume the errors in x_i as small, and $\dfrac{\Delta x_i}{x_i} \ll 1$, so that the second and higher powers of Δx_i can be ignored, Equation (1.33) gives

$$\Delta F = \sum_{i=1}^{n} \frac{\partial f}{\partial x_i} \Delta x_i$$

$$= \frac{\partial f}{\partial x_1} \Delta x_1 + \frac{\partial f}{\partial x_2} \Delta x_2 + \cdots + \frac{\partial f}{\partial x_n} \Delta x_n \qquad (1.34)$$

The relative error E_r is then given by

$$E_r = \frac{\Delta f}{f} = \frac{\partial f}{\partial x_1} \frac{\partial x_1}{f} + \frac{\partial f}{\partial x_2} \frac{\partial x_2}{f} + \cdots + \frac{\partial f}{\partial x_n} \frac{\partial x_n}{f} \qquad (1.35)$$

Replacing the function $f(h)$ with its approximation $\phi(h)$ and denoting the known error bound as $\mu(h^n)$, where n is a positive integer, we have

$$|f(h) - f(h)| \le \mu |h^n| \text{ for small } h$$

Thus, $\phi(h)$ approximates $f(h)$ with order of approximation $O(h^n)$ and we can write

$$f(h) = \phi(h) + O(h^n) \qquad (1.36)$$

EXAMPLE 1.15

Determine the maximum relative error for the function

$$F = 3x^2 y^2 + 5y^2 z^2 - 7x^2 z^2 + 38$$

For $x = y = z = 1$ and $\Delta x = -0.05$, $\Delta y = 0.001$ and $\Delta z = 0.02$.

Solution:

$$F = 3x^2y^2 + 5y^2z^2 - 7x^2z^2 + 38$$

$$\frac{\partial F}{\partial x} = 6xy^2 - 14xz^2$$

$$\frac{\partial F}{\partial y} = 6x^2y + 10yz^2$$

$$\frac{\partial F}{\partial z} = 10y^2z - 14x^2z$$

$$(\Delta F)_{max} = \left|\frac{\partial F}{\partial x}\Delta x\right| + \left|\frac{\partial F}{\partial y}\Delta y\right| + \left|\frac{\partial F}{\partial z}\Delta z\right|$$

$$= |(6xy^2 - 14xz^2)\,\Delta x| + |(6x^2y + 10yz^2)\Delta y| + |(10y^2z - 14x^2z)\Delta z| = 0.496$$

For $x = y = z = 1$ and $\Delta x = -0.05$, $\Delta y = 0.001$ and $\Delta z = 0.02$, we have the maximum relative error is given by Equation (1.34)

$$(E_r)_{max} = \frac{(\Delta F)_{max}}{F} = \frac{0.496}{39} = 0.01272$$

1.5.2 Stability and Condition

A numerical computation is said to be *numerically unstable* if the uncertainty of the input values is grossly magnified by numerical method employed.

Consider the first-order Taylor's series of a function given by

$$f(x) = f(a) + f'(a)\,(x - a) \tag{1.37}$$

The *relative error* of $f(x)$ then becomes

$$\frac{f(x) - f(a)}{f(x)} \cong \frac{f'(a)(x - a)}{f(a)} \tag{1.38}$$

The *relative error* of x becomes

$$\frac{x - a}{a} \tag{1.39}$$

A *condition number* is often defined as the ratio of the relative errors given by Equations (1.38) and (1.39) as

$$\text{Condition number} = \frac{a\,f'(a)}{f(a)} \tag{1.40}$$

The condition number given by Equation (1.40) indicates the extent to which an uncertainty in x is magnified by $f(x)$.

Condition number = 1 (function's relative error = relative error in x)

Condition number > 1 (relative error is amplified)

Condition number < 1 (relative error is attenuated) (1.41)

Condition number > very large number (the function is ill-conditioned)

EXAMPLE 1.16

Compute and interpret the condition number for

   a) $f(x) = \sin x$ for $a = 0.51\pi$

   b) $f(x) = \tan x$ for $a = 1.7$

Solution:

   a) The condition number is given by

$$\text{Condition number} = \frac{a\,f'(a)}{f(a)}$$

for $a = 0.51\pi$, $f'(a) = \cos(0.51\pi) = -0.03141$, $f(a) = \sin(0.51\pi) = 0.99951$

$$\text{Condition number} = \frac{a\,f'(a)}{f(a)} = \frac{(0.51\pi)(-0.03141)}{(0.99951)} = -0.05035$$

Since the condition number is < 1, from Equation (1.41), we conclude that the relative error is attenuated.

   b) $f(x) = \tan x$, $f'(a) = -7.6966$ for $a = 1.7$

$$f'(x) = 1/\cos^2 x, \quad f'(a) = 1/\cos^2(1.7) = 60.2377$$

$$\text{Condition number} = \frac{a\,f'(a)}{f(a)} = \frac{1.7(60.2377)}{-7.6966} = -13.305$$

Thus, the function is ill-conditioned.

1.5.3 Uncertainty in Data or Noise

Uncertainty or error in the physical data based on which the computation model is based can introduce errors in the analysis. This type of error is known as *noise*. The data can affect the accuracy of the numerical computations performed. The errors can exhibit both inaccuracy and imprecision. If the input data has d significant digits of accuracy, then the results obtained from the numerical computation should be reported in d significant digits of accuracy. For instance if $a = 5.358$ and $b = 0.06437$ both have four significant digits of accuracy, then although $a - b = 5.29363$, we should report the correct answer as $a - b = 5.293$.

TABLE 1.7 Exact arithmetic rounded to four significant digits.

No.	Exact arithmetic	Rounded to four significant digits
1.	$a - c = 237.6581 - 0.014897 = 237.643203$	237.6
2.	$b + d = 237.8389 + 137476 = 1377103.8389$	1.377×10^5
3.	$bd = (237.8389)(137476) = 32697140.62$	3.270×10^7
4.	$a/c = 237.6581/0.014897 = 15953.42015$	1.595×10^4
5.	$a - b = 237.6581 - 237.8389 = -0.1808$	-0.1808

TABLE 1.8 Calculations with a device carrying four significant digits.

No.	Calculations
1.	$A - C = 238.0 - 0.01480 = 238.0$
2.	$B + D = 238.1 + 1.375 \times 10^5 = 1.377 \times 10^5$
3.	$BD = (238.1)(1.375 \times 10^5) = 3.274 \times 10^7$
4.	$A/C = 238.0/0.01480 = 1.608 \times 10^4$
5.	$A - B = 238.0 - 238.1 = -0.1$

The following observations can be made from the results in Tables 1.7 and 1.8.

a) *Negligible addition*: Round-off error has crept into the fourth significant digit when we compare $a - c$ (rounded) to $A - C$. There is a difference in the fourth significant digit when $b + d$ (rounded) are compared to $B + D$.

b) *Error magnification*: Comparing a/c (rounded) to A/C we find a difference in the fourth significant digit when bd (rounded) and BD are compared, the two answers differ substantially by -40000.

c) *Creeping round-off*: In the calculations of $a - c$, bd, a/c, and $b + d$, we find the result of working in four significant digits as opposed to working "exactly" and then rounding would lead to a loss of precision in the fourth significant digit. These calculations show the *creeping round-off* that is the gradual loss of precision as repeated rounding errors accumulate.

d) *Subtractive calculations*: Comparing $a - b = -0.1808$ to $A - B = -0.1$, we find significant error introduced by working in fixed-precision arithmetic.

1.6 SEQUENCES

A sequence may converge to a limit in a linear fashion or in a nonlinear fashion. If the sequence is convergent, then an iterative operation produces a sequence of better and better approximate solutions.

1.6.1 Linear Convergence

Here, we consider a sequence $\{x_0, x_1, ..., x_n\}$ generated by the iteration $x_{k+1} = g(x_k)$. Table 1.9 lists k, x_k, Δx_k ($= x_{k+1} - x_k$) and $\Delta x_{k+1}/\Delta x_k$ for $g(x) = 1 + x/2$ where the *starting value* is taken as 0.85.

TABLE 1.9 Linear convergence of the iteration process for $x_{k+1} = 1 + \dfrac{x_k}{2}$

k	x_k	$\Delta x_k = x_{k+1} - x_k$	$\dfrac{\Delta x_{k+1}}{\Delta x_k}$
0	0.850000000	0.575	1/2
1	1.425000000	0.2875	1/2
2	1.712500000	0.14375	1/2
3	1.856250000	0.071875	1/2
4	1.928125000	0.0359375	1/2
5	1.964062500	0.01796875	1/2
6	1.982031250	0.008984375	1/2
7	1.991015625	0.004492188	1/2
8	1.995507813	0.002246094	1/2
9	1.997753906	0.001123047	1/2
10	1.998876953	0.000561523	1/2
11	1.999438477	0.000280762	1/2
12	1.999719238	0.000140381	1/2

Notice that the ratios of successive increments in the last column of Table 1.9 are all exactly equal to 1/2 and the convergence of the sequence to $x = 2$ is linear. We call this sequence *exactly* linear since $\Delta x_{k+1} = c_\ell \, \Delta x_k$ for all $k > 0$. Here $c_\ell = 1/2$.

The sequence $\{x_k\}$ is said to *converge linearly* provided the ratio of increments $\Delta x_{k+1}/\Delta x_k$ tends to a constant c_ℓ, where $0 < |c_\ell| < 1$.

Linear Convergence Theorem

   a) $X = g(X)$, so $x = X$ is a fixed point of the iteration $x_{k+1} = g(x_k)$.

   b) $g'(x)$ is continuous in a neighborhood of the fixed point X.   (1.42)

   c) $g'(X) \neq 0$.

Therefore,

   a) $\{x_k\}$ converges to X linearly, with $C_\ell = g'(X)$ if $0 < |g'(X)| < 1$.

   b) $\{x_k\}$ diverges linearly, with $C_\ell = g'(X)$ if $|g'(X)| > 1$.   (1.43)

   c) $\{x_k\}$ converges or diverges *slowly* if $g'(X) = \neq 1$.

If $s_0 = 0$ and $s_1 = r$, the general term in a sequence that converges exactly linearly with convergence constant C is given by

$$s_k = r\sum_{n=0}^{k-2} C^n = r\frac{C^{k-1} - 1}{C - 1} \tag{1.44}$$

The increments are then given by $\Delta s_k = s_{k+1} - s_k = rC^k$. Appropriate conditions on C and r would then guarantee *convergence*.

1.6.2 Quadratic Convergence

Consider a sequence $\{x_0, x_1, ..., x_n\}$ generated by the iteration

$$x_{k+1} = g(x_k),$$

where $g(x) = \dfrac{2.15x^2 + 2.87}{3.96x + 1.2},$

a function with fixed points.

Table 1.10 lists k, x_k, Δx_k $[= (x_{k+1} - x_k)]$, and $\Delta x_{k+1}/(\Delta x_k)^2$ for which the *starting value* is $x_0 = 2$. We observe that the sequence converges very rapidly to the point $x = 1$. The last column of Table 1.10 shows that the ratios $\Delta x_{k+1}/(\Delta x_k)^2$ are tending toward the constant $C_q = -0.4$. This confirms the quadratic convergence of the sequence $\{x_k\}$ to the point $x = 1$.

The sequence $\{x_k\}$ is said to *converge quadratically* provided the ratio of increments $\Delta x_{k+1}/(\Delta x_k)^2$ tends to a constant $C_q \neq 0$, $\neq \infty$. If $\Delta x_{k+1} = C_q(\Delta x_k)^2$ for all $k > 0$, then the sequence is said to be *exactly quadratically convergent*.

TABLE 1.10 Quadratic convergence of the iteration process for $x_{k+1} = \dfrac{2.15x^2 + 2.87}{3.96x + 1.2}$

k	x_k	$\Delta x_k = x_{k+1} - x_k$	$\Delta x_{k+1}/(\Delta x_k)^2$
0	2.000000000	−0.7423245614	−0.4276521490
1	1.257675439	−0.2356559011	−0.3930757235
2	1.022019537	−0.0218289508	−0.3999390216
3	1.000190587	−0.0001905722	−0.3999999952
4	1.000000015	−0.0000000145	
5	1.000000000		

TABLE 1.11 The structure of a sequence which is exactly quadratically convergent.

k	s_k	$\Delta s = s_{k+1} - s_k$	$\Delta s_{k+1}/(\Delta s_k)^2$
0	0	r	C
1	r	$r^2 C$	C
2	$r^2 C + r$	$r^4 C^3$	C
3	$r^4 C^3 + r^2 C + r$	$r^8 C^7$	C
4	$r^8 C^7 + r^4 C^3 + r^2 C + r$	$r^{16} C^{15}$	C
5	$r^{16} C^{15} + r^8 C^7 + r^4 C^3 + r^2 C + r$	$r^{32} C^{31}$	
6	$r^{32} C^{31} + r^{16} C^{15} + r^8 C^7 + r^4 C^3 + r^2 C + r$		

Quadratic Convergence Theorem

    a) $X = g(X)$, so $x = X$ is a fixed point of the iteration $x_{k+1} = g(x_k)$.

    b) $g''(x)$ is continuous in a neighborhood of the fixed point X.   (1.45)

    c) $g'(X) = 0$.

That is, $\{x_k\}$ converges to X quadratically, with $C_q = -\dfrac{1}{2} g''(X)$.

Table 1.11 lists s_k, $\Delta s_k = s_{k+1} - s_k$, and the ratios $\Delta s_{k+1}/(\Delta s_k)^2$ for a sequence whose convergence is exactly quadratic, with convergence constant C, and with starting values $s_0 = 0$ and $s_1 = r$.

1.6.3 Aitken's Acceleration Formula

Quadratic convergence or any convergence of order higher than 2 is faster than the linear convergence. Aitken's acceleration process is used to accelerate a linearly converging sequence to a quadratically converging sequence. Aitken's process will give better results in an iterative numerical technique with fewer number of iterative operations.

Let x be the limit of the sequence. That is,

$$\lim_{k \to \infty} x_k = x$$

If $\{x_k\}$, $k = 1$ to ∞, is a linearly convergent sequence with limit x, and $e_n = x_k - x$, then

$$\lim_{n \to \infty} \frac{|e_{k+1}|}{|e_k|^\alpha} = \lambda \quad \text{and} \quad 0 < \lambda < 1 \tag{1.46}$$

where α, ∞ is the order of convergence and λ is the asymptotic error constant. If $\alpha = 1$, convergence is linear and if $\alpha = 2$, convergence is quadratic.

Aitken's process assumes that the limiting case in Equation (1.46) occurs for all $k \geq 1$. That is,

$$e_{k+1} = \lambda e_k$$

Therefore, we can write

$$x_{k+2} = e_{k+2} + x = e_{k+1} + x \tag{1.47}$$

or $\quad x_{k+2} = \lambda(x_{k+1} - x) + x$ for all $k \geq 1$ $\tag{1.48}$

Reducing the subscript by 1, we have

$$x_{k+1} = \lambda(x_k - x) + x \tag{1.49}$$

Eliminating λ between x_{k+1} and x_{k+2} from Equations (1.48) and (1.49), we obtain

$$x = \frac{x_{k+2}x_k - x_{k+1}^2}{x_{k+2} - 2x_{k+1} + x_k}$$

$$= \frac{x_k^2 + x_k x_{k+2} - 2x_k x_{k+1} + 2x_k x_{k+1} - x_k^2 - x_{k+1}^2}{x_{k+2} - 2x_{k+1} + x_k}$$

or $\quad x = x_k - \dfrac{(x_{k+1} - x_k)^2}{x_{k+2} - 2x_{k+1} + x_k}$

The sequence $\{x_k\}$ defined by

$$x_k = x_k - \frac{(x_{k+1} - x_k)^2}{x_{k+2} - 2x_{k+1} + x_k}$$

converges more rapidly to x than the original sequence $\{x_k\}$ for $n = 1$ to ∞.

EXAMPLE 1.17

The sequence $\{x_k\}$, $n = 1$ to ∞, where $x_k = 3x^4 - 2x^3 - 2x^2 + 2.8$ converges linearly to $x = 1$ with $s_0 = 0.75$. Using Aitken's acceleration formula, obtain another sequence, which converges faster to $x = 2$.

Solution:

The results obtained using both linear convergence algorithm and Aitken's acceleration formula are shown in Table 1.12.

TABLE 1.12 Results obtained from linear convergence and Aitken's process.

k	Linear convergence			Aitken's process
	x_k	$\Delta x_k = x_{k+1} - x_k$	$\dfrac{\Delta x_{k+1}}{\Delta x_k}$	$x_k - \dfrac{(x_{k+1} - x_k)^2}{x_{k+2} - 2x_{k+1} + x_k}$
0	0.750000000	1.03046875		1.903320561
1	1.780468750	0.109765625	1/2	2.000000000
2	1.890234375	0.054882813	1/2	
3	1.945117188	0.027441406	1/2	
4	1.972558594	0.013720703	1/2	
5	1.986279297	0.006860352	1/2	
6	1.993139648	0.003430176	1/2	
7	1.996569824	0.001715088	1/2	
8	1.998284912	0.000857544	1/2	
9	1.999142456	0.000428772	1/2	
10	1.999571228	0.000214386	1/2	
11	1.999785614	0.000107193	1/2	

1.7 SUMMARY

In this chapter, we described the Taylor's theorem, as well as number representation including binary, decimal, and hexadecimal numbers. We have defined absolute and relative errors, inherent errors, round-off errors, truncation errors, machine epsilon, and error propagation. Methods for the estimation were briefly outlined. General error formulae for approximating a function, stability and condition, uncertainty in data, linear convergence, quadratic convergence, and Aitken's acceleration formulae were presented.

EXERCISES

1.1 Determine the following hyperbolic trigonometric functions to $O\,(0.9)^4$.

   (a) $\sinh\,(0.9)$

   (b) $\cosh\,(0.9)$

1.2 Determine when $f(x) = 0$, given that $f(1.7) = -1.7781$ and $f'(1.7) = 4.3257$.

1.3 Determine $f(1.2)$, given the first order differential equation

$$\frac{df}{dx} = 2x \ \text{ with } f(1) = 1.$$

1.4 **(a)** Convert $(327)_{10}$ to binary.

   (b) Convert $(0.3125)_{10}$ to binary.

1.5 Represent the number 50824.6135 in the decimal system (base-10).

1.6 Find the binary and hexadecimal values of the following numbers:

   (a) 329

   (b) 203

1.7 Convert $(75)_{10}$ to base-2.

1.8 Perform the following operations:

   (a) $(2)_8 + (5)_8$

   (b) Convert $(75)_{10}$ to base-8

   (c) Convert $(13)_8$ to base-10.

1.9 Convert $(4D3)_{16}$ to base-10.

1.10 Convert $(1001011)_2$ to base-8.

1.11 Show that the relative error E_{rxy} of the product where $x = x_e + \Delta x$ and $y = y_e + \Delta y$ is $E_{rxy} = E_{rx} + E_{ry}$. Assume $|E_{rx}| \ll 1$ and $|E_{ry}| \ll 1$.

1.12 Show that the relative error E_{rxy} of the quotient where $x = x_e + \Delta x$ and $y = y_e + \Delta y$ is $E_{rxy} = E_{rx} - E_{ry}$. Assume $|E_{rx}| \ll 1$ and $|E_{ry}| \ll 1$.

1.13   Determine the absolute and relative errors involved if $x = 2/3$ is represented in normalized decimal form with six digits by

   (a) round off

   (b) truncation

1.14   Given that 5 digit chopping is used for arithmetic calculations involving x and y where $x = 1/3$ and $y = 5/7$. Determine the absolute and relative errors involved.

1.15   If $x = 3.536$, determine the absolute error and relative error when

   (a) x is rounded

   (b) x is truncated to two decimal digits

1.16   If the number $x = 57.46235$ is rounded off to four significant figures, find the absolute error, relative error and the percentage relative error.

1.17   If the approximate value of $\pi \left(= \dfrac{22}{7} \right)$ is 3.14, determine the absolute error, relative error and relative percentage error.

1.18   Determine the true error and true percentage relative error for each case.

   (a) If the measured length of a track is approximated by 9999 cm and the true value is 10,000 cm

   (b) If the measured length of a track is approximated by 9 cm and the true value is 10 cm

1.19   The exponential function e^x can be computed using the Maclaurin series expansion as

$$e^x = 1 + x + \frac{x^2}{2!} + \frac{x^3}{3!} + \dots + \frac{x^n}{n!}$$

Include six terms in the series and compute the percentage relative error and approximate estimate of the error for each term when estimating $e^{0.5}$.

1.20 Find the relative maximum error in F which $F = \dfrac{5x^2y}{z^3}$. Given $\Delta x = \Delta y = \Delta z = 0.001$, where Δx, Δy, and Δz denote the errors in x, y, and z respectively such that $x = y = z = 1$.

1.21 Consider the trigonometric function $f(x) = \cos x$.

  (a) find the Taylor series expansion of $f(x)$ about 0.

  (b) assuming the Taylor series is truncated to $n = 6$ terms. Determine the relative error at $x = \pi/4$ due to truncation. Express it as a percentage.

  (c) determine an upper bound on the magnitude of the relative error at $x = \pi/4$ expressed as a percentage.

1.22 Suppose $f(x) = e^{-x}$ is to be expanded about the point $x = 1$ and truncated to $n = 6$ terms.
$$e^{-x} = 1 - x + \frac{x^2}{2} - \frac{x^3}{6} + \frac{x^4}{24} - \frac{x^5}{120}$$
Determine the upper bound on the magnitude of the absolute error due to truncation.

1.23 Determine the upper bound on the error for the function
$$f(x) = (x + 1)^{1/2}$$
using a polynomial approximation with third-order Taylor series (computed about $x_0 = 0$) for all $x \in [0, 1]$.

1.24 Consider the power series expansion for e^x given by
$$e^x = 1 + x + \frac{x^2}{2!} + \frac{x^3}{3!} + \cdots + \frac{x^{n-1} - 1}{(n-1)!} + \frac{x^n}{n!} e^{\xi}, \; 0 < \xi < x.$$
Determine the number of terms, n such that their sum gives the value of e^x correct to 8 decimal places at $x = 1$.

1.25 Use the Taylor's series expansion with $n = 0$ to 6 to approximate $f(x) = \cos x$ at $x_{i+1} = \pi/3$ on the premise that the value of $f(x)$ and its derivatives at $x_i = \pi/4$. Assume $h = \pi/3 - \pi/4 = \pi/12$.

1.26 Compute and interpret the condition number for

(a) $f(x) = \tan x$ for $a = \dfrac{\pi}{2} + 0.1\left(\dfrac{\pi}{2}\right)$

(b) $f(x) = \tan x$ for $a = \dfrac{\pi}{2} + 0.01\left(\dfrac{\pi}{2}\right)$

1.27 Evaluate and interpret the condition numbers for

(a) $f(x) = (x^2 - 1)^{1/2} - x$ for $x = 200$

(b) $f(x) = \dfrac{e^x + 1}{x}$ for $x = 0.01$

(c) $f(x) = \dfrac{\cos x}{1 + \sin x}$ for $x = 0.001\pi$

(d) $f(x) = e^{-2x}$ for $x = 5$

1.28 Verify whether the iteration $x_{k+1} = g(x_k)$ starting from the given x_0 converges linearly for the following functions $g(x)$.

(a) $7x^3 + x^2 - 7x - 8$, $x_0 = -0.75$

(b) $-3x^3 + 5x^2 - 4x + 1.5$, $x_0 = 0.9$

(c) $4x^4 + 5x^3 - 2x^2 - 3x - 2$, $x_0 = -0.5$

(d) $3x^3 + x^2 - 5x + 3$, $x_0 = 0.5$

(e) $-4x^3 - 8x^2 - 3x + 2$, $x_0 = -1.25$

1.29 Show that the iteration $x_{k+1} = g(x_k)$ starting from the given x_0 will not converges quadratically for the following functions $g(x)$.

(a) $g(x) = \dfrac{2x^3 + 3.5x^2 - 6}{3x^2 + 5x - 3}$, $x_0 = -4$

(b) $g(x) = \dfrac{8x^3 + 4.5x^2 - 3}{12x^2 + 8x + 1.5}$, $x_0 = -1$

(c) $g(x) = \dfrac{6x^3 + 6x^2 + 1.5}{x(9x + 11)}$, $x_0 = -2$

(d) $g(x) = \dfrac{3x^4 + 4x^3 + 6x^2 + 1}{4x^3 + 5x^2 + 10x - 1}$, $x_0 = -1$

(e) $g(x) = \dfrac{3x^4 + 9x^3 + 3x^2 + 5}{4x^3 + 14x^2 + 7x - 2}$, $x_0 = -4$

1.30 The sequence $\{s_k\}$, $n = 1$ to ∞, where $s_k = n \ln\left(1 + \dfrac{1}{n}\right)$, converges linearly to $s = 1$. Using Aitken's acceleration formula, obtain another sequence that converges faster to $s = 1$.

LINEAR SYSTEM OF EQUATIONS

2.1 INTRODUCTION

In this chapter, we present the solution of n linear simultaneous algebraic equations in n unknowns. Linear systems of equations are associated with many problems in engineering and science, as well as with applications of mathematics to the social sciences and quantitative study of business and economic problems. A system of algebraic equations has the form

$$a_{11}x_1 + a_{12}x_2 + \ldots + a_{1n}x_n = b_1$$
$$a_{12}x_1 + a_{22}x_2 + \ldots + a_{2n}x_n = b_2$$
$$\vdots \qquad \vdots \qquad \qquad \vdots$$
$$a_{n1}x_1 + a_{n2}x_2 + \ldots + a_{nn}x_n = b_n \qquad (2.1)$$

where the coefficients a_{ij} and the constants b_j are known and x_i represents the unknowns. In matrix notation, the equations are written as

$$\begin{bmatrix} a_{11} & a_{12} & \cdots & a_{1n} \\ a_{21} & a_{22} & \cdots & a_{2n} \\ \vdots & \vdots & \cdots & \vdots \\ a_{n1} & a_{n2} & \cdots & a_{nn} \end{bmatrix} \begin{bmatrix} x_1 \\ x_2 \\ \vdots \\ x_n \end{bmatrix} = \begin{bmatrix} b_1 \\ b_2 \\ \vdots \\ b_n \end{bmatrix} \qquad (2.1a)$$

or simply
$$Ax = b \qquad (2.1b)$$

A system of linear equations in n unknowns has a unique solution, provided that the determinant of the coefficient matrix is *nonsingular*, i.e., if $|A| \neq 0$. The rows and columns of a nonsingular matrix are *linearly independent* in the sense that no row (or column) is a linear combination of the other rows (or columns).

If the coefficient matrix is *singular*, the equations may have infinite number of solutions, or no solutions at all, depending on the constant vector.

Linear algebraic equations occur in almost all branches of engineering. Their most important application in engineering is in the analysis of linear systems (any system whose response is proportional to the input is deemed to be linear). Linear systems include structures, elastic solids, heat flow, seepage of fluids, electromagnetic fields and electric circuits, i.e., most topics taught in an engineering curriculum. If the system is discrete, such as a truss or an electric circuit, then its analysis leads directly to linear algebraic equations.

Summarizing the modeling of linear systems invariably gives rise to equations of the form $Ax = b$, where b is the input and x represents the response of the system. The coefficient matrix A, which reflects the characteristics of the system, is independent of the input. In other words, if the input is changed, the equations have to be solved again with a different b, but the same A. Hence, it is desirable to have an equation solving algorithm that can handle any number of constant vectors with minimal computational effort.

2.2   METHODS OF SOLUTION

There are two classes of methods for solving system of linear, algebraic equations: direct and iterative methods. The common characteristics of *direct methods* are that they transform the original equation into *equivalent equations* (equations that have the same solution) that can be solved more easily. The transformation is carried out by applying certain operations.

The solution does not contain any truncation errors but the round-off errors is introduced due to floating point operations.

Iterative or *indirect methods*, start with a guess of the solution **x**, and then repeatedly refine the solution until a certain convergence criterion is reached. Iterative methods are generally less efficient than direct methods due to the large number of operations or iterations required.

Iterative procedures are self-correcting, meaning that round-off errors (or even arithmetic mistakes) in one iteration cycle are corrected in subsequent cycles. The solution contains truncation error. A serious drawback of iterative methods is that they do not always converge to the solution. The initial guess affects only the number of iterations that are required for convergence. The indirect solution technique (iterative) is more useful to solve a set of ill-conditioned equations.

In this chapter, we will present six direct methods and two indirect (iterative) methods.

Direct Methods:

1. Matrix Inverse Method

2. Gauss Elimination Method

3. Gauss-Jordan Method

4. Cholesky's Triangularization Method

5. Crout's Method

6. Thomas Algorithm for Tridiagonal System

Indirect or Iterative Methods:

1. Jacobi's Iteration Method

2. Gauss-Seidel Iteration Method

2.3 THE INVERSE OF A MATRIX

If $\mathbf{A}$ and $\mathbf{B}$ are $m \times n$ matrices such that

$$\mathbf{AB} = \mathbf{BA} = \mathbf{I} \tag{2.2}$$

then $\mathbf{B}$ is said to be the *inverse* of $\mathbf{A}$ and is denoted by

$$\mathbf{B} = \mathbf{A}^{-1} \tag{2.2a}$$

In order to find the inverse $\mathbf{A}^{-1}$, provided the matrix $\mathbf{A}$ is given, let us consider the product,

$$\mathbf{A}adj\mathbf{A} = \begin{bmatrix} a_{11} & a_{12} & \cdots & a_{1n} \\ a_{21} & a_{22} & \cdots & a_{2n} \\ \hdashline a_{n1} & a_{n2} & \cdots & a_{nn} \end{bmatrix} \times \begin{bmatrix} |M_{11}| & -|M_{21}| & \cdots & (-1)^{1+n}|M_{n1}| \\ -|M_{12}| & |M_{22}| & \cdots & (-1)^{2+n}|M_{n2}| \\ (-1)^{1+n}|M_{1n}| & (-1)^{2+n}|M_{2n}| & \cdots & |M_{nn}| \end{bmatrix}$$

$$= \left[\sum_{j=1}^{n} (-1)^{i+j} a_{kj} |M_{ij}| \right] \tag{2.3}$$

An element of the matrix on the right side of Eq. (2.3) has the value

$$\sum_{j=1}^{n}(-1)^{i+j}a_{kj}\,|\,M_{ij}\,|=\begin{vmatrix} a_{11} & a_{12} & \cdots & a_{1n} \\ a_{21} & a_{22} & \cdots & a_{2n} \\ \hline a_{n1} & a_{n2} & \cdots & a_{nn} \end{vmatrix}=|\,a\,| \quad \text{if } i=k \qquad (2.4)$$

If $i \neq k$ the determinant possesses two identical rows, since the determinant corresponding to $i \neq k$ is obtained from the matrix $[a]$ by replacing the ith row by the kth row and keeping the kth row intact. Therefore, if $i \neq k$ the value of the element is zero.

Equation (2.3) can be written as

$$\mathbf{A}\,\text{Adj}\,\mathbf{A} = |\mathbf{A}|\,\mathbf{I} \qquad (2.5)$$

Premultiplying Eq. (2.5) throughout by $\mathbf{A}^{-1}$ and dividing the result by $|\mathbf{A}|$, we get

$$\mathbf{A}^{-1} = \frac{adj\,\mathbf{A}}{\det\,\mathbf{A}} \qquad (2.6)$$

so that the inverse of a matrix $\mathbf{A}$ is obtained by dividing its adjoint matrix by its determinant $|\mathbf{A}|$.

If $\det \mathbf{A}$ is equal to zero, then the elements of $\mathbf{A}^{-1}$ approach infinity (or are indeterminant at best), in which case the inverse $\mathbf{A}^{-1}$ is said *not to exist*, and the matrix $\mathbf{A}$ is said to be singular. The inverse of a matrix exists only if determinant is not zero, that is, the matrix must be nonsingular.

There is no direct division of matrices. The operation of division is performed by inversion; if

$$\mathbf{AB} = \mathbf{C}$$

then $$\mathbf{B} = \mathbf{A}^{-1}\mathbf{C}$$

where $\mathbf{A}^{-1}$ is called the inverse of matrix $\mathbf{A}$.

The requirements for obtaining a unique inverse of a matrix are:

1. The matrix is a square matrix.
2. The determinant of the matrix is not zero (the matrix is nonsingular).

The inverse of a matrix is also defined by the relationship:

$$\mathbf{A}^{-1}\mathbf{A} = \mathbf{I}$$

The following are the properties of an inverted matrix:

1. The inverse of a matrix is unique.
2. The inverse of the product of two matrices is equal to the product of the inverse of the two matrices in reverse order:

$$(\mathbf{AB})^{-1} = \mathbf{B}^{-1}\mathbf{A}^{-1}$$

3. The inverse of a triangular matrix is itself a triangular matrix of the same type.
4. The inverse of a symmetrical matrix is itself a symmetrical matrix.
5. The negative powers of a nonsingular matrix are obtained by raising the inverse of the matrix to positive powers.
6. The inverse of the transpose of $\mathbf{A}$ is equal to the transpose of the inverse of $\mathbf{A}$:

$$(\mathbf{A^T})^{-1} = (\mathbf{A^{-1}})^{\mathbf{T}}$$

EXAMPLE 2.1

Find the inverse of the matrix $A = \begin{bmatrix} 2 & 3 \\ 5 & 1 \end{bmatrix}$.

Solution:

If
$$A = \begin{bmatrix} 2 & 3 \\ 5 & 1 \end{bmatrix},$$

then
$$adj\,\mathbf{A} = \begin{bmatrix} 1 & -3 \\ -5 & 2 \end{bmatrix}$$

and
$$\det \mathbf{A} = 2 \times 1 - 5 \times 3 = -13$$

Hence
$$\mathbf{A}^{-1} = -\frac{1}{13}\begin{bmatrix} 1 & -3 \\ -5 & 2 \end{bmatrix} = \begin{bmatrix} -\dfrac{1}{13} & \dfrac{3}{13} \\ \dfrac{5}{13} & -\dfrac{2}{13} \end{bmatrix}$$

2.4   MATRIX INVERSION METHOD

Consider a set of three simultaneous linear algebraic equations

$$a_{11}x_1 + a_{12}x_2 + a_{13}x_3 = b_1$$

$$a_{21}x_1 + a_{22}x_2 + a_{23}x_3 = b_2$$

$$a_{31}x_1 + a_{32}x_2 + a_{33}x_3 = b_3 \tag{2.7}$$

Equation (2.7) can be expressed in the matrix form

$$\mathbf{Ax} = \mathbf{b} \tag{2.8}$$

Premultiplying by the inverse $\mathbf{A}^{-1}$, we obtain the solution of $\mathbf{x}$ as

$$\mathbf{x} = \mathbf{A}^{-1}\mathbf{b} \tag{2.9}$$

If the matrix $\mathbf{A}$ is nonsingular, that is, if det $(\mathbf{A})$ is not equal to zero, then Eq. (2.9) has a unique solution.

The solution for x_1 is

$$x_1 = \frac{1}{|\mathbf{A}|} \begin{vmatrix} b_1 & a_{12} & a_{13} \\ b_2 & a_{22} & a_{23} \\ b_3 & a_{32} & a_{33} \end{vmatrix} = \frac{1}{|\mathbf{A}|} \left\{ b_1 \begin{vmatrix} a_{22} & a_{23} \\ a_{32} & a_{33} \end{vmatrix} - b_2 \begin{vmatrix} a_{12} & a_{13} \\ a_{32} & a_{33} \end{vmatrix} + b_3 \begin{vmatrix} a_{12} & a_{13} \\ a_{22} & a_{23} \end{vmatrix} \right\}$$

$$= \frac{1}{|\mathbf{A}|} \{ b_1 C_{11} + b_2 C_{21} + b_3 C_{31} \}$$

where $\mathbf{A}$ is the determinant of the coefficient matrix $\mathbf{A}$, and C_{11}, C_{21} and C_{31} are the cofactors of $\mathbf{A}$ corresponding to element 11, 21 and 31. We can also write similar expressions for x_2 and x_3 by replacing the second and third columns by the y column respectively. Hence, the complete solution can be written in matrix form as follows:

$$\begin{Bmatrix} x_1 \\ x_2 \\ x_3 \end{Bmatrix} = \frac{1}{|\mathbf{A}|} \begin{bmatrix} C_{11} & C_{21} & C_{31} \\ C_{12} & C_{22} & C_{32} \\ C_{13} & C_{23} & C_{33} \end{bmatrix} \begin{Bmatrix} b_1 \\ b_2 \\ b_3 \end{Bmatrix} \tag{2.10}$$

or $$\{x\} = \frac{1}{|\mathbf{A}|} [C_{ji}]\{b\} = \frac{1}{|\mathbf{A}|} [adj\,\mathbf{A}]\{b\}$$

Hence $$\mathbf{A}^{-1} = \frac{1}{|\mathbf{A}|} adj\ \mathbf{A} \text{ and Adj } \mathbf{A} = \mathbf{A}^{-1} \text{ abs } [\mathbf{A}]$$ (2.11)

Although this method is quite general but it is not quite suitable for large systems because evaluation of A^{-1} by cofactors becomes very cumbersome.

EXAMPLE 2.2

Obtain the solution of the following linear simultaneous equations by the matrix inversion method.

a) $\begin{bmatrix} 1 & 3 \\ 4 & -1 \end{bmatrix} \begin{bmatrix} x_1 \\ x_2 \end{bmatrix} = \begin{bmatrix} 5 \\ 12 \end{bmatrix}$

b) $\begin{bmatrix} 1 & -1 & 3 \\ 4 & 2 & -1 \\ 1 & 3 & 1 \end{bmatrix} \begin{bmatrix} x_1 \\ x_2 \\ x_3 \end{bmatrix} = \begin{bmatrix} 5 \\ 0 \\ 5 \end{bmatrix}$

Solution:

a) $\begin{bmatrix} 1 & 3 \\ 4 & -1 \end{bmatrix} \begin{bmatrix} x_1 \\ x_2 \end{bmatrix} = \begin{bmatrix} 5 \\ 12 \end{bmatrix}$

$C_{11} = (-1)^{1+1}|-1| = -1$

$C_{12} = (-1)^{1+2}|4| = -4$

$C_{21} = (-1)^{1+}|3| = -3$

$C_{22} = (-1)^{2+2}|1| = 1$

Hence $\mathbf{C} = \begin{bmatrix} -1 & -4 \\ -3 & 1 \end{bmatrix}$

$\mathbf{C^T} = \begin{bmatrix} -1 & -3 \\ -4 & 1 \end{bmatrix}$

$\mathbf{A}^{-1} = \frac{\mathbf{C^T}}{|\mathbf{A}|} = \frac{-1}{13} \begin{bmatrix} -1 & -3 \\ -4 & 1 \end{bmatrix}$

Hence $\begin{bmatrix} x_1 \\ x_2 \end{bmatrix} = \frac{-1}{13} \begin{bmatrix} -1 & -3 \\ -4 & 1 \end{bmatrix} \begin{bmatrix} 5 \\ 12 \end{bmatrix} = \frac{-1}{13} \begin{bmatrix} -5 & -36 \\ -20 & +12 \end{bmatrix} = \frac{-1}{13} \begin{bmatrix} -41 \\ -8 \end{bmatrix}$

Therefore $\quad x_1 = \dfrac{-41}{-13} = 3.15$

and $\quad\quad x_2 = \dfrac{-8}{-13} = 0.62$

b) $\quad |A| = \begin{vmatrix} 1 & -1 & 3 \\ 4 & 2 & -1 \\ 1 & 3 & 1 \end{vmatrix} = 40$

The matrix of cofactors is given by

$$\mathbf{C} = \begin{bmatrix} 5 & -5 & 10 \\ 10 & -2 & -4 \\ -5 & 13 & 6 \end{bmatrix}$$

The transpose of $\mathbf{C}$ is the adjoint of $\mathbf{A}$ or

$$\mathbf{Adj\ A} = \mathbf{C}^T = \begin{bmatrix} 5 & 10 & -5 \\ -5 & -2 & 13 \\ 10 & -4 & 6 \end{bmatrix}$$

Hence $\quad \mathbf{A}^{-1} = \mathbf{Adj\ A}/|A| = \dfrac{1}{40}\begin{bmatrix} 5 & 10 & -5 \\ -5 & -2 & 13 \\ 10 & -4 & 6 \end{bmatrix}$

Therefore $\quad X = \mathbf{A}^{-1}Y = \dfrac{1}{40}\begin{bmatrix} 5 & 10 & -5 \\ -5 & -2 & 13 \\ 10 & -4 & 6 \end{bmatrix}\begin{bmatrix} 5 \\ 0 \\ 5 \end{bmatrix} = \dfrac{1}{40}\begin{bmatrix} 0 \\ 40 \\ 80 \end{bmatrix} = \begin{bmatrix} 0 \\ 1 \\ 2 \end{bmatrix}$

or $\quad\quad x_1 = 0, x_2 = 1, x_3 = 2.$

EXAMPLE 2.3

Find the inverse of the matrix

$$A = \begin{bmatrix} 2 & 2 & 0 \\ -2 & 1 & 1 \\ 3 & 0 & 1 \end{bmatrix}$$

and solve the system of equations $[A]\{x\} = \{b\}$ where $\{b\} = \begin{Bmatrix} 6 \\ 3 \\ 6 \end{Bmatrix}$.

Solution:

The determinant of $A = \begin{bmatrix} 2 & 2 & 0 \\ -2 & 1 & 1 \\ 3 & 0 & 1 \end{bmatrix} = 2[1(1) - 0(1)] + 2[2(1) - 0(0)] +$
$3[2(1) - 1(0)] = 12$

Since $\det \mathbf{A} = 12 \neq 0$, the given matrix is nonsingular. The cofactors corresponding to the entries in each row of $\det \mathbf{A}$ are

$$C_{11} = \begin{vmatrix} 1 & 1 \\ 0 & 1 \end{vmatrix} = 1 \qquad C_{12} = -\begin{vmatrix} -2 & 1 \\ 3 & 1 \end{vmatrix} = 5 \qquad C_{13} = \begin{vmatrix} -2 & 1 \\ 3 & 0 \end{vmatrix} = -3$$

$$C_{21} = -\begin{vmatrix} 2 & 0 \\ 0 & 1 \end{vmatrix} = -2 \quad C_{22} = \begin{vmatrix} 2 & 0 \\ 3 & 1 \end{vmatrix} = 2 \qquad C_{23} = -\begin{vmatrix} 2 & 2 \\ 3 & 0 \end{vmatrix} = 6$$

$$C_{31} = \begin{vmatrix} 2 & 0 \\ 1 & 1 \end{vmatrix} = 2 \qquad C_{32} = -\begin{vmatrix} 2 & 0 \\ -2 & 1 \end{vmatrix} = -2 \quad C_{33} = \begin{vmatrix} 2 & 2 \\ -2 & 1 \end{vmatrix} = 6$$

Hence $\quad \mathbf{A}^{-1} = \dfrac{C^T}{|\mathbf{A}|} = \dfrac{1}{12}\begin{pmatrix} 1 & -2 & 2 \\ 5 & 2 & -2 \\ -3 & 6 & 6 \end{pmatrix} = \begin{pmatrix} \frac{1}{12} & -\frac{1}{6} & \frac{1}{6} \\ \frac{5}{12} & \frac{1}{6} & -\frac{1}{6} \\ -\frac{1}{4} & \frac{1}{2} & \frac{1}{2} \end{pmatrix}$

It is easy to verify that $\mathbf{A}^{-1}\mathbf{A} = \mathbf{A}\mathbf{A}^{-1} = \mathbf{I}$.

Therefore $\begin{bmatrix} x_1 \\ x_2 \\ x_3 \end{bmatrix} = [\mathbf{A}^{-1}]b = \begin{bmatrix} \frac{1}{12} & -\frac{1}{6} & \frac{1}{6} \\ \frac{5}{12} & \frac{1}{6} & -\frac{1}{6} \\ -\frac{1}{4} & \frac{1}{2} & \frac{1}{2} \end{bmatrix}\begin{bmatrix} 6 \\ 3 \\ 6 \end{bmatrix} = \begin{bmatrix} \frac{6}{12} - \frac{3}{6} + \frac{6}{6} \\ \frac{30}{12} + \frac{3}{6} - \frac{6}{6} \\ \frac{-6}{4} + \frac{3}{2} + \frac{6}{2} \end{bmatrix} = \begin{bmatrix} \frac{6-6+12}{12} \\ \frac{30+6-12}{12} \\ \frac{-6+6+12}{4} \end{bmatrix} = \begin{bmatrix} 1 \\ 2 \\ 3 \end{bmatrix}$

2.4.1 Augmented Matrix

A system of linear equations in matrix notation takes the form $Ax = b$, where A is of order $m \times n$, x is of order $n \times 1$. The augmented matrix $[A_b]$ can be obtained by adjoining column b to matrix A. In terms of partitioned matrices, we have $[A_b] = [A : b]$.

As an example let us consider a set of linear equations

$$x + y + z = 8$$
$$x + y - z = 5$$
$$x - y + z = 2$$

Then we have,

$$[A] = \begin{bmatrix} 1 & 1 & 1 \\ 1 & 1 & -1 \\ 1 & -1 & 1 \end{bmatrix}; \quad \{b\} = \begin{Bmatrix} 8 \\ 5 \\ 2 \end{Bmatrix}$$

and

$$[A_b] = \begin{bmatrix} 1 & 1 & 1 & 8 \\ 1 & 1 & -1 & 5 \\ 1 & -1 & 1 & 2 \end{bmatrix}$$

2.5   GAUSS ELIMINATION METHOD

Consider the following system of linear simultaneous equations:

$$a_{11}x_1 + a_{12}x_2 + a_{13}x_3 = b_1 \tag{2.12}$$

$$a_{21}x_1 + a_{22}x_2 + a_{23}x_3 = b_2 \tag{2.13}$$

$$a_{31}x_1 + a_{32}x_2 + a_{33}x_3 = b_3 \tag{2.14}$$

Gauss elimination is a popular technique for solving simultaneous linear algebraic equations. It reduces the coefficient matrix into an upper triangular matrix through a sequence of operations carried out on the matrix. The vector b is also modified in the process. The solution vector {x} is obtained from a backward substitution procedure.

Two linear systems $Ax = b$ and $A'x = b'$ of equations are said to be equivalent if any solution of one is a solution of the other. Also, let $Ax = b$ is a linear nonhomogeneous system of n equations. Suppose we subject this system to the system of following operations:

1. Multiplication of one equation by a nonzero constant.

2. Addition of a multiple of one equation to another equation.

3. Interchange of two equations.

If the sequence of operations produce the new system $\mathbf{A'x} = b'$, then both the systems $\mathbf{Ax} = b$ and $A'x = b'$ are equivalent. In particular, $\mathbf{A}$ is invertible if $\mathbf{A'}$ is invertible. In the Gauss elimination method, we adopt this and the elimination process is based on this theorem.

In the Gauss elimination method, the unknowns are eliminated such that the elimination process leads to an upper triangular system and the unknowns are obtained by back substitution. It is assumed $a_{11} \neq 0$. The method can be described by the following steps:

Step 1: Eliminate x_1 from the second and third equations.

Using the first equation (2.12), the following operations are performed:

$$(2.13) - \left(\frac{a_{21}}{a_{11}}\right)(2.12) \quad \text{and} \quad (2.14) - \left(\frac{a_{31}}{a_{11}}\right)(2.12)$$

gives
$$a_{11}x_1 + a_{12}x_2 + a_{13}x_3 = b_1 \tag{2.15}$$
$$a'_{22}x_2 + a'_{23}x_3 = b'_2 \tag{2.16}$$
$$a'_{32}x_2 + a'_{33}x_3 = b'_3 \tag{2.17}$$

Equation (2.15) is called the *pivotal equation* and the coefficient a_{11} is the *pivot*.

Step 2: Eliminate x_2 from the Eq. (2.17) using Eq. (2.16) by assuming $a'_{22} \neq 0$. We perform the following operation:

$$(2.17) - \left(\frac{a'_{32}}{a'_{22}}\right)(2.16)$$

to obtain
$$a_{11}x_1 + a_{12}x_2 + a_{13}x_3 = b_1 \tag{2.18}$$
$$a'_{22}x_2 + a'_{23}x_3 = b'_2 \tag{2.19}$$

and
$$a''_{33}x_3 = b''_3 \tag{2.20}$$

Here Eq. (2.19) is called the *pivotal equation* and the coefficient a'_{22} is the *pivot*.

Step 3: To find x_1, x_2, and x_3, we apply back substitution starting from Eq. (2.20) giving x_3, then x_2 from Eq. (2.19) and x_1 from Eq. (2.18).

Pivoting:

The Gauss elimination method fails if any one of the pivots in the above equations (2.12) to (2.20) becomes zero. To overcome this difficulty, the equations are to be rewritten in a slightly different order such that the pivots are not zero.

Partial pivoting method:

Step 1: The numerically largest coefficient of x_1 is selected from all the equations are pivot and the corresponding equation becomes the first equation (2.12).

Step 2: The numerically largest coefficient of x_2 is selected from all the remaining equations as *pivot* and the corresponding equation becomes the second equation (2.16). This process is repeated till an equation into a simple variable is obtained.

Complete pivoting method:

In this method, we select at each stage the numerically largest coefficient of the complete matrix of coefficients. This procedure leads to an interchange of the equations as well as interchange of the position of variables.

EXAMPLE 2.4

Solve the following equations by the Gauss elimination method:

$$2x + 4y - 6z = -4$$
$$x + 5y + 3z = 10$$
$$x + 3y + 2z = 5$$

Solution:

$$2x + 4y - 6z = -4 \qquad\qquad (\text{E}.1)$$
$$x + 5y + 3z = 10 \qquad\qquad (\text{E}.2)$$
$$x + 3y + 2z = 5 \qquad\qquad (\text{E}.3)$$

To eliminate x from (E.2) and (E.3) using (E.1):

$$2x + 4y - 6z = -4$$
$$x + 5y + 3z = 10 \qquad\qquad 1 \times (-2)$$

$$x + 3y + 2z = 5 \qquad 1 \times (-2)$$

$$2x + 4y - 6z = -4$$

$$-2x - 10y - 6z = -20$$

$$-2x - 6y - 4z = -10$$

$$2x + 4y - 6z = -4$$

Row 1+ Row 2: $\qquad -6y - 12z = -24$ (E.6)

Row 1 + Row 3: $\qquad 2y - 10z = -14 \qquad 1 \times (-3)$ (E.5)

To eliminate y from (E.5) using (E.4):

$$2x + 4y - 6z = -4$$

$$-6y - 12z = -24$$

$$6y + 30z = 42$$

$$2x + 4y - 6z = -4$$

$$-6y - 12z = -24$$

Row 2 + Row 3: $\qquad 18z = 18 \quad \Rightarrow \boxed{z = 1}$

Evaluation of the unknowns by back substitution:

$$-6y - 12z = -24$$

$$6y = 24 - 12z \quad \Rightarrow y = \frac{24 - 12 \times 1}{6} \Rightarrow \boxed{y = 2}$$

$$2x + 4y - 6z = -4$$

$$2x = -4 - 4y + 6z \Rightarrow x = \frac{-4 - 4 \times 2 + 6 \times 1}{2} \Rightarrow \boxed{x = -3}$$

EXAMPLE 2.5

Use the method of Gaussian elimination to solve the following system of linear equations:

$$x_1 + x_2 + x_3 - x_4 = 2$$

$$4x_1 + 4x_2 + x_3 + x_4 = 11$$

$$x_1 - x_2 - x_3 + 2x_4 = 0$$

$$2x_1 + x_2 + 2x_3 - 2x_4 = 2 \qquad \text{(E.1)}$$

Solution:

In the first step, eliminate x_1 terms from second, third and fourth equations of the set of equations (E.1) to obtain:

$$x_1 + x_2 + x_3 - x_4 = 2$$
$$-3x_3 + 5x_4 = 3$$
$$-2x_2 - 2x_3 + 3x_4 = -2$$
$$-x_2 = -2 \qquad \text{(E.2)}$$

Interchanging columns in Equation (E.2) putting the variables in the order x_1, x_4, x_3 and x_2 as

$$x_1 - x_4 + x_3 + x_2 = 2$$
$$-5x_4 - 3x_3 = 3$$
$$3x_4 - 2x_3 - 2x_2 = -2$$
$$-x_2 = -2 \qquad \text{(E.3)}$$

In the second step, eliminate x_4 term in third equation of the set of equations (E.3)

$$x_1 - x_4 + x_3 + x_2 = 2$$
$$5x_4 - 3x_3 = 3$$
$$-1/5x_3 - 2x_2 = -19/5$$
$$-x_2 = -2$$

Now, by the process of back substitution, we have

$$x_2 = 2, \, x_3 = -1, \, x_4 = 0, \, x_1 = 1.$$

EXAMPLE 2.6

Using the Gaussian elimination method, solve the system of equations $[\mathbf{A}]\{x\} = \{b\}$ where

$$[\mathbf{A}] = \begin{bmatrix} 1 & 1 & 1 & 1 \\ 2 & -1 & 3 & 0 \\ 0 & 2 & 0 & 3 \\ -1 & 0 & 2 & 1 \end{bmatrix} \text{ and } \{b\} = \begin{Bmatrix} 3 \\ 3 \\ 1 \\ 0 \end{Bmatrix}$$

Solution:

The augmented matrix is

$$[\mathbf{A_b}] = \begin{bmatrix} 1 & 1 & 1 & 1 & 3 \\ 2 & -1 & 3 & 0 & 3 \\ 0 & 2 & 0 & 3 & 1 \\ 0 & 2 & 0 & 3 & 1 \\ -1 & 0 & 2 & 1 & 0 \end{bmatrix}$$

From the augmented matrix, we apply elementary transformations:

$$\begin{matrix} \text{Row } 2 - 2 \times \text{Row 1} \\ \text{Row } 4 + \text{Row 1} \end{matrix} \begin{bmatrix} 1 & 1 & 1 & 1 & 3 \\ 0 & -3 & 1 & -2 & 3 \\ 0 & 2 & 0 & 3 & 1 \\ 0 & 2 & 0 & 3 & 1 \\ 0 & 1 & 3 & 2 & 3 \end{bmatrix}$$

which gives

$$\begin{bmatrix} 1 & 1 & 1 & 1 & 3 \\ 0 & -3 & 1 & -2 & -3 \\ 0 & 0 & \frac{2}{3} & \frac{5}{3} & -1 \\ 0 & 0 & \frac{10}{3} & \frac{4}{3} & 2 \end{bmatrix} \begin{bmatrix} 1 & 1 & 1 & 1 & 3 \\ 0 & -3 & 1 & -2 & -3 \\ 0 & 0 & \frac{2}{3} & \frac{5}{3} & -1 \\ 0 & 0 & 0 & -7 & 7 \end{bmatrix}$$

Hence, by back substitution the upper triangular matrix, we obtain

$$x_4 = -1, x_3 = 1, x_2 = 2, x_1 = 1.$$

2.6 GAUSS-JORDAN METHOD

The Gauss-Jordan method is an extension of the Gauss elimination method. The set of equations $\mathbf{Ax} = b$ is reduced to a diagonal set $Ix = b'$, where I is a unit matrix. This is equivalent to $x = b'$. The solution vector is therefore obtained directly from b'. The Gauss-Jordan method implements the same series of operations as implemented by the Gauss elimination process. The main difference is that it applies these operations below as well as above the diagonal such that all off-diagonal elements of the matrix are reduced to zero. The Gauss-Jordan method also provides the inverse of the coefficient matrix

A along with the solution vector $\{x\}$. The Gauss-Jordan method is highly used due to its stability and direct procedure. The Gauss-Jordan method requires more computational effort than the Gauss elimination process.

The Gauss-Jordan method is a modification of the Gauss elimination method. The series of operations performed are quite similar to the Gauss elimination method. In the Gauss elimination method, an upper triangular matrix is derived while in the Gauss-Jordan method an identity matrix is derived. Hence, back substitutions are not required.

EXAMPLE 2.7

Solve the following equations by the Gauss-Jordan method.

$$x + 3y + 2z = 17$$

$$x + 2y + 3z = 16$$

$$2x - y + 4z = 13$$

Solution:

Consider

$$x + 3y + 2z = 17 \qquad\qquad (E.1)$$

$$x + 2y + 3z = 16 \qquad\qquad (E.2)$$

$$2x - y + 4z = 13 \qquad\qquad (E.3)$$

$$x + 3y + 2z = 17 \qquad (E.1)\,(-2) + (E.3)$$

$$x + 2y + 3z = 16 \qquad (E.2)\,(-1) + (E.1)$$

$$2x - y + 4z = 13$$

$$x + 3y + 2z = 17$$

$$y - z = 1 \qquad\qquad x(2) + (E.1)$$

$$7y = 21 \qquad\qquad \Rightarrow \boxed{y = 3}$$

$$x + 5y = 19$$

$$y - z = 1 \qquad\qquad \Rightarrow z = y - 1 = 3 - 1 \quad \Rightarrow \boxed{z = 2}$$

$$x + 5y = 19 \qquad\qquad \Rightarrow x = 19 - 5 \times 3 \quad \Rightarrow \boxed{x = 4}$$

EXAMPLE 2.8

Solve the following system of equations using the Gauss-Jordan method.

$$x - 2y = -4$$
$$5y + z = -9$$
$$4x - 3z = -10$$

Solution:

The augmented matrix is

$$\begin{bmatrix} 1 & -2 & 0 & \vdots & -4 \\ 0 & -5 & 1 & \vdots & -9 \\ 4 & 0 & -3 & \vdots & -10 \end{bmatrix}$$

Multiplying 1st row by -4 and adding the result to the 3rd row, we obtain

$$-4R_1 + R_3 \rightarrow \begin{bmatrix} 1 & -2 & 0 & \vdots & -4 \\ 0 & -5 & 1 & \vdots & -9 \\ 0 & 8 & -3 & \vdots & 6 \end{bmatrix}$$

Now, multiply the 2nd row by $-1/5$

$$-\frac{1}{5}R_2 \rightarrow \begin{bmatrix} 1 & -2 & 0 & \vdots & -4 \\ 0 & 1 & -1/5 & \vdots & 9/5 \\ 0 & 8 & -3 & \vdots & 6 \end{bmatrix}$$

Multiply the 2nd row and add the result to the 1st row. Then multiply the 2nd row by -8 and add the result to the 3rd row.

$$2R_2 + R_1 \rightarrow \begin{bmatrix} 1 & 0 & -2/5 & \vdots & -2/5 \\ 0 & 1 & -1/5 & \vdots & 9/5 \\ 0 & 0 & -7/5 & \vdots & -42/5 \end{bmatrix}$$

Multiply 3rd row by $-5/7$

$$-\frac{5}{7}R_3 \rightarrow \begin{bmatrix} 1 & 0 & -2/5 & \vdots & -2/5 \\ 0 & 1 & -1/5 & \vdots & 9/5 \\ 0 & 0 & 1 & \vdots & 6 \end{bmatrix}$$

Multiply the 3rd row by 2/5 and add the result to the 1st row. Then multiply the 3rd row by 1/5 and add the result to the 2nd row.

$$\begin{array}{c} \frac{2}{5}R_3 + R_1 \rightarrow \\ \\ \frac{1}{5}R_3 + R_2 \rightarrow \end{array} \left[\begin{array}{ccc|c} 1 & 0 & 0 & 2 \\ 0 & 1 & 0 & 3 \\ 0 & 0 & 1 & 6 \end{array} \right]$$

Hence, the last matrix above represents the system with $x = 2$, $y = 3$ and $z = 6$.

EXAMPLE 2.9

Solve the following set of equations by the Gauss-Jordan method.

$$2x_1 + x_2 - 3x_3 = 11$$
$$4x_1 - 2x_2 + 3x_3 = 8$$
$$-2x_1 + 2x_2 - x_3 = -6$$

Solution:

The augmented matrix for the given set of equations is

$$\left[\begin{array}{cccc} 2 & 1 & -3 & 11 \\ 4 & -2 & 3 & 8 \\ -2 & 2 & -1 & -6 \end{array} \right]$$

Step 1: Divide Row 1 by 2

$$\left[\begin{array}{cccc} 1 & \dfrac{1}{2} & -\dfrac{3}{2} & \dfrac{11}{2} \\ 4 & -2 & 3 & 8 \\ -2 & 2 & -1 & -6 \end{array} \right]$$

$$\begin{array}{c} \\ \text{Row } 2 - 4 \times \text{Row } 1 \\ \text{Step 2:} \qquad \text{Row } 3 - 2 \times \text{Row } 1 \\ \\ \end{array} \left[\begin{array}{cccc} 1 & \dfrac{1}{2} & -\dfrac{3}{2} & \dfrac{11}{2} \\ 0 & -4 & 9 & -14 \\ 0 & 3 & -4 & 5 \end{array} \right]$$

Step 3: Divide Row 2 by –4
$$\begin{bmatrix} 1 & \dfrac{1}{2} & -\dfrac{3}{2} & \dfrac{11}{2} \\ 0 & 1 & -\dfrac{9}{4} & \dfrac{7}{2} \\ 0 & 3 & -4 & 5 \end{bmatrix}$$

Step 4:
Row $1 - 1/2 \times$ Row 2
Row $3 - 3 \times$ Row 2
$$\begin{bmatrix} 1 & 0 & -\dfrac{3}{8} & \dfrac{15}{4} \\ 0 & 1 & -\dfrac{9}{4} & \dfrac{7}{2} \\ 0 & 3 & \dfrac{11}{4} & -\dfrac{11}{2} \end{bmatrix}$$

Step 5: Divide Row 3 by 11/4
$$\begin{bmatrix} 1 & 0 & -\dfrac{3}{8} & \dfrac{15}{4} \\ 0 & 1 & -\dfrac{9}{4} & \dfrac{7}{2} \\ 0 & 0 & 1 & -2 \end{bmatrix}$$

Step 6:
Row $1 + 3/8 \times$ Row 3
Row $2 + 9/4 \times$ Row 3
$$\begin{bmatrix} 1 & 0 & 0 & 3 \\ 0 & 1 & 0 & -1 \\ 0 & 0 & 1 & -2 \end{bmatrix}$$

Hence the solution is $x_1 = 3$, $x_2 = -1$, $x_3 = -2$.

EXAMPLE 2.10

Solve

$$2x_1 + 6x_2 + x_3 = 7$$

$$x_1 + 2x_2 - x_3 = -1$$

$$5x_1 + 7x_2 - 4x_3 = 9$$

Using (a) Gaussian elimination and (b) Gauss-Jordan elimination.

Solution:

a) Using row operations on the augmented matrix of the system,

$$\begin{pmatrix} 2 & 6 & 1 & | & 7 \\ 1 & 2 & -1 & | & -1 \\ 5 & 7 & -4 & | & 9 \end{pmatrix} \xrightarrow{R_{12}} \begin{pmatrix} 1 & 2 & -1 & | & -1 \\ 2 & 6 & 1 & | & 7 \\ 5 & 7 & -4 & | & 9 \end{pmatrix} \xrightarrow[-5R_1+R_3]{-2R_1+R_2} \begin{pmatrix} 1 & 2 & -1 & | & -1 \\ 0 & 2 & 3 & | & 9 \\ 0 & -3 & 1 & | & 14 \end{pmatrix}$$

$$\xrightarrow{\frac{1}{2}R_2} \begin{pmatrix} 1 & 2 & -1 & | & -1 \\ 0 & 1 & \frac{3}{2} & | & \frac{9}{2} \\ 0 & -3 & 1 & | & 14 \end{pmatrix} \xrightarrow{3R_2+R_3} \begin{pmatrix} 1 & 2 & -1 & | & -1 \\ 0 & 1 & \frac{3}{2} & | & \frac{9}{2} \\ 0 & 0 & \frac{11}{2} & | & \frac{55}{2} \end{pmatrix}$$

$$\xrightarrow{\frac{2}{11}R_3} \begin{pmatrix} 1 & 2 & -1 & | & -1 \\ 0 & 1 & \frac{3}{2} & | & \frac{9}{2} \\ 0 & 0 & 1 & | & 5 \end{pmatrix} \tag{E.1}$$

The last matrix is in row-echelon form and represents the system

$$x_1 + 2x_2 - x_3 = -1 \tag{E.2}$$

$$x_2 + \frac{3}{2}x_3 = \frac{9}{2} \tag{E.3}$$

$$x_3 = 5 \tag{E.4}$$

Substituting $x_3 = 5$ into Eq. (E.3) gives $x_2 = -3$. Substituting both of these values back into Eq. (E.2) finally yields $x_1 = 10$.

b) We start with the last matrix in Eq. (E.1) above. Since the first entries in the second and third rows are 1's, we must, in turn, make the remaining entries in the second and third columns 0's:

$$\begin{pmatrix} 1 & 2 & -1 & | & -1 \\ 0 & 1 & \frac{3}{2} & | & \frac{9}{2} \\ 0 & 0 & 1 & | & 5 \end{pmatrix} \xrightarrow{-2R_2+R_1} \begin{pmatrix} 1 & 0 & -4 & | & -10 \\ 0 & 1 & \frac{3}{2} & | & \frac{9}{2} \\ 0 & 0 & 1 & | & 5 \end{pmatrix} \xrightarrow[-\frac{3}{2}R_3+R_2]{4R_3+R_1} \begin{pmatrix} 1 & 0 & 0 & | & 10 \\ 0 & 1 & 0 & | & -3 \\ 0 & 0 & 1 & | & 5 \end{pmatrix}$$

$$\tag{E.5}$$

The last matrix in Eq. (E.5) is now in reduced row-echelon form. It is evident that the solution of the system is $x_1 = 10$, $x_2 = -3$, $x_3 = 5$.

LU Decomposition: It is possible to show that any square matrix **A** can be expressed as a product of a lower triangular matrix L and an upper triangular matrix U.

$$\mathbf{A} = LU$$

For instance

$$\begin{bmatrix} a_{11} & a_{12} & a_{13} \\ a_{21} & a_{22} & a_{23} \\ a_{31} & a_{32} & a_{33} \end{bmatrix} = \begin{bmatrix} L_{11} & 0 & 0 \\ L_{21} & L_{22} & 0 \\ L_{31} & L_{32} & L_{33} \end{bmatrix} \begin{bmatrix} U_{11} & U_{12} & U_{13} \\ 0 & U_{22} & U_{23} \\ 0 & 0 & U_{33} \end{bmatrix}$$

The process of computing L and U for a given **A** is known as *LU Decomposition or LU Factorization. LU* decomposition is not unique (the combinations of L and U for a prescribed **A** are endless), unless certain constraints are placed on L or U. These constraints distinguish one type of decomposition from another. Two commonly used decompositions are given below:

1. Cholesky's decomposition: Constraints are $L = U^T$

2. Crout's decomposition: Constrains are $U_{ii} = 1, i = 1, 2, ..., n$.

After decomposing the matrix **A**, it is easier to solve the equations $Ax = b$.

We can rewrite the equations as

$$LUx = b$$

or denoting Ux = y, the above equation becomes

$$Ly = b$$

This equation $Ly = b$ can be solved for y by forward substitution. Then $Ux = y$ will yield x by the backward substitution process. The advantage of the LU decomposition method over the Gauss elimination method is that once **A** is decomposed, we can solve $\mathbf{A}x = b$ for as many constant vectors b as we please. Also, the forward and backward substitutions operations are much less time consuming than the decomposition process.

2.7   CHOLESKY'S TRIANGULARIZATION METHOD

Cholesky's decomposition method is faster than the LU decomposition. There is no need for pivoting. If the decomposition fails, the matrix is not positive definite.

Consider the system of linear equations:

$$a_{11}x_1 + a_{12}x_2 + a_{13}x_3 = b_1$$

$$a_{21}x_1 + a_{22}x_2 + a_{23}x_3 = b_2$$

$$a_{31}x_1 + a_{32}x_2 + a_{33}x_3 = b_3 \qquad (2.21)$$

The above system can be written as

$$\mathbf{A}x = b \qquad (2.22)$$

where $\quad \mathbf{A} = \begin{bmatrix} a_{11} & a_{12} & a_{13} \\ a_{21} & a_{22} & a_{23} \\ a_{31} & a_{32} & a_{33} \end{bmatrix}, \ x = \begin{bmatrix} x_1 \\ x_2 \\ x_3 \end{bmatrix}, \ b = \begin{bmatrix} b_1 \\ b_2 \\ b_3 \end{bmatrix}$

Let $\quad \mathbf{A} = LU\ldots \qquad (2.23)$

where $\quad L = \begin{bmatrix} 1 & 0 & 0 \\ l_{21} & 1 & 0 \\ l_{31} & l_{32} & 1 \end{bmatrix}$ and $U = \begin{bmatrix} u_{11} & u_{12} & u_{13} \\ 0 & u_{22} & u_{23} \\ 0 & 0 & u_{33} \end{bmatrix}$

Equation (2.21) can be written as

$$LUX = b \qquad (2.24)$$

If we write $\qquad\qquad\qquad UX = V \qquad (2.25)$

Equation (2.24) becomes

$$LV = b \qquad (2.26)$$

Equation (2.26) is equivalent to the system

$$v_1 = b_1$$

$$l_{21}v_1 + v_2 = b_2$$

$$l_{31}v_1 + l_{32}v_2 + v_3 = b_3 \qquad (2.27)$$

The above system can be solved to find the values of v_1, v_2, and v_3 which give us the matrix V.

$$UX = V$$

then becomes

$$u_{11}x_1 + u_{12}x_2 + u_{13}x_3 = v_1$$

$$u_{22}x_2 + u_{23}x_3 = v_2$$

$$u_{33}x_3 = v_3 \tag{2.28}$$

which can be solved for x_3, x_2, and x_1 by the backward substitution process.

In order to compute the matrices L and U, we write Eq. (2.23) as

$$\begin{bmatrix} 1 & 0 & 0 \\ l_{21} & 1 & 0 \\ l_{31} & l_{32} & 1 \end{bmatrix} \begin{bmatrix} u_{11} & u_{12} & u_{13} \\ 0 & u_{22} & u_{23} \\ 0 & 0 & u_{33} \end{bmatrix} = \begin{bmatrix} a_{11} & a_{12} & a_{13} \\ a_{21} & a_{22} & a_{23} \\ a_{31} & a_{32} & a_{33} \end{bmatrix} \tag{2.29}$$

Multiplying the matrices on the left and equating the corresponding elements of both sides, we obtain

$$u_{11} = a_{11}, u_{12} = a_{12}, u_{13} = a_{13} \tag{2.30}$$

$$\left. \begin{aligned} l_{21}u_{11} = a_{21} \Rightarrow l_{21} = \frac{a_{21}}{a_{11}} \\ l_{31}u_{11} = a_{31} \Rightarrow l_{31} = \frac{a_{31}}{a_{11}} \end{aligned} \right\} \tag{2.31}$$

$$\left. \begin{aligned} l_{21}u_{12} + u_{22} = a_{22} \Rightarrow u_{22} = a_{22} - \frac{a_{21}}{a_{11}}a_{12} \\ l_{21}u_{13} + u_{23} = a_{23} \Rightarrow u_{23} = a_{23} - \frac{a_{21}}{a_{11}}a_{13} \end{aligned} \right\} \tag{2.32}$$

$$l_{31}u_{12} + l_{32}u_{22} = a_{32} \Rightarrow l_{32} = \frac{1}{u_{22}}\left[a_{32} - \frac{a_{31}}{a_{11}}a_{12} \right] \tag{2.33}$$

and $\quad l_{31}u_{13} + l_{32}u_{23} + u_{33} = a_{33} \tag{2.34}$

The value of u_{33} can be computed from Eq. (2.34).

To obtain the elements of L and U, we first find the first row of U and the first column of L. Then, determine the second row of U and the second column of L. Finally, we compute the third row of U.

Cholesky's triangularization method is also known as Crout's triangularization method or method of factorization.

EXAMPLE 2.11

Solve the following equations by Cholesky's triangularization method.

$$2x + y + 4z = 12$$

$$8x - 3y + 2z = 20$$

$$4x + 11y - z = 33$$

Solution:

We have
$$A = \begin{bmatrix} 2 & 1 & 4 \\ 8 & -3 & 2 \\ 4 & 11 & -1 \end{bmatrix}, \ X = \begin{bmatrix} x \\ y \\ z \end{bmatrix}, \ B = \begin{bmatrix} 12 \\ 20 \\ 33 \end{bmatrix}$$

Let
$$\begin{bmatrix} 1 & 0 & 0 \\ l_{21} & 1 & 0 \\ l_{31} & l_{32} & 1 \end{bmatrix} \begin{bmatrix} u_{11} & u_{12} & u_{13} \\ 0 & u_{22} & u_{23} \\ 0 & 0 & u_{33} \end{bmatrix} = \begin{bmatrix} 2 & 1 & 4 \\ 8 & -3 & 2 \\ 4 & 11 & -1 \end{bmatrix}$$

Multiplying and equating we get:

$$l \times u_{11} = 2 \Rightarrow \boxed{u_{11}} = 2$$

$$l \times u_{12} = 1 \Rightarrow \boxed{u_{12}} = 1$$

$$l \times u_{13} = 4 \Rightarrow \boxed{u_{13}} = 4$$

$$l_{21} \times u_{11} = 8 \Rightarrow \boxed{l_{21}} = \frac{8}{u_{11}} = \frac{8}{2} = 4$$

$$l_{21} \times u_{12} + u_{22} = -3 \Rightarrow \boxed{u_{22}} = -3 - l_{21} \times u_{12} = -3 - 4 \times 1 = -7$$

$$l_{21} \times u_{13} + u_{23} = 2 \Rightarrow \boxed{u_{23}} = 2 - l_{21} \times u_{13} = 2 - 4 \times 4 = -14$$

$$l_{31} \times u_{11} = 4 \Rightarrow \boxed{l_{31}} = \frac{4}{u_{11}} = \frac{4}{2} = 2$$

$$l_{31} \times u_{12} + l_{32} \times u_{22} = 11 \Rightarrow \boxed{l_{32}} = \frac{11 - l_{31} \times u_{12}}{u_{22}} = \frac{11 - 2 \times 1}{-7} = -\frac{9}{7}$$

$$l_{31} \times u_{13} + l_{32} \times u_{23} + l \times u_{33} = -1 \Rightarrow \boxed{u_{33}} = -1 - l_{31} \times u_{13} - l_{32} \times u_{23}$$

$$= -1 - 2 \times 4 - \left(-\frac{9}{7}(-14) \right) = -27$$

We get: $A = \begin{bmatrix} 1 & 0 & 0 \\ 4 & 1 & 0 \\ 2 & -\dfrac{9}{7} & 1 \end{bmatrix} \begin{bmatrix} 2 & 1 & 4 \\ 0 & -7 & -14 \\ 0 & 0 & -27 \end{bmatrix}$

and the given system can be written as:

$$\begin{bmatrix} 1 & 0 & 0 \\ 4 & 1 & 0 \\ 2 & -\dfrac{9}{7} & 1 \end{bmatrix} \begin{bmatrix} 2 & 1 & 4 \\ 0 & -7 & -14 \\ 0 & 0 & -27 \end{bmatrix} \begin{bmatrix} x \\ y \\ z \end{bmatrix} = \begin{bmatrix} 12 \\ 20 \\ 33 \end{bmatrix}$$

Writing: $LV = B$, we get

$$\begin{bmatrix} 1 & 0 & 0 \\ 4 & 1 & 0 \\ 2 & -\dfrac{9}{7} & 1 \end{bmatrix} \begin{bmatrix} V_1 \\ V_2 \\ V_3 \end{bmatrix} = \begin{bmatrix} 12 \\ 20 \\ 33 \end{bmatrix}$$

which gives

$$\boxed{V_1} = 12 \Rightarrow \boxed{V_2}$$

$$4V_1 + V_2 = 20 \Rightarrow \boxed{V_2} = 20 - 4 \times 12 = -28$$

$$2V_1 - \frac{9}{7}V_2 + V_3 = 33 \Rightarrow \boxed{V_3} = 33 + \frac{9}{7}(-28) - 2 \times 12 = -27$$

The solution to the original system is given by:

$$UX = V$$

$$\begin{bmatrix} 2 & 1 & 4 \\ 0 & -7 & -14 \\ 0 & 0 & -27 \end{bmatrix} \begin{bmatrix} x \\ y \\ z \end{bmatrix} = \begin{bmatrix} 12 \\ -28 \\ -27 \end{bmatrix}$$

$$2x + y + 4z = 12$$

$$-7y - 14z = -28$$

$$-27z = -27 \Rightarrow \boxed{z = 1}$$

$$7y = 28 - 14 \times 1 \Rightarrow y = \frac{14}{7} \Rightarrow \boxed{y = 2}$$

$$2x = 12 - y - 4z = 12 - 2 - 4 \times 1 \Rightarrow x = \frac{6}{2} \Rightarrow \boxed{x = 3}$$

EXAMPLE 2.12

Solve the system of equations using Cholesky's factorizations.

$$x_1 + x_2 + x_3 - x_4 = 2$$

$$x_1 - x_2 - x_3 + 2x_4 = 0$$

$$4x_1 + 4x_2 + x_3 + x_4 = 11$$

$$2x_1 + x_2 + 2x_3 - 2x_4 = 2$$

Solution:

The set of equations can be written in the matrix form $[A]\{x\} = \{b\}$

$$\begin{bmatrix} 1 & 1 & 1 & -1 \\ 1 & -1 & -1 & 2 \\ 4 & 4 & 1 & 1 \\ 2 & 1 & 2 & -2 \end{bmatrix} \begin{Bmatrix} x_1 \\ x_2 \\ x_3 \\ x_4 \end{Bmatrix} = \begin{Bmatrix} 2 \\ 0 \\ 11 \\ 2 \end{Bmatrix}$$

Let us decompose $[A]$ in the form

$$[A] = [L]\,[U]$$

where $\quad [L] = \begin{bmatrix} 1 & 0 & 0 & 0 \\ l_{21} & 1 & 0 & 0 \\ l_{31} & l_{32} & 1 & 0 \\ l_{41} & l_{42} & l_{43} & 1 \end{bmatrix}$ and $[U] = \begin{bmatrix} u_{11} & u_{12} & u_{13} & u_{14} \\ 0 & u_{22} & u_{23} & u_{24} \\ 0 & 0 & u_{33} & u_{34} \\ 0 & 0 & 0 & u_{44} \end{bmatrix}$

The product of $[L][U]$ gives

$$[L][U] = \begin{bmatrix} u_{11} & u_{12} & u_{13} & u_{14} \\ l_{21}u_{11} & l_{21}u_{12} + u_{22} & l_{21}u_{13} + u_{23} & l_{21}u_{14} + u_{24} \\ l_{31}u_{11} & l_{31}u_{12} + l_{32}u_{22} & l_{31}u_{13} + l_{32}u_{23} + u_{33} & l_{31}u_{14} + l_{32}u_{24} + u_{34} \\ l_{41}u_{11} & l_{41}u_{12} + l_{42}u_{22} & l_{41}u_{13} + l_{42}u_{23} + l_{43}u_{33} & l_{41}u_{14} + l_{42}u_{24} + l_{43}u_{34} + u_{44} \end{bmatrix}$$

Equating the elements of this matrix to the $[A]$ matrix yields the following equations

$$u_{11} = 1 \quad l_{21}u_{11} = 1 \qquad l_{31}u_{11} = 4 \qquad\qquad l_{41}u_{11} = 2$$

$$u_{12} = 1 \quad l_{21}u_{12} + u_{22} = -1 \quad l_{31}u_{12} + l_{32}u_{22} = 4 \qquad l_{41}u_{12} + l_{42}u_{22} = 1$$

$$u_{13} = 1 \quad l_{21}u_{13} + u_{23} = -1 \quad l_{31}u_{13} + l_{32}u_{23} + u_{33} = 1 \quad l_{41}u_{13} + l_{42}u_{23} + l_{23}u_{33} = 2$$

$$u_{14} = -1 \quad l_{21}u_{14} + u_{24} = 2 \quad l_{31}u_{14} + l_{32}u_{24} + u_{34} = 1 \quad l_{41}u_{14} + l_{42}u_{24} + l_{43}u_{34} + u_{44} = -2$$

By solving these sixteen equations we get

$$[L] = \begin{bmatrix} 1 & 0 & 0 & 0 \\ 1 & 1 & 0 & 0 \\ 4 & 0 & 1 & 0 \\ 2 & \frac{1}{2} & -\frac{1}{3} & 1 \end{bmatrix} \text{and } [U] = \begin{bmatrix} 1 & 1 & 1 & -1 \\ 0 & -2 & -2 & 3 \\ 0 & 0 & -3 & 5 \\ 0 & 0 & 0 & \frac{1}{6} \end{bmatrix}$$

To solve $[A]\{x\} = \{b\}$ we have to solve the two systems

$$[L]\{Y\} = \{b\}$$

$$[U]\{x\} = \{Y\}$$

i.e.,
$$\begin{bmatrix} 1 & 0 & 0 & 0 \\ 1 & 1 & 0 & 0 \\ 4 & 0 & 1 & 0 \\ 2 & \frac{1}{2} & \frac{1}{3} & 1 \end{bmatrix} \begin{Bmatrix} y_1 \\ y_2 \\ y_3 \\ y_4 \end{Bmatrix} = \begin{Bmatrix} 2 \\ 0 \\ 11 \\ 2 \end{Bmatrix}$$

which gives by forward substitution

$$y_1 = 2, y_2 = -2, y_3 = 3, y_4 = 0$$

and hence $[U]\{x\} = \{y\}$ becomes

$$\begin{bmatrix} 1 & 1 & 1 & -1 \\ 0 & -2 & -2 & 3 \\ 0 & 0 & -3 & 5 \\ 0 & 0 & 0 & \frac{1}{6} \end{bmatrix} \begin{Bmatrix} x_1 \\ x_2 \\ x_3 \\ x_4 \end{Bmatrix} = \begin{Bmatrix} 2 \\ -2 \\ 3 \\ 0 \end{Bmatrix}$$

Then by back substitution we obtain

$$x_4 = 0, x_3 = -1, x_2 = 2, x_1 = 1.$$

EXAMPLE 2.13

Solve the system of linear equations using Cholesky's factorization method.

$$2x - 6y + 8z = 24$$
$$5x + 4y - 3z = 2$$
$$3x + y + 2z = 16$$

Solution:

$$\begin{bmatrix} 1 & 0 & 0 \\ l_{21} & 1 & 0 \\ l_{31} & l_{32} & 1 \end{bmatrix} \begin{bmatrix} u_{11} & u_{12} & u_{13} \\ 0 & u_{22} & u_{23} \\ 0 & 0 & u_{33} \end{bmatrix} = \begin{bmatrix} 2 & -6 & 8 \\ 5 & 4 & -3 \\ 3 & 1 & 2 \end{bmatrix}$$

$$\begin{bmatrix} u_{11} & u_{12} & u_{13} \\ l_{21}u_{11} & l_{21}u_{12}+u_{22} & l_{21}u_{13}+u_{23} \\ l_{31}u_{11} & l_{31}u_{12}+l_{32}u_{22} & l_{31}u_{13}+l_{32}u_{23}+u_{33} \end{bmatrix} = \begin{bmatrix} 2 & -6 & 8 \\ 5 & 4 & -3 \\ 3 & 1 & 2 \end{bmatrix}$$

$$u_{11} = 2, \qquad u_{12} = -6, \qquad u_{13} = 8$$

$$l_{21} = \frac{5}{u_{11}} = 2.5$$

$$l_{31} = \frac{3}{u_{11}} = 1.5$$

$$u_{22} = 4 - l_{21}u_{12} = 19$$

$$u_{23} = -3 - l_{21}u_{13} = -23$$

$$l_{32} = \frac{1 - l_{31}u_{12}}{u_{22}} = \frac{10}{19}$$

$$l_{33} = 2 - l_{31}u_{13} - l_{32}u_{23} = \frac{40}{19}$$

$$L = \begin{bmatrix} 1 & 0 & 0 \\ 2.5 & 1 & 0 \\ 1.5 & \frac{10}{19} & 1 \end{bmatrix}, \ U = \begin{bmatrix} 2 & -6 & 8 \\ 0 & 19 & -23 \\ 0 & 0 & \frac{40}{19} \end{bmatrix}$$

$$LV = B \Rightarrow \begin{bmatrix} 1 & 0 & 0 \\ 2.5 & 1 & 0 \\ 1.5 & \frac{10}{19} & 1 \end{bmatrix} \begin{bmatrix} v_1 \\ v_2 \\ v_3 \end{bmatrix} = \begin{bmatrix} 24 \\ 2 \\ 16 \end{bmatrix}$$

$\Rightarrow \qquad v_1 = 24$

$v_2 = 2 - 2.5 \times 24 = -58$

$v_3 = 16 - 1.5 \times 24 - \dfrac{10}{19}(-58) = \dfrac{200}{19}$

$$UX = V \Rightarrow \begin{bmatrix} 2 & -6 & 8 \\ 0 & 19 & -23 \\ 0 & 0 & \frac{40}{19} \end{bmatrix} \begin{bmatrix} x \\ y \\ z \end{bmatrix} = \begin{bmatrix} 24 \\ -58 \\ \dfrac{200}{19} \end{bmatrix}$$

$2x - 6y + 8z = 24$ \hfill (E.1)

$19y - 23z = -58$ \hfill (E.2)

$\dfrac{40}{19}z = \dfrac{200}{19} \Rightarrow \boxed{z = 5}$ \hfill (E.3)

From Eqs. (E.2), and (E.3), we have

$$\boxed{y = 3}$$

From Eqs. (E.1), (E.2), and (E.3), we get

$$\boxed{x = 1}$$

2.8 CROUT'S METHOD

This method is based on the fact that every square matrix A can be expressed as the product of a lower triangular matrix and an upper triangular matrix, provided all the principle minors of A are nonsingular. Also, such a factorization, if exists, is unique.

This method is also called *triangularization* or *factorization method*. Here, we factorize the given matrix as $A = LU$, where L is a lower triangular matrix with unit diagonal elements and U is an upper triangular matrix. Then,

$$A^{-1} = (LU)^{-1} = U^{-1}L^{-1}$$

Consider the system

$$a_{11}x_1 + a_{12}x_2 + a_{13}x_3 = b_1$$

$$a_{21}x_1 + a_{22}x_2 + a_{23}x_3 = b_2$$

$$a_{31}x_1 + a_{32}x_2 + a_{33}x_3 = b_3 \qquad (2.35)$$

The above system can be written as

$$Ax = b$$

Let

$$A = LU \tag{2.36}$$

where $L = \begin{bmatrix} l_{11} & 0 & 0 \\ l_{21} & l_{22} & 0 \\ l_{31} & l_{32} & l_{33} \end{bmatrix}$ and $U = \begin{bmatrix} 1 & u_{12} & u_{13} \\ 0 & 1 & u_{23} \\ 0 & 0 & 1 \end{bmatrix}$ \tag{2.37}

Here, L is a lower triangular matrix and U is an upper triangular matrix with diagonal elements equal to unity.

$$A = LU \Rightarrow A^{-1} = U^{-1}L^{-1} \tag{2.38}$$

Now $A = LU \Rightarrow \begin{bmatrix} a_{11} & a_{12} & a_{13} \\ a_{21} & a_{22} & a_{23} \\ a_{31} & a_{32} & a_{33} \end{bmatrix} = \begin{bmatrix} l_{11} & 0 & 0 \\ l_{21} & l_{22} & 0 \\ l_{31} & l_{32} & l_{33} \end{bmatrix} \begin{bmatrix} 1 & u_{12} & u_{13} \\ 0 & 1 & u_{23} \\ 0 & 0 & 1 \end{bmatrix}$

or $\begin{bmatrix} a_{11} & a_{12} & a_{13} \\ a_{21} & a_{22} & a_{23} \\ a_{31} & a_{32} & a_{33} \end{bmatrix} = \begin{bmatrix} l_{11} & l_{11}u_{12} & l_{11}u_{13} \\ l_{21} & l_{21}u_{12} + l_{22} & l_{21}u_{13} + l_{22}u_{23} \\ l_{31} & l_{31}u_{12} + l_{32} & l_{31}u_{13} + l_{32}u_{23} + l_{33} \end{bmatrix}$

Equating the corresponding elements, we obtain

$$l_{11} = a_{11} \qquad\qquad l_{21} = a_{21} \qquad\qquad l_{31} = a_{31} \tag{2.39}$$

$$l_{11}u_{12} = a_{12} \qquad\qquad l_{11}u_{13} = a_{13} \tag{2.40}$$

$$l_{21}u_{12} + l_{22} = a_{22} \qquad l_{31}u_{12} + l_{32} = a_{32} \tag{2.41}$$

$$l_{21}u_{13} + l_{22}u_{23} = a_{23} \tag{2.42}$$

and $\quad l_{31}u_{13} + l_{32}u_{23} + l_{33} = a_{33}$ \tag{2.43}

from (2.40) we find

$$u_{12} = a_{12}/l_{11} = a_{12}/a_{11}$$

from (2.41) we obtain

$$l_{22} = a_{22} - l_{21}u_{12} \tag{2.44}$$

$$l_{32} = a_{32} - l_{31}u_{12} \tag{2.45}$$

Equation (2.42) gives

$$u_{23} = (a_{23} - l_{21}u_{23})/l_{22} \qquad (2.46)$$

from the relation (2.43) we get

$$l_{33} = a_{33} - l_{31}u_{13} - l_{32}u_{23} \qquad (2.47)$$

Thus, we have determined all the elements of L and U.

From Eqs. (2.36) and (2.37) we have

$$LUx = b \qquad (2.48)$$

Let

$$UX = V$$

where

$$V = \begin{bmatrix} v_1 \\ v_2 \\ \vdots \\ v_n \end{bmatrix}$$

From Eq. (2.48) we have $LV = b$, which on forward substitution yields V.
From $UX = V$, we find x (by backward substitution).

EXAMPLE 2.14

Solve the following set of equations by Crout's method:

$$2x + y + 4z = 12$$
$$8x - 3y + 2z = 20$$
$$4x + 11y - z = 33$$

Solution:

We have $\mathbf{A} = \begin{bmatrix} 2 & 1 & 4 \\ 8 & -3 & 2 \\ 4 & 11 & -1 \end{bmatrix}$, $X = \begin{bmatrix} x \\ y \\ z \end{bmatrix}$, $B = \begin{bmatrix} 12 \\ 20 \\ 33 \end{bmatrix}$

$$\mathbf{AX} = B$$

Let

$$\mathbf{A} = LU$$

$$L = \begin{bmatrix} l_{11} & 0 & 0 \\ l_{21} & l_{22} & 0 \\ l_{31} & l_{32} & l_{33} \end{bmatrix} \qquad U = \begin{bmatrix} 1 & u_{12} & u_{13} \\ 0 & 1 & u_{23} \\ 0 & 0 & 1 \end{bmatrix}$$

$$\begin{bmatrix} 2 & 1 & 4 \\ 8 & -3 & 2 \\ 4 & 11 & -1 \end{bmatrix} = \begin{bmatrix} l_{11} & 0 & 0 \\ l_{21} & l_{22} & 0 \\ l_{31} & l_{32} & l_{33} \end{bmatrix} \begin{bmatrix} 1 & u_{12} & u_{13} \\ 0 & 1 & u_{23} \\ 0 & 0 & 1 \end{bmatrix}$$

$$\begin{bmatrix} 2 & 1 & 4 \\ 8 & -3 & 2 \\ 4 & 11 & -1 \end{bmatrix} = \begin{bmatrix} l_{11} & l_{11}u_{12} & l_{11}u_{13} \\ l_{21} & l_{21}u_{12} + l_{22} & l_{21}u_{13} + l_{22}u_{23} \\ l_{31} & l_{31}u_{12} + l_{32} & l_{31}u_{13} + l_{32}u_{23} + l_{33} \end{bmatrix}$$

$$l_{11}u_{12} = 1 \Rightarrow \boxed{u_{12}} = \frac{1}{2}$$

$$l_{11}u_{12} = 4 \Rightarrow \boxed{u_{13}} = \frac{4}{2} = 2$$

$$l_{22} + l_{21}u_{12} = -3 \Rightarrow \boxed{l_{22}} = -3 - 8\left(\frac{1}{2}\right) = -7$$

$$l_{32} + l_{21}u_{12} = -3 \Rightarrow \boxed{l_{32}} = 11 - 4\left(\frac{1}{2}\right) = 9$$

$$l_{21}u_{13} + l_{22}u_{23} = 2 \Rightarrow \boxed{u_{23}} = \frac{2 - 8 \times 2}{-7} = 2$$

$$l_{31}u_{13} + l_{32}u_{23} + l_{33} = -1 \Rightarrow \boxed{l_{33}} = -1 - 4 \times 2 - 9 \times 2 = -27$$

$$L = \begin{bmatrix} 2 & 0 & 0 \\ 8 & -7 & 0 \\ 4 & 9 & -27 \end{bmatrix} \text{ and } U = \begin{bmatrix} 1 & \frac{1}{2} & 2 \\ 0 & 1 & 2 \\ 0 & 0 & 1 \end{bmatrix}$$

$$LV = B$$

$$\begin{bmatrix} 2 & 0 & 0 \\ 8 & -7 & 0 \\ 4 & 9 & -27 \end{bmatrix} \begin{bmatrix} v_1 \\ v_2 \\ v_3 \end{bmatrix} = \begin{bmatrix} 12 \\ 20 \\ 33 \end{bmatrix}$$

$$2v_1 = 12 \Rightarrow \boxed{v_1} = 6$$

$$8v_1 - 7v_2 = 20 \Rightarrow \boxed{v_2} = \frac{-20 + 8 \times 6}{7} = 4$$

$$4v_1 + 9v_2 - 27v_3 = 33 \Rightarrow \boxed{v_3} = \frac{-33 + 4 \times 6 + 9 \times 4}{27} = 1$$

$$V = \begin{bmatrix} V_1 \\ V_2 \\ V_3 \end{bmatrix} = \begin{bmatrix} 6 \\ 4 \\ 1 \end{bmatrix}; \qquad Ux = V$$

$$\begin{bmatrix} 1 & \frac{1}{2} & 2 \\ 0 & 1 & 2 \\ 0 & 0 & 1 \end{bmatrix} \begin{bmatrix} x \\ y \\ z \end{bmatrix} = \begin{bmatrix} 6 \\ 4 \\ 1 \end{bmatrix}$$

$$x + \frac{1}{2}y + 2z = 6$$

$$y + 2z = 4$$

$$\boxed{z = 1}$$

$$y = 4 - 2x1 \Rightarrow \boxed{y = 2}$$

$$x = 6 - \frac{1}{2} \times 2 - 2 \times 1 \Rightarrow \boxed{x = 3}$$

EXAMPLE 2.15

Solve the following set of equations by using Crout's method:

$$2x_1 + x_2 + x_3 = 7$$

$$x_1 + 2x_2 + x_3 = 8$$

$$x_1 + x_2 + 2x_3 = 9$$

Solution:

$$\mathbf{A} = \begin{bmatrix} 2 & 1 & 1 \\ 1 & 2 & 1 \\ 1 & 1 & 2 \end{bmatrix}, \ x = \begin{bmatrix} x \\ y \\ z \end{bmatrix}, \ \mathbf{B} = \begin{bmatrix} 7 \\ 8 \\ 9 \end{bmatrix}$$

Let $\qquad \mathbf{A} = LU$

$$L = \begin{bmatrix} l_{11} & 0 & 0 \\ l_{21} & l_{22} & 0 \\ l_{31} & l_{32} & l_{33} \end{bmatrix} \qquad U = \begin{bmatrix} 1 & u_{12} & u_{13} \\ 0 & 1 & u_{23} \\ 0 & 0 & 1 \end{bmatrix}$$

$$\begin{bmatrix} 2 & 1 & 1 \\ 1 & 2 & 1 \\ 1 & 1 & 2 \end{bmatrix} = \begin{bmatrix} l_{11} & l_{11}u_{12} & l_{11}u_{13} \\ l_{21} & l_{21}u_{12} + l_{22} & l_{21}u_{13} + l_{22}u_{23} \\ l_{31} & l_{31}u_{12} + l_{32} & l_{31}u_{13} + l_{32}u_{23} + l_{33} \end{bmatrix}$$

$$l_{11} = 2, \qquad l_{21} = 1, \qquad l_{31} = 1$$

$$u_{12} = \frac{1}{2}, \qquad u_{13} = \frac{1}{2}$$

$$l_{22} = 2 - l_{21}u_{12} = 2 - 1 \times \frac{1}{2} = \frac{3}{2}$$

$$l_{32} = 1 - l_{31}u_{12} = 1 - 1 \times \frac{1}{2} = \frac{1}{2}$$

$$u_{23} = \frac{1 - l_{21}u_{13}}{l_{22}} = \frac{1}{3}$$

$$l_{33} = 2 - l_{31}u_{13} - l_{32}u_{23} = 2 - \frac{1}{2} - \frac{1}{2} \times \frac{1}{3} = \frac{4}{3}$$

$$L = \begin{bmatrix} 2 & 0 & 0 \\ 1 & 3/2 & 0 \\ 1 & 1/2 & 4/3 \end{bmatrix}, U = \begin{bmatrix} 1 & 1/2 & 1/2 \\ 0 & 1 & 1/3 \\ 0 & 0 & 1 \end{bmatrix}$$

$$\mathbf{A}x = B, \qquad LU.x = B, \qquad Ux = V$$

$$LV = B \Rightarrow \begin{bmatrix} 2 & 0 & 0 \\ 1 & 3/2 & 0 \\ 1 & 1/2 & 4/3 \end{bmatrix} \begin{bmatrix} v_1 \\ v_2 \\ v_3 \end{bmatrix} = \begin{bmatrix} 7 \\ 8 \\ 9 \end{bmatrix}$$

$$\left. \begin{array}{l} 2v_1 = 7 \Rightarrow v_1 = 3.5 \\[2mm] v_1 + \dfrac{3}{2}v_2 = 8 \Rightarrow v_2 = 3 \\[2mm] v_1 + \dfrac{1}{2}v_2 + \dfrac{4}{3}v_3 = 9 \Rightarrow v_3 = 3 \end{array} \right\} \Rightarrow V = \begin{bmatrix} 3.5 \\ 3 \\ 3 \end{bmatrix}$$

$$Ux = V \Rightarrow \begin{bmatrix} 1 & 1/2 & 1/2 \\ 0 & 1 & 1/3 \\ 0 & 0 & 1 \end{bmatrix} \begin{bmatrix} x_1 \\ x_2 \\ x_3 \end{bmatrix} = \begin{bmatrix} 3.5 \\ 3 \\ 3 \end{bmatrix}$$

$$x_1 + \frac{1}{2}x_2 + \frac{1}{2}x_3 = 3.5 \tag{E.1}$$

$$x_2 + \frac{1}{3}x_3 = 3 \tag{E.2}$$

$$\boxed{x_3 = 3} \tag{E.3}$$

From Eqs. (E.2) and (E.3), we have

$$\boxed{x_2 = 2}$$

From Eq. (E.1), we get

$$\boxed{x_1 = 1}$$

2.9   THOMAS ALGORITHM FOR TRIDIAGONAL SYSTEM

Consider the system of linear simultaneous algebraic equations given by

$$\mathbf{A}x = b$$

where $\mathbf{A}$ is a tridiagonal matrix, $x = [x_1, x_2, ..., x_n]^T$ and $b = [b_1, b_2, ..., b_n]^T$. Hence, we consider a 4×4 tridiagonal system of equations given by

$$\begin{bmatrix} a_{12} & a_{13} & 0 & 0 \\ a_{21} & a_{22} & a_{23} & 0 \\ 0 & a_{31} & a_{32} & a_{33} \\ 0 & 0 & a_{41} & a_{42} \end{bmatrix} \begin{bmatrix} x_1 \\ x_2 \\ x_3 \\ x_4 \end{bmatrix} = \begin{bmatrix} b_1 \\ b_2 \\ b_3 \\ b_4 \end{bmatrix} \tag{2.48a}$$

Equation (2.48a) can be written as

$$a_{12}x_1 + a_{13}x_2 = b_1$$

$$a_{21}x_1 + a_{22}x_2 + a_{23}x_3 = b_2$$

$$a_{31}x_2 + a_{32}x_3 + a_{33}x_4 = b_3$$

$$a_{41}x_3 + a_{42}x_4 = b_4 \tag{2.48b}$$

The system of equations given by Eq. (2.48b) is solved using the Thomas algorithm which is described in three steps as shown below:

Step 1: Set $y_1 = a_{12}$ and compute

$$y_i = a_{i2} - \frac{a_{i1}a_{(i-1)3}}{y_{i-1}} \qquad\qquad i = 2, 3, \dots, n$$

Step 2: Set $z_1 = \dfrac{b_1}{a_{12}}$ and compute

$$z_i = \frac{b_i - a_{i1}z_{i-1}}{y_i} \qquad\qquad i = 2, 3, \dots, n$$

Step 3: $x_i = z_i - \dfrac{a_{i3}x_{i+1}}{y_i} \qquad\qquad i = n-1, n-2, \dots, 1$, where $x_n = z_n$

EXAMPLE 2.16

Solve the following equations by using the Thomas algorithm.

$$3x_1 - x_2 = 5$$
$$2x_1 - 3x_2 + 2x_3 = 5$$
$$x_2 + 2x_3 + 5x_4 = 10$$
$$x_3 - x_4 = 1$$

Solution:

Here
$$\begin{bmatrix} 3 & -1 & 0 & 0 \\ 2 & -3 & 2 & 0 \\ 0 & 1 & 2 & 5 \\ 0 & 0 & 1 & -1 \end{bmatrix}\begin{bmatrix} x_1 \\ x_2 \\ x_3 \\ x_4 \end{bmatrix} = \begin{bmatrix} 5 \\ 5 \\ 10 \\ 1 \end{bmatrix}$$

$$[a_2, a_3, a_4] = [2, 1, 1]$$
$$[b_1, b_2, b_3, b_4] = [3, -3, 2, -1]$$
$$[c_1, c_2, c_3] = [-1, 2, 5]$$

Step 1: Set $y_1 = b_1$ and compute

$$y_i = b_i - \frac{a_i c_{i-1}}{y_{i-1}}, \qquad\qquad i = 2, 3, \dots, n$$

$$y_1 = 3$$

$$i = 2, \boxed{y_2} = b_2 - \frac{a_2 c_1}{y_1} = -3 - \frac{2(-1)}{3} = -\frac{7}{3}$$

$$i = 3, \boxed{y_3} = b_3 - \frac{a_3 c_2}{y_2} = 2 - \frac{1 \times 2}{-\dfrac{7}{3}} = \frac{20}{7}$$

$$i = 4, \boxed{y_4} = b_4 - \frac{a_4 c_3}{y_3} = -1 - \frac{1.5}{\dfrac{20}{7}} = -\frac{55}{20}$$

Step 2: Set $z_1 = \dfrac{d_1}{b_1} = \dfrac{5}{3}$, $z_i = \dfrac{d_i - a_i z_{i-1}}{y_i}$ $\qquad i = 2, 3, \dots, n$

$$i = 2, \boxed{z_2} = \frac{d_2 - a_2 z_1}{y_2} = \frac{5 - 2 \times \dfrac{5}{3}}{-\dfrac{7}{3}} = -\frac{5}{7}$$

$$i = 3, \boxed{z_3} = \frac{d_3 - a_3 z_2}{y_3} = \frac{10 - 1\left(-\dfrac{5}{7}\right)}{\dfrac{20}{7}} = \frac{75}{20}$$

$$i = 4, \boxed{z_4} = \frac{d_4 - a_4 z_3}{y_4} = \frac{1 - 1 \times \dfrac{75}{20}}{-\dfrac{55}{20}} = 1$$

Step 3: Set $x_i = z_i - \dfrac{c_i x_{i+1}}{y_i}$, $\qquad i = n-1, n-2, \dots, 1;\quad x_n = z_n$

$$\boxed{x_4} = z_4 = \boxed{1}$$

$$i = 3, \boxed{x_3} = z_3 - \frac{c_3 x_4}{y_3} = \frac{75}{20} - \frac{5 \times 1}{\dfrac{20}{7}} = \boxed{2}$$

$$i = 2, \boxed{x_2} = z_2 - \frac{c_2 x_3}{y_2} = -\frac{5}{7} - \frac{2 \times 2}{-\frac{7}{3}} = \boxed{1}$$

$$i = 1, \boxed{x_1} = z_1 - \frac{c_1 x_2}{y_1} = \frac{5}{3} - \frac{(-1) \times 1}{3} = \boxed{2}$$

EXAMPLE 2.17

Solve the following set of tridiagonal set of algebraic equations using the Thomas method.

$$x_1 + 4x_2 = 10$$
$$2x_1 + 10x_2 - 4x_3 = 7$$
$$x_2 + 8x_3 - x_4 = 6$$
$$x_3 - 6x_4 = 4$$

Solution:

$$b_1 x_1 + c_1 x_2 = d_1$$
$$a_2 x_1 + b_2 x_2 + c_2 x_3 = d_2$$
$$a_3 x_2 + b_3 x_3 + c_3 x_4 = d_3$$
$$a_4 x_3 + b_4 x_4 = d_4$$

$$a_2 = 2, a_3 = 1, a_4 = 1$$
$$b_1 = 1, b_2 = 10, b_3 = 8, b_4 = -6$$
$$c_1 = 4, c_2 = -4, c_3 = -1$$
$$d_1 = 10, d_2 = 7, d_3 = 6, d_4 = 4$$

Step 1: Set $y_1 = b_1$,

$$y_i = b_i - \frac{a_i c_{i-1}}{y_{i-1}}, \quad i = 2, 3, \ldots, n$$

$$y_1 = 1$$

$$y_2 = 10 - \frac{2 \times 4}{1} = 2$$

$$y_3 = \frac{8 - 1(-4)}{2} = 8 + 2 = 10; \qquad y_4 = -6 - \frac{1 \times (-1)}{10} \Rightarrow$$

$$y_4 = \frac{-60 + 1}{10} = -\frac{59}{10}$$

Step 2: Set $z_1 = \dfrac{d_1}{b_1}$, $z_i = \dfrac{d_i - a_i z_{i-1}}{y_i}$ $\qquad i = 2, 3, \ldots, n$

$$z_1 = \frac{10}{1} = 10$$

$$z_2 = \frac{7 - 2.10}{2} = -\frac{13}{2}$$

$$z_3 = \frac{6 - 1(-13/2)}{10} = \frac{6 + 13/2}{10} = \frac{25}{20}$$

$$z_4 = \frac{4 - 1 \times 25/20}{-59/10} = -\frac{55}{118}$$

Step 3: Set $x_{11} = z_{11}$, $x_i = z_i - \dfrac{c_i x_{i+1}}{y_i}$, $\qquad i = n - 1, n - 2, \ldots, 1$

$$\boxed{x_4} = -\frac{55}{118} = -0.4661$$

$$\boxed{x_3} = \frac{25}{20} - \frac{(-1)(-55/118)}{10} = 1.2034$$

$$\boxed{x_2} = -\frac{13}{2} - \frac{(-4)1.203}{2} = -4.0932$$

$$\boxed{x_1} = 10 - \frac{4(-4.094)}{1} = 26.3729$$

2.10 JACOBI'S ITERATION METHOD

This method is also known as *the method of simultaneous displacements.* Consider the system of linear equations

$$a_{11}x_1 + a_{12}x_2 + a_{13}x_3 = b_1$$

$$a_{21}x_1 + a_{22}x_2 + a_{23}x_3 = b_2$$

$$a_{31}x_1 + a_{32}x_2 + a_{33}x_3 = b_3 \qquad (2.49)$$

Here, we assume that the coefficients a_{11}, a_{22}, and a_{33} are the largest coefficients in the respective equations so that

$$|a_{11}| > |a_{12}| + |a_{13}|$$
$$|a_{22}| > |a_{21}| + |a_{23}|$$
$$|a_{33}| > |a_{31}| + |a_{32}| \qquad (2.50)$$

Jacobi's iteration method is applicable only if the conditions given in Eq. (2.50) are satisfied.

Now, we can write Eq. (2.49)

$$x_1 = \frac{1}{a_{11}}(b_1 - a_{12}x_2 - a_{13}x_3)$$

$$x_2 = \frac{1}{a_{22}}(b_2 - a_{21}x_1 - a_{23}x_3) \qquad (2.51)$$

$$x_3 = \frac{1}{a_{33}}(b_3 - a_{31}x_1 - a_{32}x_2)$$

Let the initial approximations be x_1^0, x_2^0, and x_3^0 respectively. The following iterations are then carried out.

Iteration 1: The first improvements are found as

$$x_{11} = \frac{1}{a_{11}}\left(b_1 - a_{12}x_2^0 - a_{13}x_3^0\right)$$

$$x_{21} = \frac{1}{a_{22}}\left(b_2 - a_{21}x_1^0 - a_{23}x_3^0\right)$$

$$x_{31} = \frac{1}{a_{33}}\left(b_3 - a_{31}x_1^0 - a_{32}x_2^0\right) \qquad (2.52)$$

Iteration 2: The second improvements are obtained as

$$x_{12} = \frac{1}{a_{11}}(b_1 - a_{12}x_{21} - a_{13}x_{31})$$

$$x_{22} = \frac{1}{a_{22}}(b_2 - a_{21}x_{11} - a_{23}x_{31})$$

$$x_{32} = \frac{1}{a_{33}}(b_3 - a_{31}x_{11} - a_{32}x_{21}) \qquad (2.53)$$

The above iteration process is continued until the values of x_1, x_2, and x_3 are found to a preassigned degree of accuracy. That is, the procedure is continued until the relative error between two consecutive vector norm is satisfactorily small. In Jacobi method, it is a general practice to assume $x_1^0 = x_2^0 = x_3^0 = 0$. The method can be extended to a system of n linear simultaneous equations in n unknowns.

EXAMPLE 2.18

Solve the following equations by using Jacobi method.

$$15x + 3y - 2z = 85$$

$$2x + 10y + z = 51$$

$$x - 2y + 8z = 5$$

Solution:

In the above equations:

$$|15| > |3| + |-2|$$

$$|10| > |2| + |1|$$

$$|8| > |1| + |-2|$$

then Jacobi method is applicable. We rewrite the given equations as follows:

$$x = \frac{1}{a_1}(d_1 - b_1 y - c_1 z) = \frac{1}{15}(85 - 3y + 2z)$$

$$y = \frac{1}{b_2}(d_2 - a_2 x - c_2 z) = \frac{1}{10}(51 - 2x - z)$$

$$z = \frac{1}{c_3}(d_3 - a_3 x - b_3 y) = \frac{1}{8}(5 - x + 2y)$$

Let the initial approximations be:

$$x^0 = y^0 = z^0 = 0$$

Iteration 1:

$$\boxed{x_1} = \frac{d_1}{a_1} = \frac{85}{15} = \frac{17}{3}$$

$$\boxed{y_1} = \frac{d_2}{b_2} = \frac{51}{10}$$

$$\boxed{z_1} = \frac{d_3}{c_3} = \frac{5}{8}$$

Iteration 2:

$$x_2 = \frac{1}{a_1}(d_1 - b_1 y_1 - c_1 z_1) = \frac{1}{15}\left(85 - 3 \times \frac{51}{10} - (-2) \times \frac{5}{8}\right)$$

$$\boxed{x_2} = 4.73$$

$$y_2 = \frac{1}{b_2}(d_2 - a_2 x_1 - c_2 z_1) = \frac{1}{10}\left(51 - 2 \times \frac{17}{3} - 1 \times \frac{5}{8}\right)$$

$$\boxed{y_2} = 3.904$$

$$z_2 = \frac{1}{c_3}(d_3 - a_3 x_1 - b_3 y_1) = \frac{1}{8}\left(5 - 1 \times \frac{17}{3} - (-2) \times \frac{51}{10}\right)$$

$$\boxed{z_2} = 1.192$$

Iteration 3:

$$\boxed{x_3} = \frac{1}{15}(85 - 3 \times 3.904 + 2 \times 1.192) = 5.045$$

$$\boxed{y_3} = \frac{1}{10}(51 - 2 \times 4.73 - 1 \times 1.192) = 4.035$$

$$\boxed{z_3} = \frac{1}{8}(5 - 1 \times 4.173 + 2 \times 3.904) = 1.010$$

Iteration 4:

$$\boxed{x_4} = \frac{1}{15}(85 - 3 \times 4.035 + 2 \times 1.010) = 4.994$$

$$\boxed{y_4} = \frac{1}{10}(51 - 2 \times 5.045 - 1 \times 1.010) = 3.99$$

$$\boxed{z_4} = \frac{1}{8}(5 - 1 \times 5.045 + 2 \times 4.035) = 1.003$$

Iteration 5:

$$\boxed{x_5} = \frac{1}{15}(85 - 3 \times 3.99 + 2 \times 1.003) = 5.002$$

$$\boxed{y_5} = \frac{1}{10}(51 - 2 \times 4.994 - 1 \times 1.003) = 4.001$$

$$\boxed{z_5} = \frac{1}{8}(5 - 14.994 + 2 \times 3.99) = 0.998$$

Iteration 6:

$$\boxed{x_6} = \frac{1}{15}(85 - 3 \times 4.001 + 2 \times 0.998) = 5.0$$

$$\boxed{y_6} = \frac{1}{10}(51 - 2 \times 5.002 - 1 \times 0.998) = 4.0$$

$$\boxed{z_6} = \frac{1}{8}(5 - 1 \times 5.002 + 2 \times 4.001) = 1.0$$

Iteration 7:

$$\boxed{x_7} = \frac{1}{15}(85 - 3 \times 4 + 2 \times 1) = 5.0$$

$$\boxed{y_7} = \frac{1}{10}(51 - 2 \times 5 - 1 \times 1) = 4.0$$

$$\boxed{z_7} = \frac{1}{8}(5 - 1 \times 5 + 2 \times 4) = 1.0$$

EXAMPLE 2.19

Use the Jacobi iterative scheme to obtain the solutions of the system of equations correct to three decimal places.

$$x + 2y + z = 0$$

$$3x + y - z = 0$$

$$x - y + 4z = 3$$

Solution:

Rearrange the equations in such a way that all the diagonal terms are dominant.

$$3x + y - z = 0$$
$$x + 2y + z = 0$$
$$x - y + 4z = 3$$

Computing for x, y, and z we get

$$x = (z - y)/3$$
$$y = (-x - z)/2$$
$$z = (3 + y - x)/4$$

The iterative equation can be written as

$$x^{(r+1)} = (z^{(r)} - y^{(r)})/3$$
$$y^{(r+1)} = (-x^{(r)} - z^{(r)})/2$$
$$z^{(r+1)} = (3 - x^{(r)} + y^{(r)})/4$$

The initial vector is not specified in the problem. Hence, we choose

$$x^{(0)} = y^{(0)} = z^{(0)} = 1$$

Then, the first iteration gives

$$x^{(1)} = (z^{(0)} - y^{(0)})/3 = (1 - 1)/3 = 0$$
$$y^{(1)} = (-x^{(0)} - z^{(0)})/2 = (-1 - 1)/2 = -1.0$$
$$z^{(1)} = (3 - x^{(0)} + y^{(0)})/4 = (3 - 1 + 1)/4 = 0.750$$

similarly, second iteration yields

$$x^{(2)} = (z^{(1)} - y^{(1)})/3 = (0.75 + 1.0)/3 = 0.5833$$
$$y^{(2)} = (-x^{(1)} - z^{(1)})/2 = (-0 - 0.75)/2 = -0.3750$$
$$z^{(2)} = (3 - x^{(1)} + y^{(1)})/4 = (3 - 0 - 0)/4 = 0.500$$

Subsequent iterations result in the following:

$x^{(3)} = 0.29167$	$y^{(3)} = -0.47917$	$z^{(3)} = 0.51042$
$x^{(4)} = 0.32986$	$y^{(4)} = -0.40104$	$z^{(4)} = 0.57862$

$$x^{(5)} = 0.32595 \qquad y^{(5)} = -0.45334 \qquad z^{(5)} = 0.56728$$

$$x^{(6)} = 0.34021 \qquad y^{(6)} = -0.44662 \qquad z^{(6)} = 0.55329$$

$$x^{(7)} = 0.3333 \qquad y^{(7)} = -0.44675 \qquad z^{(7)} = 0.55498$$

$$x^{(8)} = 0.33391 \qquad y^{(8)} = -0.44414 \qquad z^{(8)} = 0.55498$$

$$x^{(9)} = 0.33304 \qquad y^{(9)} = -0.44445 \qquad z^{(9)} = 0.5555$$

so to three decimal places the approximate solution

$$x = 0.333 \qquad\qquad y = -0.444 \qquad\qquad z = 0.555$$

EXAMPLE 2.20

Use the Jacobi iterative scheme to obtain the solution of the system of equations correct to two decimal places.

$$\begin{bmatrix} 5 & -2 & 1 \\ 1 & 4 & -2 \\ 1 & 2 & 4 \end{bmatrix} = \begin{bmatrix} 4 \\ 3 \\ 17 \end{bmatrix}$$

Solution:

The Jacobi method is applicable only if the conditions given by Eq. (2.50) are satisfied.

Here $\quad |5| > |-2| + |1| \quad$ or $\quad 5 > 3$

$\qquad\quad |4| > |1| + |-1| \quad$ or $\quad 4 > 3$

$\qquad\quad |4| > |1| + |2| \quad$ or $\quad 4 > 3$

Clearly, the iterative approach will converse. Hence, writing the set of equations in the form of (2.51), we have

$$\begin{Bmatrix} x \\ y \\ z \end{Bmatrix}_{k+1} = \begin{Bmatrix} 0.8 \\ 0.75 \\ 4.25 \end{Bmatrix} - \begin{bmatrix} 0 & -0.4 & 0.2 \\ 0.25 & 0 & -0.5 \\ 0.25 & 0.5 & 0 \end{bmatrix} \begin{Bmatrix} x \\ y \\ z \end{Bmatrix}_{k} \qquad\qquad \text{(E.1)}$$

Assuming the initial approximation $\begin{Bmatrix} x \\ y \\ z \end{Bmatrix}_{0} = \begin{Bmatrix} 0 \\ 0 \\ 0 \end{Bmatrix}$ and substituting into Eq. (E.1)

gives our first approximation to the solution. Hence

$$
\begin{Bmatrix} x \\ y \\ z \end{Bmatrix}_2 = \begin{Bmatrix} 0.8 \\ 0.75 \\ 4.25 \end{Bmatrix} - \begin{bmatrix} 0 & -0.4 & 0.2 \\ 0.25 & 0 & -0.5 \\ 0.25 & 0.5 & 0 \end{bmatrix} \begin{Bmatrix} 0.8 \\ 0.75 \\ 4.25 \end{Bmatrix}_1 \qquad (\text{E.2})
$$

The process is continued until successive values of each vector are very close in magnitude. Here, the eleven iterations obtained accurate to two decimal places are shown below in Table 2.1.

TABLE 2.1

Variable	1	2	3	4	5	6	7	8	9	10	11
x	0.8	0.25	1.14	1.24	1.02	0.92	0.98	1.02	1.01	0.99	1
y	0.75	2.68	2.53	1.89	1.79	1.99	2.07	2.02	1.98	1.99	2
z	4.25	3.68	2.85	2.70	2.99	3.10	3.02	2.97	2.98	3.01	3

Hence, the solution is given by $x = 1$, $y = 2$, and $z = 3$.

2.11 GAUSS-SEIDEL ITERATION METHOD

The Gauss-Seidel method is applicable to *predominantly diagonal systems*. A predominantly diagonal system has large diagonal elements. The absolute value of the diagonal element in each case is larger than the sum of the absolute values of the other elements in that row of the matrix **A**. For such predominantly diagonal systems, the Gauss-Seidel method always converges to the correct solution, irrespective of the choice of the initial estimates. Since the most recent approximations of the variables are used while proceeding to the next step, the convergence of the Gauss-Seidel method is twice as fast as in Jacobi's method. The Gauss-Seidel and Jacobi's methods converge for any choice of the initial approximations, if in each equation of the system, the absolute value of the largest coefficient is greater than the sum of the absolute values of the remaining coefficients. In other words,

$$
\sum_{\substack{i=1 \\ j \neq 1}}^{n} \frac{|a_{ij}|}{|a_{ii}|} \leq 1, \qquad i = 1, 2, 3, \dots, n
$$

where the inequality holds in case of at least one equation. Convergence is assured in the Gauss-Seidel method if the matrix A is diagonally dominant and *positive definite*. If it is not in a diagonally dominant form, it should be connected to a diagonally dominant form by row exchanger, before starting the Gauss-Seidel iterative scheme.

The Gauss-Seidel method is also an iterative solution procedure which is an improved version of Jacobi's method. The method is also known as the *method of successive approximations*.

Consider the system of linear simultaneous equations

$$a_{11}x_1 + a_{12}x_2 + a_{13}x_3 = b_1$$

$$a_{21}x_1 + a_{22}x_2 + a_{23}x_3 = b_2$$

$$a_{31}x_1 + a_{32}x_2 + a_{33}x_3 = b_3 \tag{2.54}$$

If the absolute value of the largest coefficient in each equation is greater than the sum of the absolute values of all the remaining coefficients, then the Gauss-Seidel iteration method will converge. If this condition is not satisfied, then the Gauss-Seidel method is not applicable. Here, in Eq. (2.54), we assume the coefficient a_{11}, a_{22}, and a_{33} are the largest coefficients.

We can rewrite Eq. (2.54) as

$$x_1 = \frac{1}{a_{11}}(b_1 - a_{12}x_2 - a_{13}x_3)$$

$$x_2 = \frac{1}{a_{22}}(b_2 - a_{21}x_1 - a_{23}x_3)$$

$$x_3 = \frac{1}{a_{33}}(b_3 - a_{31}x_1 - a_{32}x_2) \tag{2.55}$$

Let the initial approximations be x_1^0, x_2^0, and x_3^0 respectively. The following iterations are then carried out.

Iteration 1: The first improvements of x_1, x_2, and x_3 are obtained as

$$x_{11} = \frac{1}{a_{11}}\left(b_1 - a_{12}\dot{x}_2 - a_{13}x_3^0\right)$$

$$x_{21} = \frac{1}{a_{22}}\left(b_2 - a_{21}x_{11} - a_{23}x_3^0\right)$$

$$x_{31} = \frac{1}{a_{33}}\left(b_3 - a_{31}x_{11} - a_{32}x_{21}\right) \tag{2.56}$$

Iteration 2: The second improvements of x_1, x_2, and x_3 are obtained as

$$x_{12} = \frac{1}{a_{11}}(b_1 - a_{12}x_{11} - a_{13}x_{31})$$

$$x_{22} = \frac{1}{a_{22}}(b_2 - a_{21}x_{12} - a_{23}x_{31})$$

$$x_{32} = \frac{1}{a_{33}}(b_3 - a_{31}x_{12} - a_{32}x_{22}) \tag{2.57}$$

The above iteration process is continued until the values of x_1, x_2, and x_3 are obtained to a preassigned or desired degree of accuracy. In general, the initial approximations are assumed as $x_1^0 = x_2^0 = x_3^0 = 0$. The Gauss-Seidel method generally converges for any initial values of x_1^0, x_2^0, x_3^0. The convergence rate of the Gauss-Seidel method is found to be twice to that of the Jacobi's method. Like Jacobi's method, the Gauss-Seidel method can also be extended to n linear simultaneous algebraic equations in n unknowns.

EXAMPLE 2.21

Solve the following equations by the Gauss-Seidel method.

$$8x + 2y - 2z = 8$$

$$x - 8y + 3z = -4$$

$$2x + y + 9z = 12$$

Solution:

In the above equations:

$$|8| > |2| + |-2|$$

$$|-8| > |1| + |3|$$

$$|9| > |2| + |1|$$

So, the conditions of convergence are satisfied and we can apply the Gauss-Seidel method. Then we rewrite the given equations as follows:

$$x_1 = \frac{1}{a_1}(d_1 - b_1 y^0 - c_1 z^0)$$

$$y_1 = \frac{1}{b_2}(d_2 - a_2 x_1 - c_2 z^0)$$

$$z_1 = \frac{1}{c_3}(d_3 - a_3 x_1 - b_3 y_1)$$

Let the initial approximations be:

$$x^0 = y^0 = z^0 = 0$$

Iteration 1:

$$\boxed{x_1} = \frac{d_1}{a_1} = \frac{8}{8} = 1.0$$

$$\boxed{y_1} = \frac{1}{b_2}(d_2 - a_2 x_1) = \frac{1}{-8}(-4 - 1 \times 1.0) = 0.625$$

$$\boxed{z_1} = \frac{1}{c_3}(d_3 - a_3 x_1 - b_3 y_1) = \frac{1}{9}(12 - 2. = 2 \times 1.0 - 1 \times 0.625) = 1.042$$

Iteration 2:

$$\boxed{x_2} = \frac{1}{a_1}(d_1 - b_1 y_1 - c_1 z_1) = \frac{1}{8}(8 - 2 \times 0.625 - (-2) \times 1.042) = 1.104$$

$$\boxed{y_2} = \frac{1}{b_2}(d_2 - a_2 x_2 - c_2 z_1) = \frac{1}{-8}(-4 - 1 \times 1.104 - 3 \times 1.042) = 1.029$$

$$\boxed{z_2} = \frac{1}{c_3}(d_3 - a_3 x_2 - b_3 y_2) = \frac{1}{9}(12 - 2 \times 1.104 - 1 \times 1.029) = 0.974$$

Iteration 3:

$$\boxed{x_3} = \frac{1}{a_1}(d_1 - b_1 y_2 - c_1 z_2) = \frac{1}{8}(8 - 2 \times 1.029 - (-2) \times 0.974) = 0.986$$

$$\boxed{y_3} = \frac{1}{b_2}(d_2 - a_2 x_3 - c_2 z_2) = \frac{1}{-8}(-4 - 1 \times 0.986 - 3 \times 0.974) = 0.989$$

$$\boxed{z_3} = \frac{1}{c_3}(d_3 - a_3 x_3 - b_3 y_3) = \frac{1}{9}(12 - 2 \times 0.986 - 1 \times 0.989) = 1.004$$

Iteration 4:

$$x_4 = \frac{1}{8}(8 - 2 \times 0.989 - (-2) \times 1.004) = 1.004$$

$$y_4 = \frac{1}{-8}(-4 - 1 \times 1.004 - 3 \times 1.004) = 1.002$$

$$z_4 = \frac{1}{9}(12 - 2 \times 1.004 - 1 \times 1.002) = 0.999$$

Iteration 5:

$$x_5 = \frac{1}{8}(8 - 2 \times 1.002 - (-2) \times 0.999) = 0.999$$

$$y_5 = \frac{1}{-8}(-4 - 1 \times 0.999 - 3 \times 0.999) = 1.0$$

$$z_5 = \frac{1}{9}(12 - 2 \times 0.999 - 1 \times 1.0) = 1.0$$

Iteration 6:

$$x_6 = \frac{1}{8}(8 - 2 \times 1 + 2 \times 1) = \boxed{1.0}$$

$$y_6 = \frac{1}{-8}(-4 - 1 \times 1.0 - 3 \times 1.0) = \boxed{1.0}$$

$$z_6 = \frac{1}{9}(12 - 2 \times 1.0 - 1 \times 1.0) = \boxed{1.0}$$

EXAMPLE 2.22

Using the Gauss-Seidel method, solve the system of equations correct to three decimal places.

$$x + 2y + z = 0$$

$$3x + y - z = 0$$

$$x - y + 4z = 3$$

Solution:

Rearranging the given equations to give dominant diagonal elements, we obtain

$$3x + y - z = 0$$

$$x + 2y + z = 0$$

$$x - y + 4z = 3 \qquad \text{(E.1)}$$

Equation (E.1) can be rewritten as

$$x = (z - y)/3$$

$$y = -(x + z)/2$$

$$z = (3 + x + y)/2 \qquad \text{(E.2)}$$

Writing Eq. (E.2) in the form of the Gauss-Seidel iterative scheme, we get

$$x^{(r+1)} = (z^{(r)} - y^{(r)})/3$$

$$y^{(r+1)} = -(x^{(r+1)} - z^{(r)})/2$$

$$z^{(r+1)} = (3 - x^{(r+1)} + y^{(r+1)})/4$$

We start with the initial value

$$x^{(0)} = y^{(0)} = z^{(0)} = 1$$

The iteration scheme gives

$$x^{(1)} = (z^{(0)} - y^{(0)})/3 = (1 - 1)/3 = 0$$

$$y^{(1)} = (-x^{(1)} - z^{(0)})/2 = (0 - 1)/2 = -0.5$$

$$z^{(1)} = (3 - x^{(1)} + y^{(1)})/4 = (3 - 0 - 0)/4 = 0.625$$

The second iteration gives

$$x^{(2)} = (z^{(1)} - y^{(1)})/3 = (0.625 + 0.5)/3 = 0.375$$

$$y^{(2)} = (-x^{(2)} - z^{(1)})/2 = (-0.375 - 0.625)/2 = -0.50$$

$$z^{(2)} = (3 - x^{(2)} + y^{(2)})/4 = (3 - 0.375 - 0.5)/4 = 0.53125$$

Subsequent iterations result in

$$x^{(3)} = 0.34375 \qquad y^{(3)} = -0.4375 \qquad z^{(3)} = 0.55469$$

$$x^{(4)} = 0.33075 \qquad y^{(4)} = -0.44271 \qquad z^{(4)} = 0.55664$$

$$x^{(5)} = 0.33312 \qquad y^{(5)} = -0.44488 \qquad z^{(5)} = 0.5555$$

$$x^{(6)} = 0.33346 \qquad y^{(6)} = -0.44448 \qquad z^{(6)} = 0.55552$$

Hence, the approximate solution is as follows:

$$x = 0.333 \qquad y = -0.444 \qquad z = 0.555$$

EXAMPLE 2.23

Solve the following equations by the Gauss-Seidel method.

$$4x - y + z = 12$$
$$-x + 4y - 2z = -1$$
$$x - 2y + 4z = 5$$

Solution:

The iteration formula is

$$x_i \leftarrow \frac{1}{A_{ii}}\left[b_i - \sum_{\substack{j=1 \\ j \neq i}}^{n} A_{ij} x_j\right], \qquad i = 1, 2, \ldots, n$$

Hence, $x = \dfrac{1}{4}(12 + y - z)$

$$y = \frac{1}{4}(-1 + x + 2z)$$

$$z = \frac{1}{4}(5 - x + 2y)$$

Choosing the starting values $x = y = z = 0$, we have the first iteration

$$x = \frac{1}{4}(12 + 0 - 0) = 3$$

$$y = \frac{1}{4}[-1 + 3 + 2(0)] = 0.5$$

$$z = \frac{1}{4}(5 - 3 + 2(0.5)] = 0.75$$

The second iteration gives

$$x = \frac{1}{4}(12 + 0.5 - 0.75) = 2.9375$$

$$y = \frac{1}{4}[-1 + 2.9375 + 2(0.75)] = 0.8594$$

$$z = \frac{1}{4}(5 - 2.9375 + 2(0.8594)] = 0.9453$$

The third iteration yields

$$x = \frac{1}{4}[12 + 0.8594 - 0.9453] = 2.9785$$

$$y = \frac{1}{4}[-1 + 2.9785 + 2(0.9453)] = 0.9673$$

$$z = \frac{1}{4}(5 - 2.9785 + 2(0.9673)] = 0.9890$$

After five more iterations, we obtain the final values for x, y, and z as $x = 3$, $y = 1$ and $z = 1$.

2.12 SUMMARY

A matrix is a rectangular array of elements, in rows and columns. The elements of a matrix can be numbers, coefficients, terms or variables. This chapter provided the relevant and useful elements of matrix analysis for the solution of linear simultaneous algebraic equations. Topics covered include matrix definitions, matrix operations, determinants, matrix inversion, trace, transpose, and system of algebraic equations and solution. The solution of n linear simultaneous algebraic equations in n unknowns is presented. There are two classes of methods of solving a system of linear algebraic equations: direct and iterative methods. Direct methods transform the original equation into equivalent equations that can be solved more easily. Iterative or indirect methods start with a guess of the solution x, and then repeatedly refine the solution until a certain convergence criterion is reached. Six direct methods (matrix inversion method, Gauss elimination method, Gauss-Jordan method, Cholesky's triangularization method, Crout's method, and the Thomas algorithm for tridiagonal system) are presented. Two indirect or iterative methods (Jacobi's iteration method and the Gauss-Seidel iteration method) are presented.

The LU decomposition method is closely related to the Gauss elimination method. LU decomposition is computationally very effective if the coefficient matrix remains the same but the right-hand side vector changes. Cholesky's decomposition method can be used when the coefficient matrix A is symmetric and positive definite. The Gauss-Jordan method is a very stable method for solving linear algebraic equations. The Gauss-Seidel iterative substitution technique is very suitable for predominantly diagonal systems. It requires a guess of the solution.

EXERCISES

2.1 Determine the inverse of the following matrices:

(a) $A = \begin{bmatrix} -1 & 1 & 2 \\ 3 & -1 & 1 \\ -1 & 3 & 4 \end{bmatrix}$

(b) $A = \begin{bmatrix} 1 & 2 & 0 \\ 3 & -1 & -2 \\ 1 & 0 & -3 \end{bmatrix}$

(c) $A = \begin{bmatrix} 10 & 3 & 10 \\ 8 & -2 & 9 \\ 8 & 1 & -10 \end{bmatrix}$

(d) $A = \begin{bmatrix} 1 & 2 & 3 \\ 4 & 5 & 6 \\ 7 & 0 & 0 \end{bmatrix}$

(e) $A = \begin{bmatrix} 1 & 1 & 1 \\ 1 & 2 & 2 \\ 1 & 0 & 3 \end{bmatrix}$

(f) $\mathbf{A} = \begin{bmatrix} 1 & 0 & 3 \\ 2 & 1 & -1 \\ 1 & -1 & 1 \end{bmatrix}$

2.2 Solve the following set of simultaneous linear equations by the matrix inverse method.

(a) $2x + 3y - z = -10$

$-x + 4y + 2z = -4$

$2x - 2y + 5z = 35$

(b) $10x + 3y + 10z = 5$

$8x - 2y + 9z = 2$

$8x + y - 10z = 35$

(c) $2x + 3y - z = 1$

$-x + 2y + z = 8$

$x - 3y - 2z = -13$

(d) $2x - y + 3z = 4$

$x + 9y - 2z = -8$

$4x - 8y + 11z = 15$

(e) $x_1 - x_2 + 3x_3 - x_4 = 1$

$x_2 - 3x_3 + 5x_4 = 2$

$x_1 - x_3 + x_4 = 0$

$x_1 + 2x_2 - x_4 = -5$

(f) $x_1 + 2x_2 + 3x_3 + 4x_4 = 8$

$2x_1 - 2x_2 - x_3 - x_4 = -3$

$x_1 - 3x_2 + 4x_3 - 4x_4 = 8$

$2x_1 + 2x_2 - 3x_3 + 4x_4 = -2$

2.3   Solve the following set of simultaneous linear equations using the method of Gaussian elimination.

(a) $2x + y - 3z = 11$

$4x - 2y + 3z = 8$

$-2x + 2y - z = -6$

(b) $6x + 3y + 6z = 30$

$2x + 3y + 3z = 17$

$x + 2y + 2z = 11$

(c) $2x_1 + x_2 + x_3 = 4$

$3x_2 - 3x_3 = 0$

$-x_2 + 2x_3 = 1$

(d) $x_1 + 2x_2 + 3x_3 + 4x_4 = 8$

$2x_1 - 2x_2 - x_3 - x_4 = -3$

$x_1 - 3x_2 + 4x_3 - 4x_4 = 8$

$2x_1 + 2x_2 - 3x_3 + 4x_4 = -2$

(e) $2x_1 + x_2 + x_3 - x_4 = 10$

$x_1 + 5x_2 - 5x_3 + 6x_4 = 25$

$-7x_1 + 3x_2 - 7x_3 - 5x_4 = 5$

$x_1 - 5x_2 + 2x_3 + 7x_4 = 11$

(f) $x_1 + x_2 + x_3 + x_4 = 3$

$2x_1 - x_2 + 3x_3 = 3$

$2x_2 + 3x_4 = 1$

$-x_1 + 2x_3 + x_4 = 0$

2.4   Solve the following set of simultaneous linear equations by using the Gauss-Jordan method.

(a) $4x - 3y + 5z = 34$

$2x - y - z = 6$

$x + y + 4z = 15$

(b) $2x - y + z = -1$

$3x + 3y + 9z = 0$

$3x + 3y + 5z = 4$

(c) $x + y - z = 1$

$x + 2y - 2z = 0$

$-2x + y + z = 1$

(d) $x - y = 2$

$-2x + 2y - z = -1$

$y - 2z = 6$

(e) $x + y + z = 3$

$2x + 3y + z = 6$

$x - y - z = -3$

(f) $4x_1 - 2x_2 - 3x_3 + 6x_4 = 12$

$-5x_1 + 7x_2 + 6.5x_3 - 6x_4 = -6.5$

$x_1 + 7.5x_2 + 6.25x_3 + 5.5x_4 = 16$

$-12x_1 + 22x_2 + 15.5x_3 - x_4 = 17$

2.5   Solve the following set of simultaneous linear equations by using Cholesky's factorization method.

(a) $2x - y = 3$

$-x + 2y - z = -3$

$-y + z = 2$

(b) $x + y + z = 7$

$3x + 3y + 4z = 23$

$2x + y + z = 10$

(c) $x + 0.5y = 1$

$0.5x + y + 0.5z = 2$

$0.5y + z = 3$

(d) $2x + 3y + z = 9$

$x + 2y + 3z = 6$

$3x + y + 2z = 8$

(e) $x - 2y + z = 2$

$5x + y - 3z = 0$

$3x + 4y + z = 9$

(f) $12x_1 - 6x_2 - 6x_3 + 1.5x_4 = 1$

$-6x_1 + 4x_2 + 3x_3 + 0.5x_4 = 2$

$-6x_1 + 3x_2 + 6x_3 + 1.5x_4 = 3$

$-1.5x_1 + 0.5x_2 + 1.5x_3 + x_4 = 4$

2.6 Solve the following set of simultaneous linear equations using Crout's method.

(a) $2x + y = 7$

$x + 2y = 5$

(b) $3x + 2y + 7z = 4$

$2x + 3y + z = 5$

$3x - 4y + z = 7$

(c) $x + y + z = 9$

$2x - 3y + 4z = 13$

$3x + y + 5z = 40$

(d) $3x + y = -1$

$2x + 4y + z = 7$

$2y + 5z = 9$

(e) $2x + y - z = 6$

$x - 3y + 5z = 11$

$-x + 5y + 4z = 13$

(f) $2x_1 - x_2 = 1$

$-x_1 + 2x_2 - x_3 = 0$

$-x_2 + 2x_3 - x_4 = 0$

$-x_3 + 2x_4 = 1$

2.7 Solve the following tridiagonal system of equations using the Thomas algorithm.

(a) $2x_1 + x_2 = 3$

$-x_1 + 2x_2 + x_3 = 6$

$3x_2 + 2x_3 = 12$

(b) $2x_1 + x_2 = 4$

$3x_1 + 2x_2 + x_3 = 8$

$x_2 + 2x_3 + 2x_4 = 8$

$x_3 + 4x_4 = 9$

(c) $3x_1 - x_2 = 2$

$2x_1 - 3x_2 + 2x_3 = 1$

$x_2 + 2x_3 + 5x_4 = 13$

$x_3 - x_4 = -1$

(d) $2x_1 + x_2 = 3$

$x_1 + 3x_2 + x_3 = 3$

$x_2 + x_3 + 2x_4 = 4$

$2x_3 + 3x_4 = 4$

(e) $2x_1 + x_2 = 1$

$3x_1 + 2x_2 + x_3 = 2$

$x_2 + 2x_3 + 2x_4 = -1$

$x_3 + 4x_4 = -3$

(f) $2x_1 - x_2 = 1$

$x_1 + 3x_2 + x_3 = 3$

$x_2 + x_3 + 2x_4 = 0$

$2x_3 + 3x_4 = -1$

2.8 Solve the following set of simultaneous linear equations using Jacobi's method.

(a) $2x - y + 5z = 15$

$2x + y + z = 7$

$x + 3y + z = 10$

(b) $20x + y - 2z = 17$

$3x + 20y - z = -18$

$2x - 3y + 20z = 25$

(c) $5x + 2y + z = 12$

$x + 4y + 2z = 15$

$x + 2y + 5z = 20$

(d) $10x - y + 2z = 6$

$-x + 11y + z = 22$

$2x - y + 10z = -10$

(e) $8x + 2y - 2z = 8$

$x - 8y + 3z = -4$

$2x + y + 9z = 12$

(f) $10x_1 + x_2 + 2x_3 = 6$

$\quad -x_1 + 11x_2 - x_3 + 3x_4 = 25$

$\quad 2x_1 - x_2 + 10x_3 - x_4 = -1$

$\quad 3x_2 - x_3 + 8x_4 = 15$

2.9 Solve the following system of simultaneous linear equations using the Gauss-Seidel method.

(a) $4x - 3y + 5z = 34$

$\quad 2x - y - z = 6$

$\quad z + y + 4z = 15$

(b) $2x - y + 5z = 15$

$\quad 2x + y + z = 7$

$\quad x + 3y + z = 10$

(c) $15x + 3y - 2z = 85$

$\quad 2x + 10y + z = 51$

$\quad x - 2y + 8z = 5$

(d) $10x_1 - 2x_2 - x_3 - x_4 = 3$

$\quad -2x_1 + 10x_2 - x_3 - x_4 = 15$

$\quad -x_1 - x_2 + 10x_3 - 2x_4 = 27$

$\quad -x_1 - x_2 - 2x_3 + 10x_4 = -9$

(e) $4x_1 + 2x_2 = 4$

$\quad 2x_1 + 8x_2 + 2x_3 = 0$

$\quad 2x_2 + 8x_3 + 2x_3 = 0$

$\quad 2x_3 + 4x_4 = 0$

(f) $4x_1 + 2x_2 = 4$

$\quad 2x_1 + 8x_2 + 2x_3 = 0$

$\quad 2x_2 + 8x_3 + 2x_3 = 0$

$\quad 2x_3 + 4x_4 = 14$

3

SOLUTION OF ALGEBRAIC AND TRANSCENDENTAL EQUATIONS

3.1   INTRODUCTION

One of the most common problems encountered in engineering analysis is given a function $f(x)$, find the values of x for which $f(x) = 0$. The solution (values of x) are known as the *roots* of the equation $f(x) = 0$, or the *zeroes* of the function $f(x)$.

The roots of equations may be real or complex. In general, an equation may have any number of (real) roots or no roots at all. For example, $\sin x - x = 0$ has a single root, namely, $x = 0$, whereas $\tan x - x = 0$ has an infinite number of roots $(x = 0, \pm 4.493, \pm 7.725, \ldots)$. There are two types of methods available to find the roots of algebraic and transcendental equations of the form $f(x) = 0$.

1. **Direct Methods:** Direct methods give the exact value of the roots in a finite number of steps. We assume that there are no round-off errors. Direct methods determine all the roots at the same time.

2. **Indirect or Iterative Methods:** Indirect or iterative methods are based on the concept of successive approximations. The general procedure is to start with one or more initial approximation to the root and obtain a sequence of iterates (x_k) which in the limit converges to the actual or true solution to the root. Indirect or iterative methods determine one or two roots at a time.

The indirect or iterative methods are further divided into two categories: bracketing and open methods. The bracketing methods require the limits between which the root lies, whereas the open methods require the initial estimation of the solution. The bisection and false position methods are two known examples of bracketing methods. Among the open methods, the Newton-Raphson and the method of successive approximation is most commonly used. The most popular method for solving a nonlinear equation is the Newton-Raphson method, and this method has a high rate of convergence to a solution.

In this chapter, we present the following indirect or iterative methods with illustrative examples:

- bisection method
- method of false position (regula falsi method)
- Newton-Raphson method (Newton's method)
- successive approximation method

3.2 BISECTION METHOD

After a root of $f(x) = 0$ has been bracketed in the interval (a, b), The bisection method can be used to close in on it. The bisection method accomplishes this by successfully halving the interval until it becomes sufficiently small. The bisection method is also known as the *interval halving method*. The bisection method is not the fastest method available for finding roots of a function, but it is the most reliable method. Once a has been bracketed, the bisection method will always close in on it.

We assume that $f(x)$ is a function that is real-valued and that x is a real variable. Suppose that $f(x)$ is continuous on an interval $a \leq x \leq b$ and that $f(a)$ $f(b) < 0$. When this is the case, $f(x)$ will have opposite signs at the end points of the interval (a, b). As shown in Figure 3.1 (a) and (b), if $f(x)$ is continuous and has a solution between the points $x = a$ and $x = b$, then either $f(a) > 0$ and $f(b) < 0$ or $f(a) < 0$ and $f(b) > 0$. In other words, if there is a solution between $x = a$, and $x = b$, then $f(a) f(b) < 0$.

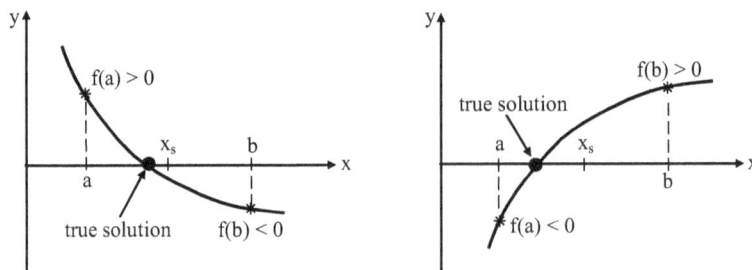

FIGURE 3.1 Solution of f(x) = 0 between x = a and x = b.

The method of finding a solution with the bisection method is illustrated in Figure 3.2. It starts by finding points a and b that define an interval where a solution exists. The midpoint of the interval x_{s_1} is then taken as the first estimate for the numerical solution. The true solution is either in the portion between points a and x_{s_1} or in the portion between points x_{s_1} and b. If the solution obtained is not accurate enough, a new interval that contains the true solution is defined. The new interval selected is the half of the original interval that contains the true solution, and its midpoint is taken as the new (second) estimate of the numerical solution. The procedure is repeated until the numerical solution is accurate enough according to a certain criterion that is selected.

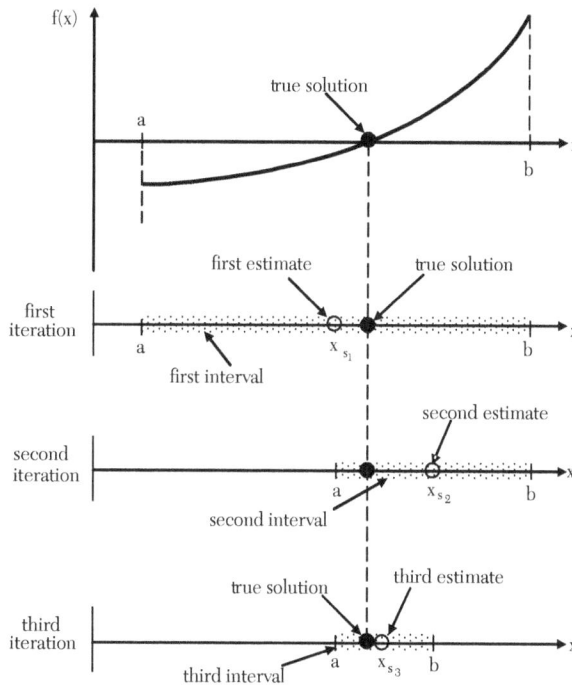

FIGURE 3.2 Bisection method.

The procedure or algorithm for finding a numerical solution with the bisection method is given in the following section.

Algorithm for the bisection method

1. Compute the first estimate of the numerical solution x_{s_1} by

$$x_{s_1} = \frac{a+b}{2}$$

2. Determine whether the true solution is between a and x_{s_1} or between x_{s_1} and b by checking the sign of the product

$$f(a)f(x_{s_1}):$$

If $f(a)f(x_{s_1}) < 0$, the true solution is between a and x_{s_1}.

If $f(a)f(x_{s_1}) > 0$, the true solution is between x_{s_1} and b.

If $b - c \leq \in$, then accept c as the root and stop. $\in$ is the error tolerance, $\in > 0$.

3. Choose the subinterval that contains the true solution (a to x_{s_1} or x_{s_1} to b) as the new interval (a, b), and go back to step 1.

    Steps 1 through 3 are repeated until a specified tolerance or error bound is attained.

3.2.1 Error Bounds

Let a_n, b_n, and c_n denote the nth computed values of a, b, and x_{s_1} respectively. Then, we have

$$b_{n+1} - a_{n+1} = \frac{1}{2}(b_n - a_n) \qquad n \geq 1 \qquad (3.1)$$

also

$$b_n - a_n = \frac{1}{2^{n-1}}(b - a) \qquad n \geq 1 \qquad (3.2)$$

where $(b - a)$ denotes the length of the original interval with which we started. Since the root x_s is in either the interval (a_n, c_n) or (c_n, b_n), we know that

$$|x_s - c_n| \leq c_n - a_n = b_n - c_n = \frac{1}{2}(b_n - a_n) \qquad (3.3)$$

This is the error bound for c_n that is used in step 2 of the algorithm described earlier.

From Equations (3.2) and (3.3), we obtain the further bound

$$|x_s - c_n| \leq \frac{1}{2^n}(b - a) \qquad (3.4)$$

Equation (3.4) shows that the iterate c_n converges to x_s as $n \rightarrow \infty$.

To find out how many iterations will be necessary, suppose we want to have

$$|x_s - c_n| \leq \in$$

This will be satisfied if

$$\frac{1}{2^n}(b-a) \leq \in \tag{3.5}$$

Taking logarithms of both sides of Equation (3.5), and simplifying the resulting expression, we obtain

$$n \geq \frac{\log\left(\dfrac{b-a}{\in}\right)}{\log 2} \tag{3.6}$$

There are several advantages to the bisection method. The method is guaranteed to converge, and always converges to an answer, provided a root was bracketed in the interval (a, b) to start with. In addition, the error bound, given in Equation (3.4), is guaranteed to decrease by one half with each iteration. The method may fail when the function is tangent to the axis and does not cross the x-axis at $f(x) = 0$. The disadvantage of the bisection method is that it generally converges more slowly than most other methods. For functions $f(x)$ that have a continuous derivative, other methods are usually faster. These methods may not always converge. When these methods do converge, they are almost always much faster than the bisection method.

EXAMPLE 3.1

Use the bisection method to find a root of the equation $x^3 - 4x - 8.95 = 0$ accurate to three decimal places.

Solution:

Here, $\quad f(x) = x^3 - 4x - 8.95 = 0$

$$f(2) = 2^3 - 4(2) - 8.95 = -8.95 < 0$$

$$f(3) = 3^3 - 4(3) - 8.95 = 6.05 > 0$$

Hence, a root lies between 2 and 3.

$f'(x) = 3x^2 - 4 > 0$ for x in the interval $(2, 3)$. Hence, we have $a = 2$ and $b = 3$. The results of the algorithm for the bisection method are shown in Table 3.1.

TABLE 3.1 Bisection method results.

n	a	b	x_{s_1}	$b - x_{s_1}$	$f(x_{s_1})$
1	2	3	2.5	0.5	−3.25
2	2.5	3	2.75	0.25	0.84688
3	2.5	2.75	2.625	0.125	−1.36211
4	2.75	2.625	2.6875	−0.0625	−0.28911
5	2.75	2.6875	2.71875	−0.03125	0.27092
6	2.6875	2.71875	2.70313	0.01563	−0.01108
7	2.71875	2.70313	2.71094	−0.00781	0.12942
8	2.71875	2.71094	2.71484	−0.00391	0.20005
9	2.71094	2.71484	2.71289	0.00195	0.16470
10	2.71094	2.71289	2.71191	0.00098	0.14706
11	2.71094	2.71191	2.71143	0.00049	0.13824

Hence, the root is 2.711 accurate to three decimal places.

EXAMPLE 3.2

Find one root of $e^x - 3x = 0$ correct to two decimal places using the method of bisection.

Solution:

Here, $f(x) = e^x - 3x$

$$f(1.5) = e^{1.5} - 3(1.5) = -0.01831$$

$$f(1.6) = e^{1.6} - 3(1.6) = 0.15303$$

$f'(x) = e^x - 3 > 0$ for x in the interval $(1.5, 1.6)$. Hence, a root lies in the interval $(1.5, 1.6)$. Therefore, here we have $a = 1.5$ and $b = 1.6$. The results of the algorithm for the bisection method are shown in Table 3.2.

TABLE 3.2 Bisection method results.

n	a	b	x_{s_1}	$b - x_{s_1}$	$f(x_{s_1})$
1	1.5	1.6	1.55	0.05	0.06147
2	1.5	1.55	1.525	0.025	0.02014
3	1.5	1.525	1.5125	0.0125	0.00056
4	1.5	1.5125	1.50625	0.00625	−0.00896
5	1.50625	1.5125	1.50938	0.00313	−0.00422
6	1.50938	1.5125	1.51094	0.00156	−0.00184

Hence the root of $f(x) = 0$ is $x = 1.51$ accurate up to two decimal places.

EXAMPLE 3.3

Determine the largest root of $f(x) = x^6 - x - 1 = 0$ accurate to within $\epsilon = 0.001$. Use the bisection method.

Solution:

Here $f(x) = x^6 - x - 1 = 0$

$$f(1) = 1^6 - 1 - 1 = -1$$

$$f(2) = 2^6 - 2 - 1 = 61$$

Since $f(1)f(2) < 0$, $f(x) = 0$ has at least one root on the interval. The results of the algorithm for the bisection method are shown in Table 3.3.

TABLE 3.3 Bisection method results.

n	a	b	x_{s_1}	$b - x_{s_1}$	$f(x_{s_1})$
1	1	2	1.5	0.5	8.89063
2	1	1.5	1.25	0.25	1.56470
3	1	1.25	1.25	0.125	−0.09771
4	1.125	1.25	1.1875	0.0625	0.61665
5	1.125	1.1875	1.15625	0.03125	0.23327
6	1.125	1.15625	1.14063	0.01563	0.06158
7	1.125	1.14063	1.13281	0.00781	−0.01958
8	1.13281	1.14063	1.13672	0.00391	0.02062
9	1.13281	1.13672	1.13477	0.00195	0.00043
10	1.13281	1.13477	1.13379	0.00098	−0.00960

3.3 METHOD OF FALSE POSITION

The method of false position (also called *the regula falsi method,* and *the linear interpolation method*) is another well-known bracketing method. It is very similar to the bisection method with the exception that it uses a different strategy to conclude its new root estimate. Rather than bisecting the interval (a, b), it locates the root by joining $f(a_1)$ and $f(b_1)$ with a straight line. The intersection of this line with the x-axis represents an improved estimate of the root.

Here again, we assume that within a given interval (a, b), $f(x)$ is continuous and the equation has a solution. As shown in Figure 3.3, the method starts by finding an initial interval (a_1, b_1) that brackets the solution. $f(a_1)$ and $f(b_1)$

are the values of the function at the end points a_1 and b_1. These end points are connected by a straight line, and the first estimate of the numerical solution, x_{s_1}, is the point where the straight line crosses the axis. For the second iteration, a new interval (a_2, b_2) is defined. The new interval is either (a_1, x_{s_1}) where a_1 is assigned to a_2 and x_{s_1} to b_2 or (x_{s_1}, b_1) where x_{s_1} is assigned to a_2 and b_1 to b_2. The end points of the second interval are connected with a straight line, and the point where this new line crosses the x-axis is the second estimate of the solution, x_{s_1}. A new subinterval (a_3, b_3) is selected for the third iteration and the iterations will be continued until the numerical solution is accurate enough.

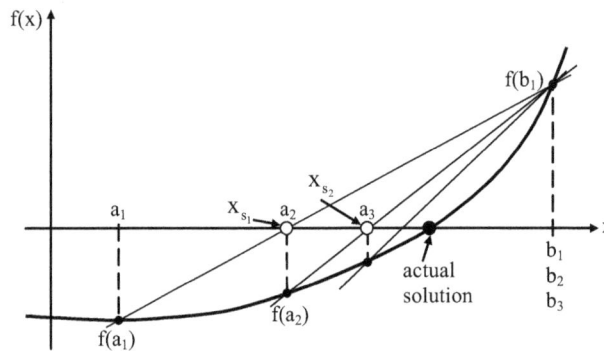

FIGURE 3.3 Method of false position.

The equation of a straight line that connects points $(b, f(b))$ to point $(a, f(a))$ is given by

$$y = \frac{f(b) - f(a)}{b - a}(x - b) + f(b) \tag{3.7}$$

The points x_s where the line intersects the x-axis is determined by substituting $y = 0$ in Equation (3.7) and solving the equation for x.

Hence
$$x_s = \frac{a\,f(b) - b\,f(a)}{f(b) - f(a)} \tag{3.8}$$

The procedure (or algorithm) for finding a solution with the method of false position is given below:

Algorithm for the method of false position

1. Define the first interval (a, b) such that a solution exists between them. Check $f(a)f(b) < 0$.

2. Compute the first estimate of the numerical solution x_s using Eq. (3.8).

3. Find out whether the actual solution is between a and x_{s_1} or between x_{s_1} and b. This is accomplished by checking the sign of the product $f(a)f(x_{s_1})$.
   If $f(a)f(x_{s_1}) < 0$, the solution is between a and x_{s_1}.
   If $f(a)f(x_{s_1}) > 0$, the solution is between x_{s_1} and b.

4. Select the subinterval that contains the solution (a to x_{s_1}, or x_{s_1} to b) which is the new interval (a, b) and go back to step 2. Steps 2 through 4 are repeated until a specified tolerance or error bound is attained. The method of false position always converges to an answer, provided a root is initially bracketed in the interval (a, b).

EXAMPLE 3.4

Using the false position method, find a root of the function $f(x) = e^x - 3x^2$ to an accuracy of five digits. The root is known to lie between 0.5 and 1.0.

Solution:

We apply the method of false position with $a = 0.5$ and $b = 1.0$. Equation (3.8) is

$$x_s = \frac{a\,f(b) - b\,f(a)}{f(b) - f(a)}$$

The calculations based on the method of false position are shown in Table 3.4.

TABLE 3.4

n	a	b	f(a)	f(b)	x_{s_1}	$f(x_{s_1})$	ξ Relative error
1	0.5	1	0.89872	−0.28172	0.88067	0.08577	—
2	0.88067	1	0.08577	−0.28172	0.90852	0.00441	0.03065
3	0.90852	1	0.00441	−0.28172	0.90993	0.00022	0.00155
4	0.90993	1	0.00022	−0.28172	0.91000	0.00001	0.00008
5	0.91000	1	0.00001	−0.28172	0.91001	0	3.7952×10^{-6}

The relative error after the fifth step is $\left(\dfrac{0.91001 - 0.91}{0.91001} \right) = 3.7952 \times 10^{-6}$. The root is 0.91 accurate to five digits.

EXAMPLE 3.5

Find a real root of $\cos x - 3x + 5 = 0$, correct to four decimal places using the method of false position method.

Solution:

Here $f(x) = \cos x - 3x + 5 = 0$

$$f(0) = \cos 0 - 3(0) + 5 = 5 > 0$$

$$f(\pi/2) = \frac{\cos \pi}{2} - 3\left(\frac{\pi}{2}\right) + 5 = \frac{-3\pi}{2} + 5 < 0$$

Therefore, a root of $f(x) = 0$ lies between 0 and $\pi/2$. We apply the method of false position with $a = 0$ and $b = \pi/2$. Equation (3.8) is

$$x_s = \frac{a\,f(b) - b\,f(a)}{f(b) - f(a)}$$

The calculations based on the method of false position are shown in Table 3.5.

TABLE 3.5

n	a	b	f(a)	f(b)	x_{s_1}	$f(x_{s_1})$	ξ
1	0	1.5708	6	0.28761	1.64988	−0.02866	—
2	1.64988	1.5708	−0.02866	0.28761	1.64272	−0.00001	−0.00436
3	1.64272	1.5708	−0.00001	0.28761	1.64271	0	-1.97337×10^{-6}

The relative error after the third step is

$$x = \frac{1.64271 - 1.64272}{1.64271} = -1.97337 \times 10^{-6}$$

The root is 1.6427 accurate to four decimal places.

EXAMPLE 3.6

Using the method of false position, find a real root of the equation $x^4 - 11x + 8 = 0$ accurate to four decimal places.

Solution:

Here $f(x) = x^4 - 11x + 8 = 0$

$$f(1) = 1^4 - 11(1) + 8 = -2 < 0$$

$$f(2) = 2^4 - 11(2) + 8 = 4 > 0$$

Therefore, a root of $f(x) = 0$ lies between 1 and 2. We apply the method of false position with $a = 1$ and $b = 2$. Equation (3.8) is

$$x_s = \frac{a\,f(b) - b\,f(a)}{f(b) - f(a)}$$

The calculations based on the method of false position are summarized in Table 3.6.

TABLE 3.6

n	a	b	f(a)	f(b)	x_{s_1}	$f(x_{s_1})$	ξ
1	1	2	−2	2	1.5	−3.4375	—
2	1.5	2	−3.4375	2	1.81609	−1.9895	0.17405
3	1.81609	2	−1.09895	2	1.88131	−0.16758	3.4666×10^{-2}
4	1.88131	2	−0.16758	2	1.89049	−0.02232	4.85383×10^{-3}
5	1.89049	2	−0.02232	2	1.89169	−0.00292	6.3902×10^{-4}
6	1.89169	2	−0.00292	2	1.89185	−0.00038	8.34227×10^{-5}
7	1.89185	2	−0.00038	2	1.89187	−0.00005	1.08786×10^{-5}

The relative error after the seventh step is

$$\xi = \frac{1.89187 - 1.89185}{1.89187} = 1.08786 \times 10^{-5}$$

Hence, the root is 1.8918 accurate to four decimal places.

3.4   NEWTON-RAPHSON METHOD

The Newton-Raphson method is the best-known method of finding roots of a function $f(x)$. The method is simple and fast. One drawback of this method is that it uses the derivative $f'(x)$ of the function as well as the function $f(x)$. Hence, the Newton-Raphson method is usable only in problems where

$f'(x)$ can be readily computed. The Newton-Raphson method is also called *Newton's method*. Here, again we assume that $f(x)$ is continuous and differentiable and the equation is known to have a solution near a given point. Figure 3.4 illustrates the procedure used in the Newton-Raphson method. The solution process starts by selecting point x_1 as the first estimate of the solution. The second estimate x_2 is found by drawing the tangent line to $f(x)$ at the point $(x_1, f(x_1))$ and determining the intersection point of the tangent line with the x-axis. The next estimate x_3 is the intersection of the tangent line to $f(x)$ at the point $(x_2, f(x_2))$ with the x-axis, and so on. The slope, $f'(x_1)$, of the tangent at point $(x_1, f(x_1))$ is written as

$$f'(x_1) = \frac{f(x_1) - 0}{x_1 - x_2} \tag{3.9}$$

Rewriting Equation (3.9) for x_2 gives

$$x_2 = x_1 - \frac{f(x_1)}{f'(x_1)} \tag{3.10}$$

Equation (3.10) can be generalized for determining the next solution x_{i+1} from the current solution x_i as

$$x_{i+1} = x_i - \frac{f(x_i)}{f'(x_i)} \tag{3.11}$$

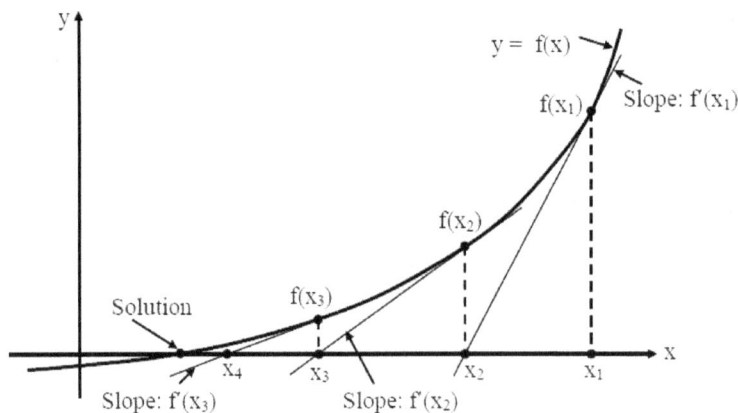

FIGURE 3.4 The Newton-Raphson method.

The solution is obtained by repeated application of the iteration formula given by Eq. (3.11) for each successive value of "i".

Algorithm for the Newton-Raphson Method:

1. Select a point x_1 as an initial guess of the solution.

2. For $i = 1, 2, \ldots$, until the error is smaller than a specified value, compute x_{i+1} by using Eq. (3.11).

Two error estimates that are generally used in the Newton-Raphson method are given below:

The iterations are stopped when the estimated relative error $\left|\dfrac{x_{i+1} - x_i}{x_i}\right|$ is smaller than a specified value $\in$.

$$\left|\frac{x_{i+1} - x_i}{x_i}\right| \leq \in \qquad (3.12)$$

The iterations are stopped when the absolute value of $f(x_i)$ is smaller than some number δ:

$$|f(x_i)| \leq \delta \qquad (3.13)$$

The Newton-Raphson method, when successful, works well and converges fast. Convergence problems occur when the value of $f'(x)$ is close to zero in the vicinity of the solution, where $f(x) = 0$. The Newton-Raphson method generally converges when $f(x)$, $f'(x)$ and $f''(x)$ are all continuous, if $f'(x)$ is not zero at the solution and if the starting value x_1 is near the actual solution.

3.4.1 Convergence of the Newton-Raphson Method

The Newton-Raphson iteration formula is given by

$$x_{i+1} = x_n - \frac{f(x_i)}{f'(x_i)} = \phi(x_n) \qquad (3.13a)$$

The general form of Eq. (3.13a) is given by

$$x = \phi(x) \qquad (3.13b)$$

The Newton-Raphson iteration method given by Eq. (3.13b) converges if $|\phi'(x)| < 1$.

Here
$$\phi(x) = x - \frac{f(x_i)}{f'(x_i)}$$

Hence
$$\phi'(x) = 1 - \left[\frac{[f'(x)]^2 - f(x)f''(x)}{[f'(x)]^2} \right] = \frac{f(x)f''(x)}{[f'(x)]^2}$$

or
$$|\phi'(x)| = \left| \frac{f(x)f''(x)}{[f'(x)]^2} \right|$$

Hence, the Newton-Raphson method converges if

$$\left| \frac{f(x)f''(x)}{[f'(x)]^2} \right| < 1$$

or
$$|f(x)f''(x)| < [f'(x)]^2 \qquad (3.13c)$$

If α denotes the actual root of $f(x) = 0$, then we can select a small interval in which $f(x)$, $f'(x)$ and $f''(x)$ are all continuous and the condition given by Eq. (3.13c) is satisfied. Therefore, the Newton-Raphson method always converges provided the initial approximation x_0 is taken very close to the actual root α.

3.4.2 Rate of Convergence of the Newton-Raphson Method

Let α denote he exact value of the root of $f(x) = 0$, and let x_i, x_{i+1}, be two successive approximations to the actual root α. If $\in_i$ and $\in_{i+1}$ are the corresponding errors, we have

$$x_i = \alpha + \in_i \text{ and } x_{i+1} = \alpha + \varepsilon_{i+1}$$

by Newton-Raphson's iterative formula

$$\alpha + \in_{i+1} = \alpha + \in_i - \frac{f(\alpha + \in_i)}{f'(\alpha + \in_i)}$$

$$\in_{i+1} - \in_i = -\frac{f(\alpha + \in_i)}{f'(\alpha + \in_i)}$$

or
$$\in_{i+1} = \in_i - \frac{f(\alpha) + \in_i f'(\alpha) + \left(\dfrac{\in_i^2}{2} \right) f''(\alpha) + \dots}{f'(\alpha) + \in_i f''(\alpha) + \dots}$$

$$= \in_i - \frac{\in_i f'(\alpha) + \frac{\in_i^2}{2} f''(\alpha) + \ldots}{f'(\alpha) + \in_i f''(\alpha) + \ldots} \qquad \text{(since f } (\alpha) = 0\text{)}$$

$$= \in_i - \frac{\in_i \left[f'(\alpha) + \frac{\in_i}{2} f''(\alpha) + \ldots \right]}{f'(\alpha) + \in_i f''(\alpha) + \ldots} = \frac{1}{2} \left[\frac{\in_i^2 f''(\alpha)}{f'(\alpha) + \in_i f''(\alpha) + \ldots} \right]$$

$$= \frac{1}{2} \left[\frac{\in_i^2 f''(\alpha)}{f'(\alpha) \left(1 + \in_i \frac{f''(\alpha)}{f'(\alpha)} \right) + \ldots} \right] = \in_{i+1} \approx \frac{f''(\alpha)}{2 f'(\alpha)} \qquad (3.13\text{d})$$

Equation (3.13d) shows that the error at each stage is proportional to the sequence of the error in the previous stage. Hence, the Newton-Raphson method has a quadratic convergence.

EXAMPLE 3.7

Use the Newton-Raphson method to find the real root near 2 of the equation $x^4 - 11x + 8 = 0$ accurate to five decimal places.

Solution:

Here $\quad f(x) = x^4 - 11x + 8$

$\qquad f'(x) = 4x^3 - 11$

$\qquad x_0 = 2$

and $\quad f(x_0) = f(2) = 2^4 - 11(2) + 8 = 2$

$\qquad f'(x_0) = f'(2) = 4(2)^3 - 11 = 21$

Therefore, $x_1 = x_0 - \dfrac{f(x_0)}{f'(x_0)} = 2 - \dfrac{2}{21} = 1.90476$

$$x_2 = x_1 - \frac{f(x_1)}{f'(x_1)} = 1.90476 - \frac{(1.90476)^4 - 11(1.90476) + 8}{4(1.90476)^3 - 11} = 1.89209$$

$$x_3 = x_2 - \frac{f(x_2)}{f'(x_2)} = 1.89209 - \frac{(1.89209)^4 - 11(1.89209) + 8}{4(1.89209)^3 - 11} = 1.89188$$

$$x_4 = x_3 - \frac{f(x_3)}{f'(x_3)} = 1.89188 - \frac{(1.89188)^4 - 11(1.89188) + 8}{4(1.89188)^3 - 11} = 1.89188$$

Hence the root of the equation is 1.89188.

EXAMPLE 3.8

Using the Newton-Raphson method, find a root of the function $f(x) = e^x - 3x^2$ to an accuracy of five digits. The root is known to lie between 0.5 and 1.0. Take the starting value of x as $x_0 = 1.0$.

Solution:

Start at $x_0 = 1.0$ and prepare a table as shown in Table 3.7, where $f(x) = e^x - 3x^2$ and $f'(x) = e^x - 6x$. The relative error

$$\xi = \left| \frac{x_{i+1} - x_i}{x_i + 1} \right|$$

The Newton-Raphson iteration method is given by

$$x_{i+1} = x_i - \frac{f(x_i)}{f'(x_i)}$$

TABLE 3.7

i	x_i	$f(x_i)$	$f'(x_i)$	x_{i+1}	ξ
0	1.0	−0.28172	−3.28172	0.91416	0.09391
1	0.91416	−0.01237	−2.99026	0.91002	0.00455
2	0.91002	−0.00003	−2.97574	0.91001	0.00001
3	0.91001	0	−2.97570	0.91001	6.613×10^{-11}

EXAMPLE 3.9

Evaluate $\sqrt{29}$ to five decimal places by the Newton-Raphson iterative method.

Solution:

Let $x = \sqrt{29}$ then $x^2 - 29 = 0$.

We consider $f(x) = x^2 - 29 = 0$ and $f'(x) = 2x$

The Newton-Raphson iteration formula gives

$$x_{i+1} = x_i - \frac{f(x_i)}{f'(x_i)} = x_i - \frac{x_i^2 - 29}{2x_i} = \frac{1}{2}\left(x_i + \frac{29}{x_i}\right) \tag{E.1}$$

Now $f(5) = 25 - 29 = -4 < 0$ and $f(6) = 36 - 29 = 7 > 0$.

Hence, a root of $f(x = 0)$ lies between 5 and 6.

Taking $x_0 = 3.3$, Equation (E.1) gives

$$x_1 = \frac{1}{2}\left(5.3 + \frac{29}{5.3}\right) = 5.38585$$

$$x_2 = \frac{1}{2}\left(5.38585 + \frac{29}{5.38585}\right) = 5.38516$$

$$x_3 = \frac{1}{2}\left(5.38516 + \frac{29}{5.38516}\right) = 5.38516$$

Since $x_2 = x_3$ up to five decimal places, $\sqrt{29} = 5.38516$.

3.4.3 Modified Newton-Raphson Method

Here, the iteration scheme is written as

$$x_{i+1} = x_i - \frac{f(x_i)}{f'(x_i + a(x_i)f(x_i))} = \phi(x_i) \text{ (say)} \tag{3.13e}$$

or $\quad \phi(x) = x - \dfrac{f(x)}{f'(x + a(x)f(x))}$

where $a(x)$ is a smooth function.

Consider

$$\phi'(x) = 1 - \frac{f'(x)}{f'(x + a(x)f(x))} + \frac{f(x)f''(x + a(x)f(x))(1 + a'(x)f(x) + a(x)f'(x))}{[f'(x + a(x)f(x))]^2} \tag{3.13f}$$

and

$$\phi''(x) = \frac{f''(x)}{f'(x + a(x)f(x))} + 2\frac{f'(x)f''(x + a(x)f(x))[1 + a'(x)f(x) + a(x)f'(x)]}{[f'(x + a(x)f(x))]^2}$$

$$-2\frac{f(x)[f''(x + a(x)f(x))]^2[1 + a'(x)f(x) + a(x)f'(x)]^2}{[f'(x + a(x)f(x))]^2}$$

$$+\frac{f(x)f'''(x+a(x)f(x))[1+a'(x)f(x)+a(x)f'(x)]^2}{[f'(x+a(x)f(x))]^2}$$

$$+\frac{[f(x)]^2 f''(x+a(x)f(x))a''(x)}{[f'(x+a(x)f(x))]^2}$$

$$+\frac{f(x)f''(x+a(x)f(x))[2a'(x)f'(x)+a(x)f''(x)]}{[f'(x+a(x)f(x))]^2} \qquad (3.13g)$$

If ξ is the root of the equation $f(x) = 0$, then $f(\xi) = 0$ and therefore $\phi(\xi) = \xi$ and $\phi'(\xi) = 0$.

Now, from Eq. (3.13g)

$$\phi''(\xi) = -\frac{f''(\xi)}{f'(\xi)} + \frac{2f'(\xi)f''(\xi)[1+a(\xi)f'(\xi)]}{[f'(\xi)]^2} = \frac{f''(\xi)}{f'(\xi)}[1+2a(\xi)f'(\xi)] \qquad (3.13h)$$

If $\qquad a(\xi) = -\dfrac{1}{2f'(\xi)}$ then $\phi''(\xi) = 0$.

Therefore, the iteration scheme in the modified Newton-Raphson method is given by

$$x_{i+1} = x_i - \frac{f(x_i)}{f'[x_i + a(x_i) + f(x_i)]} \qquad (3.13i)$$

where $\qquad a(x_i) = -\dfrac{1}{2f'(x_i)}$

Equation (3.13i) can also be written as

$$x_{i+1} = x_i - \frac{f(x_i)f'(x_i)}{[f'(x_i)]^2 - f(x_i)f''(x_i)} \qquad (3.13j)$$

In addition, we have

$$\phi(\xi) = \xi, \; \phi'(\xi) = 0 \text{ and } \phi'(\xi) = 0 \qquad (3.13k)$$

3.4.4 Rate of Convergence of the Modified Newton-Raphson Method

Let ξ be the root of the equation $f(x) = 0$. In addition, let

$$\in_i = x_i - \xi$$

Hence
$$x_{i+1} = \phi(x_i) = \phi(\in_i + \xi)$$

or
$$\in_{i+1} + \xi = \phi(\xi) + \in_i \phi'(\xi) + \frac{\in_i^2}{2!} \phi'(\xi) + \frac{\in_i^3}{3!} \phi'''(\xi) + ...$$

or
$$\in_{i+1} = \frac{\in_i^3}{3!} \phi'''(\xi) + O(\in_i^4) \qquad (3.13l)$$

If we neglect the terms $\in_i^4$ and higher powers of $\in_i^4$, Equation (3.13l) reduces to

$$\in_{i+1} = A \in_i^3$$

in which
$$A = \frac{1}{3!} \phi'''(\xi) \qquad (3.13m)$$

Equation (3.13m) shows that the rate of convergence of the modified Newton-Raphson method is cubic.

EXAMPLE 3.10

Repeat Example 3.7 using the modified Newton-Raphson method.

Solution:

$$f(x) = x^4 - 11x + 8$$
$$f'(x) = 4x^3 - 11$$
$$f''(x) = 12x^2$$

The modified Newton-Raphson's formula is

$$x_{i+1} = x_i - \frac{f(x_i)f'(x_i)}{[f'(x_i)]^2 - f(x_i)f''(x_i)}$$

The calculations are shown in Table 3.8.

TABLE 3.8

n	x_i	$f(x_i)$	$f'(x_i)$	$f''(x_i)$	x_{i+1}
0	2	2	21	48	1.878261
1	1.878261	−0.21505	15.50499	42.33437	1.891624
2	1.891624	−0.00405	16.07476	42.93891	1.891876
3	1.891876	-1.4×10^{-6}	16.08557	42.95034	1.891876

Hence, the root is 1.891876.

3.5   SUCCESSIVE APPROXIMATION METHOD

Suppose we are given an equation $f(x) = 0$ whose roots are to be determined. The equation can be written as

$$x = f(x) \tag{3.14}$$

Let $x = x_0$ be an initial approximation to the desired root α. Then, the first approximation x_1 is given by

$$x_1 = \phi(x_0)$$

The second approximation $x_2 = \phi(x_1)$. The successive approximations are then given by $x_3 = \phi(x_2)$, $x_4 = \phi(x_3)$, ..., $x_n = \phi(x_{n-1})$.

The sequence of approximations of x_1, x_2, ..., x_n always converge to the root of $x = \phi(x)$ and it can be shown that if $|\phi'(x)| < 1$, when x is sufficiently close to the exact value c of the root and $x_n \to c$ as $n \to \infty$. The convergence of $x_{i+1} = \phi(x_n)$, for $|\phi'(x)| < 1$ is shown in Figure 3.5. The following theorem presents the convergence criteria for the iterative sequence of solution for the successive approximation method.

Theorem 3.5:   Let α be a root of $f(x) = 0$ which is equivalent to $x = \phi(x)$, $\phi(x)$ is continuously differentiable function in an interval I containing the root $x = \alpha$, if $|\phi'(x)| < 1$, then the sequence of approximations x_0, x_1, x_2, ..., x_n will converge to the root α provided the initial approximation $x_0 \in I$.

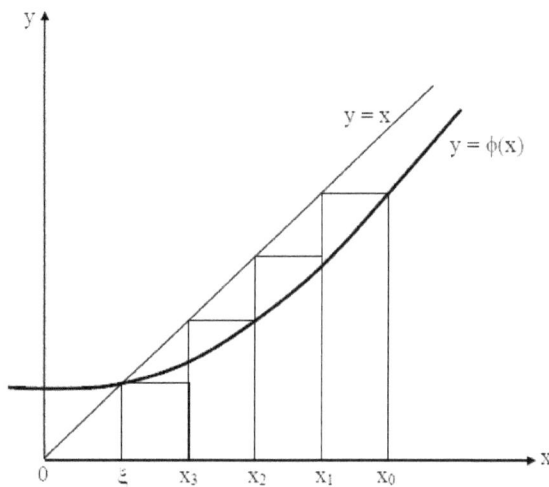

FIGURE 3.5 Converge of $x_{i+1} = \phi(x_n)$, for $|\phi'(x)| < 1$.

Proof: Let α be the actual root of $x = \phi(x)$, then we can write

$$\alpha = \phi(\alpha) \tag{3.15}$$

Let $x = x_0$ be an initial approximation to the root, then

$$x_1 = \phi(x_0) \tag{3.16}$$

From Equations (3.15) and (3.16), we obtain

$$\alpha - x_1 = \phi(\alpha) - \phi(x_0)$$

By using the Lagrange's mean value theorem, we can write

$$\alpha - x_1 = (\alpha - x_0)\phi'(\xi_0) \qquad \text{for } x_0 < \xi_0 < \alpha$$

Similarly, $\quad \alpha - x_2 = (\alpha - x_1)\phi'(\xi_1) \qquad \text{for } x_1 < \xi_1 < \alpha$

$$\alpha - x_3 = (\alpha - x_2)\phi'(\xi_2) \qquad \text{for } x_2 < \xi_2 < \alpha$$

and so on, or

$$\alpha - x_n = (\alpha - x_{n-1})\phi'(\xi_{n-1}) \qquad \text{for } x_{n-1} < \xi_{n-1} < \alpha$$

Multiplying the above equations, we obtain

$$\alpha - x_n = (\alpha - x_0)\phi'(\xi_0)\phi'(\xi_1)\phi'(\xi_2), \ldots, \phi'(\xi_{n-1})$$

If $|\phi'(x_i)| \leq \kappa < 1$, for all I, then

$$|x_n - \alpha| \leq |x_0 - \alpha||\phi'(\xi_0)|\phi'(\xi_1)|\ldots|\phi'(\xi_{n-1})| \leq \kappa, \kappa.\ldots\kappa|x_0 - \alpha| \leq \kappa^n|x_0 - \alpha|$$

As $\kappa < 1$ therefore $k^n \to 0$, as $n \to \infty$, and thus we have $x_n \to \alpha$, provided $x_0 \in I$. Hence, the theorem is proved.

3.5.1   Error Estimate in the Successive Approximation Method

Let $\in_n = x_n - \xi$, the error estimate at the nth iteration, then $\lim\limits_{n \to \infty}(\in_{n+1} / \in_n) = \phi'(\xi)$ is satisfied. We know that $|x_{n+1} - \xi| = |\phi(x_n) - \phi(\xi)| = |\phi'(\xi_n)||x_n - \xi|, \xi_n \in (x_n, \xi)$ by the mean value theorem. Hence

$$\in_{n+1} = \in_n|\phi'(\xi_n)| \Rightarrow \lim\limits_{n \to \infty}(\in_{n+1} / \in_n) = \phi'(\xi)$$

Therefore, we obtain the order of convergence as linear. But if $\phi'(\xi) = 0$ and $\phi''(\xi) \neq 0$, then the Taylor series expansion of ϕ in a neighborhood of ξ is given by

$$\phi(x_n) = \phi(\xi) + (x_n - \xi)\phi'(\xi) + \frac{(x_n - \xi)^2}{\angle 2}\phi''(\xi) + \dots$$

which shows that

$$\in_{n+1} = \in_n \phi'(\xi) - \frac{\in_n^2}{2}\phi''(\xi) + \frac{\in_n^3}{6}\phi''(\xi) - \dots.$$

on using $x_{n+1} = \phi(x_n)$ and $\in_n = |x_{n+1} - \xi|$.

Hence, $\in_{n+1} = -\frac{\in_n^2}{2}\phi''(\xi)$, on neglecting the terms containing cubes and higher power of $\in_n$. This is a quadratic convergence.

EXAMPLE 3.11

Find a real root of $x^3 - 2x - 3 = 0$, correct to three decimal places using the successive approximation method.

Solution:

Here $f(x) = x^3 - 2x - 3 = 0$ (E.1)

Also $f(1) = 1^3 - 2(1) - 3 = -4 < 0$

and $f(2) = 2^3 - 2(2) - 3 = 1 > 0$

Therefore, the root of Equation (E.1) lies between 1 and 2. Since $f(1) < f(2)$, we can take the initial approximation $x_0 = 1$. Now, Eq. (E.1) can be rewritten as

$$x^3 = 2x + 3$$

or $x = (2x + 3)^{1/3} = \phi(x)$

The successive approximations of the root are given by

$$x_1 = \phi(x_0) = (2x_0 + 3)^{1/3} = [2(1) + 3]^{1/3} = 1.25992$$

$$x_2 = \phi(x_1) = (2x_1 + 3)^{1/3} = [2(1.25992) + 3]^{1/3} = 1.31229$$

$$x_3 = \phi(x_2) = (2x_2 + 3)^{1/3} = [2(1.31229) + 3]^{1/3} = 1.32235$$

$$x_4 = \phi(x_3) = (2x_3 + 3)^{1/3} = [2(1.32235) + 3]^{1/3} = 1.32427$$

$$x_5 = \phi(x_4) = (2x_4 + 3)^{1/3} = [2(1.32427) + 3]^{1/3} = 1.32463$$

Hence, the real root of $f(x) = 0$ is 1.324 correct to three decimal places.

EXAMPLE 3.12

Find a real root of $2x - \log_{10} x - 9$ using the successive approximation method.

Solution:

Here $\quad f(x) = 2x - \log_{10} x - 9$ $\hfill$ (E.1)

$\quad\quad f(4) = 2(4) - \log_{10}(4) - 9 = 8 - 0.60206 - 9 = -1.60206$

$\quad\quad f(5) = 2(5) - \log_{10}(5) - 9 = 10 - 0.69897 - 9 = 0.30103$

Therefore, the root of Equation (E.1) lies between 4 and 5. Rewriting Equation (E.1) as

$$x = \frac{1}{2}(\log_{10} x + 9) = \phi(x)$$

We start with $x_0 = 4$.

$$x_1 = \phi(x_0) = \frac{1}{2}(\log_{10}4 + 9) = 4.80103$$

$$x_2 = \phi(x_1) = \frac{1}{2}(\log_{10}4.80103 + 9) = 4.84067$$

$$x_3 = \phi(x_2) = \frac{1}{2}(\log_{10}4.84067 + 9) = 4.84245$$

$$x_4 = \phi(x_3) = \frac{1}{2}(\log_{10}4.84245 + 9) = 4.84253$$

$$x_5 = \phi(x_4) = \frac{1}{2}(\log_{10}4.84253 + 9) = 4.84254$$

Hence, $x = 4.8425$ is the root of Eq. (E.1) correct to four decimal places.

EXAMPLE 3.13

Find a real root of $\cos x - 3x + 5 = 0$. Correct to four decimal places using the successive approximation method.

Solution:

Here, we have

$$f(x) = \cos x - 3x + 5 = 0 \qquad\qquad\qquad \text{(E.1)}$$

$$f(0) = \cos(0) - 3(0) + 5 = 5 > 0$$

$$f(\pi/2) = \cos(\pi/2) - 3(\pi/2) + 5 = -3\pi/2 + 5 < 0$$

Also $f(0)f(\pi/2) < 0$

Hence, a root of $f(x) = 0$ lies between 0 and $\pi/2$.

The given Eq. (E.1) can be written as

$$x = \frac{1}{3}[5 + \cos x]$$

Here $\phi(x) = \frac{1}{3}[5 + \cos x]$ and $\phi'(x) = -\frac{\sin x}{3}$

$$|\phi'(x)| = \left|\frac{\sin x}{3}\right| < 1 \text{ in } (0, \pi/2)$$

Hence, the successive approximation method applies.

Let $x_0 = 0$

$$x_1 = \phi(x_0) = \frac{1}{3}[5 + \cos 0] = 2$$

$$x_2 = \phi(x_1) = \frac{1}{3}[5 + \cos(2)] = 1.52795$$

$$x_3 = \phi(x_2) = \frac{1}{3}[5 + \cos(1.52795)] = 1.68094$$

$$x_4 = \phi(x_3) = \frac{1}{3}[5 + \cos(1.68094)] = 1.63002$$

$$x_5 = \phi(x_4) = \frac{1}{3}[5 + \cos(1.63002)] = 1.64694$$

$$x_6 = \phi(x_5) = \frac{1}{3}[5 + \cos(1.64694)] = 1.64131$$

$$x_7 = \phi(x_6) = \frac{1}{3}[5 + \cos(1.64131)] = 1.64318$$

$$x_8 = \phi(x_7) = \frac{1}{3}[5 + \cos(1.64318)] = 1.64256$$

$$x_9 = \phi(x_8) = \frac{1}{3}[5 + \cos(1.64256)] = 1.64277$$

$$x_{10} = \phi(x_9) = \frac{1}{3}[5 + \cos(1.64277)] = 1.64270$$

Hence, the root of the equation is 1.6427 correct to four decimal places.

3.6 SECANT METHOD

The secant method is very similar to the Newton-Raphson method. The main disadvantage of the Newton-Raphson method is that the method requires the determination of the derivatives of the function at several points. Often, the calculation of these derivatives takes too much time. In some cases, a closed-form expression for $f'(x)$ may be difficult to obtain or may not be available.

To remove this drawback of the Newton-Raphson method, the derivatives of the function being approximated by finite differences instead of being calculated analytically. In particular, the derivative $f'(x)$ is approximated by the backward difference

$$f'(x_i) = \frac{f(x_i) - f(x_{i-1})}{x_i - x_{i-1}} \tag{3.17}$$

where x_i and x_{i-1} are two approximations to the root but does not require the condition $f(x_i).f(x_{i-1}) < 0$.

Now, from the Newton-Raphson method, we have

$$x_{i+1} = x_i - \frac{f(x_i)}{f'(x_i)} = x_i - \frac{f(x_i)(x_i - x_{i-1})}{f(x_i) - f(x_{i-1})} \tag{3.18}$$

It should be noted here that from Equation (3.18) this method requires two initial guess values x_0 and x_1 for the root. The secant method is illustrated geometrically as shown in Figure 3.6, where a secant is drawn connecting $f(x_{i-1})$ and $f(x_i)$. The point where it intersects the x-axis is x_{i+1}. Another secant is drawn connecting $f(x_i)$ and $f(x_{i+1})$ to obtain x_{i+2} and the process continues.

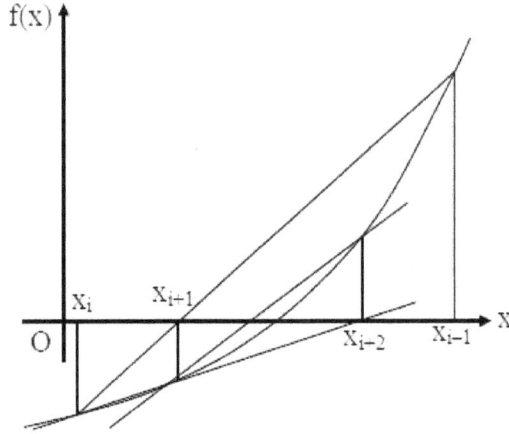

FIGURE 3.6 The secant method.

3.6.1 Convergence of the Secant Method

The formula for the secant method can be written as

$$x_{n+1} = x_n - \frac{(x_n - x_{n-1})}{f(x_n) - f(x_{n-1})}$$ (3.19)

Let ξ be the exact root of the equation $f(x) = 0$ and $f(\xi) = 0$. The error at the nth iteration is given by

$$\in_n = x_n - \xi$$ (3.20)

Now Eq. (3.19) becomes

$$\in_{n+1} = \in_n - \frac{(\in_n - \in_{n-1}) f(\in_n + \xi)}{f(\in_n + \xi) - f(\in_{n-1} + \xi)}$$

$$= \in_n - \frac{(\in_n - \in_{n-1})[f(\xi) + \in_n f'(\xi) + (\in_n^2 /2) f''(\xi) + ...]}{(\in_n - \in_{n-1}) f'(\xi) + \frac{1}{2}(\in_n^2 - \in_{n-1}^2) f''(\xi) + ...}$$

$$= \in_n - \left[\in_n + \frac{\in_n^2 f''(\xi)}{2 f'(\xi)} + ... \right] \left[1 + \frac{1}{2}(\in_n + \in_{n-1}) \frac{f''(\xi)}{f'(\xi)} + ... \right]^{-1}$$

$$= \frac{1}{2} \epsilon_n \epsilon_{n-1} \frac{f''(\xi)}{f'(\xi)} + O(\epsilon_n^2 \epsilon_{n-1} + \epsilon_n \epsilon_{n-1}^2) \tag{3.21}$$

Equation (3.21) can be expressed as

$$\epsilon_{n+1} = c \epsilon_n \epsilon_{n-1} \tag{3.22}$$

where $\quad c = \dfrac{1}{2} \dfrac{f''(\xi)}{f'(\xi)} \tag{3.23}$

Equation (3.23) is a non-linear difference equation which can be solved by letting $\epsilon_{n+1} = A \epsilon_n^p$ or $\epsilon_n = A \epsilon_{n-1}^p$ and that gives

$$\epsilon_{n-1} = \epsilon_n^{1/p} A^{-1/p} \tag{3.24}$$

Hence $\quad A \epsilon_n^p = c \epsilon_n \epsilon_n^{1/p} A^{-1/p}$

or $\quad \epsilon_n^p = c A^{-(1+1/p)} \epsilon_n^{1+1/p} \tag{3.25}$

Now by equating the power of ϵ_n both sides of Eq. (3.25), we obtain

$$p = 1 + \frac{1}{p}$$

or $\quad \mathrm{p} = \dfrac{1}{2} \left(1 \pm \sqrt{5} \right) \tag{3.26}$

Therefore taking the positive sign in Eq. (3.26), we get

$$p = 1.618$$

and $\quad \epsilon_{n+1} = A \epsilon_n^{1.618} \tag{3.27}$

Hence, the rate of convergence of the secant method is 1.618, which is lesser than the Newton-Raphson method. The second method evaluates the function only once in each iteration, whereas the Newton-Raphson method evaluates two functions f and f' in each iteration. Therefore, the second method is more efficient than the Newton-Raphson method.

EXAMPLE 3.14

Find a root of the equation $x^3 - 8x - 5 = 0$ using the secant method.

Solution:

$$f(x) = x^3 - 8x - 5 = 0$$
$$f(3) = 3^3 - 8(3) - 5 = -2$$
$$f(4) = 4^3 - 8(4) - 5 = -27$$

Therefore, one root lies between 3 and 4. Let the initial approximations be $x_0 = 3$, and $x_1 = 3.5$. Then, x_2 is given by

$$x_2 = \frac{x_0 f(x_1) - x_1 f(x_0)}{f(x_1) - f(x_0)}$$

The calculations are summarized in Table 3.9.

TABLE 3.9 Secant method.

x_0	$f(x_0)$	x_1	$f(x_1)$	x_2	$f(x_2)$
3	−2	3.5	9.875	3.08421	−0.33558
3.5	9.875	3.08421	−0.33558	3.09788	−0.05320
3.08421	−0.33558	3.09788	−0.05320	3.10045	0.00039
3.08788	−0.05320	3.10045	0.00039	3.10043	0
3.10045	0.00039	3.10043	0	3.10043	0

Hence, a root is 3.1004 correct up to five significant figures.

EXAMPLE 3.15

Determine a root of the equation $\sin x + 3 \cos x - 2 = 0$ using the secant method. The initial approximations x_0 and x_1 are 2 and 1.5.

Solution:

The formula for x_2 is given by

$$x_2 = \frac{x_0 f(x_1) - x_1 f(x_0)}{f(x_1) - f(x_0)}$$

The calculations are summarized in Table 3.10.

TABLE 3.10 Secant method

x_0	$f(x_0)$	x_1	$f(x_1)$	x_2	$f(x_2)$
2	−2.33914	1.5	−0.79029	1.24488	−0.09210
1.5	−0.79029	1.24488	−0.09210	1.21122	−0.00833
1.24488	−0.09210	1.21122	−0.00833	1.20788	−0.00012
1.21122	−0.00833	1.20788	−0.00012	1.20783	0
1.20788	−0.00012	1.20783	0	1.20783	0

Hence, a root is 1.2078 correct up to five significant figures.

EXAMPLE 3.16

Repeat Example 3.14 with initial approximations of $x_0 = -2$ and $x_1 = -1.5$.

Solution:

x_2 is given by

$$x_2 = \frac{x_0 f(x_1) - x_1 f(x_0)}{f(x_1) - f(x_0)}$$

The calculations are summarized in Table 3.11.

TABLE 3.11 Secant method.

x_0	$f(x_0)$	x_1	$f(x_1)$	x_2	$f(x_2)$
−2	−4.15774	−1.5	−2.78528	−0.48529	0.18715
−1.5	−2.78528	−0.48529	0.18715	−0.54918	0.03687
−0.48529	0.18715	−0.54918	0.03687	−0.56488	−0.00129
−0.54918	0.03687	−0.56485	−0.00129	−0.56432	0.00001
−0.56485	−0.00129	−0.56432	0.00001	−0.56433	0
−0.56432	0.00001	−0.56433	0	−0.56433	0

Hence, a root is −0.5643 correct up to five significant figures.

3.7 MULLER'S METHOD

Muller's method is an iterative method and free from the evaluation of derivative as in the Newton-Raphson method. It requires three starting points (x_{n-2}, f_{n-2}), (x_{n-1}, f_{n-1}), and (x_n, f_n). A parabola is constructed that passes through these points then the quadratic formula is employed to find a root of the quadratic for the next approximation. In other words, we assume that x_n is the best approximation to the root, and consider the parabola through the three starting values as shown in Figure 3.7. We denote $f(x_{n-2}) = f_{n-2}$, $f(x_{n-1}) = f_{n-1}$ and $f(x_n) = f_n$.

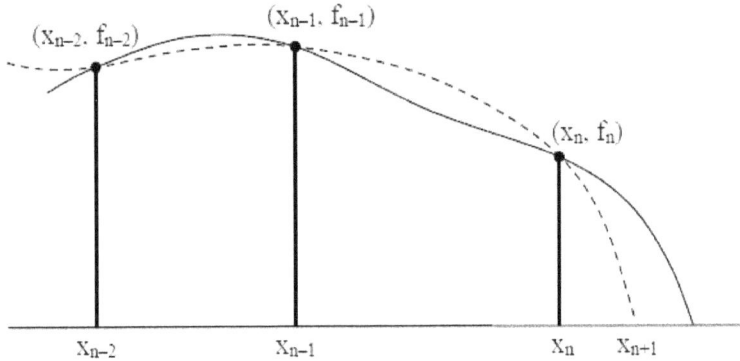

FIGURE 3.7 Muller's method.

Let the quadratic polynomial be

$$f(x) = ax^2 + bx + c \tag{3.28}$$

If Equation (3.28) passes through the points (x_{n-2}, f_{n-2}), (x_{n-1}, f_{n-1}), and (x_n, f_n), then

$$ax_{n-2}^2 + bx_{n-2} + c = f_{n-2}$$
$$ax_{n-1}^2 + bx_{n-1} + c = f_{n-1} \tag{3.29}$$
$$ax_n^2 + bx_n + c = f_n$$

Eliminating a, b, c from Equation (3.29), we obtain the following determinant

$$\begin{vmatrix} f(x) & x^2 & x & 1 \\ f_{n-2} & x_{n-2}^2 & x_{n-2} & 1 \\ f_{n-1} & x_{n-1}^2 & x_{n-1} & 1 \\ f_n & x_n^2 & x_n & 1 \end{vmatrix} = 0 \tag{3.30}$$

By expanding this determinant in Eq. (3.30) the function $f(x)$ can be written as

$$f(x) = \frac{(x - x_{n-1})(x - x_n)}{(x_{n-2} - x_{n-1})(x_{n-2} - x_n)} f_{n-2} + \frac{(x - x_{n-2})(x - x_n)}{(x_{n-1} - x_{n-2})(x_{n-1} - x_n)} f_{n-1}$$

$$+ \frac{(x - x_{n-2})(x - x_{n-1})}{(x_n - x_{n-2})(x_n - x_{n-1})} f_n \qquad (3.31)$$

Equation (3.31) is a quadratic polynomial passing through the three given points.

Let $\qquad h = x - x_n, h_n = x_n - x_{n-1}$ and $h_{n-1} = x_{n-1} - x_{n-2}$.

Now, Equation (3.31) becomes

$$\frac{h(h + h_n)}{h_{n-1}(h_{n-1} + h_n)} f_{n-2} - \frac{h(h + h_n + h_{n-1})}{h_n h_{n-1}} f_{n-1} + \frac{(h + h_n)(h + h_n + h_{n-1})}{h_n(h_n + h_{n-1})} f_n = 0 \quad (3.32)$$

Noting $f(x) = 0.$

Let $\qquad \lambda = \dfrac{h}{h_n}, \; \lambda_n = \dfrac{h_n}{h_{n-1}}$ and $\delta_n = 1 + \lambda_n.$

The Equation (3.32) now reduces to the following form:

$$\lambda^2(f_{n-1}\lambda_n^2 f_{n-1}\lambda_n\delta_n + f_n\lambda_n)\delta_n^{-1} + \lambda\{f_{n-2}\lambda_n^2 - f_{n-2}\delta_n^2 + f_n(\lambda_n + \delta_n)\}\delta_n^{-1} + f_n = 0 \quad (3.33)$$

or $\qquad \lambda^2 c_n + \lambda g_n + \delta_n f_n = 0 \qquad\qquad\qquad (3.34)$

where $\quad g_n = \lambda_n^2 f_{n-2} - \delta_n^2 f_{n-1} + (\lambda_n + \delta_n)f_n$

$\qquad\quad c_n = \lambda_n(\lambda_n f_{n-2} - \delta_n f_{n-1} + f_n)$

Equation (3.34) can be written as

$$\delta_n f_n\left(\frac{1}{\lambda^2}\right) + \frac{g_n}{\lambda} + c_n = 0 \qquad (3.35)$$

Solving Equation (3.35) for $1/\lambda$, we obtain

$$\lambda = -\frac{2\delta_n f_n}{g_n \pm \sqrt{g_n^2 - 4\delta_n f_n c_n}} \qquad (3.36)$$

The sign in the denominator of Equation (3.36) is ± according as $g_n > 0$ or $g_n < 0$.

Hence $\quad \lambda = \dfrac{x - x_n}{x_n - x_{n-1}}$ or $x = x_n + (x_n - x_{n-1})\lambda$ $\hfill$ (3.37)

Now, replacing x on left-hand side by x_{n+1} in Equation (3.37), we obtain

$$x_{n+1} = x_n + (x_n - x_{n-1})\lambda \hfill (3.38)$$

Equation (3.38) is the Muller's formula for the root and the procedure for finding a root is summarized in Table 3.12.

TABLE 3.12 Muller's method.

$$h_n = x_n - x_{n-1}, \ \lambda_n = \frac{h_n}{h_{n-2}}, \ \delta_n = 1 + \lambda_n$$

$$g_n = \lambda_n^2 f_{n-2} - \delta_n^2 f_{n-1} + (\lambda_n + \delta_n)f_n$$

$$c_n = \lambda_n(\lambda_n f_{n-2} - \delta_n f_{n-1} + f_n)$$

$$\lambda = -\frac{2\delta_n f_n}{g_n \pm \sqrt{g_n^2 - 4\delta_n f_n c_n}}$$

$$x_{n+1} = x_n + (x_n - x_{n-1})\lambda$$

$$x_{n-1} = x_n + (x_n - x_{n-1})\lambda$$

EXAMPLE 3.17

Find a root of the equation $x^3 - 3x - 7 = 0$ using Muller's method where the root lies between 2 and 3.

Solution:

Let $x_0 = 2$, $x_1 = 2.5$ and $x_2 = 3$. The calculations are shown in Tables 3.13 and 3.14.

TABLE 3.13 Muller's method.

n	x_{n-2}	x_{n-1}	x_n	h_n	h_{n-1}	λ_n	δ_n
2	2	2.5	3	0.5	0.5	1	2
3	2.5	3	2.4272	−0.5728	0.5	−1.14559	−0.14559
4	3	2.4272	2.42599	−0.00122	−0.5728	0.00213	1.00213
5	2.4272	2.42599	2.42599	0	−0.00122	−0.0029	0.99710

TABLE 3.14 Muller's method.

n	f_{n-2}	f_{n-1}	f_n	g_n	c_n	λ	x_{n+1}
2	–5	1.125	11	23.5	3.75	–1.14559	2.42720
3	1.125	11	0.01781	1.22026	–0.37867	0.00213	2.42599
4	11	0.01781	–0.00005	–0.01789	0.00001	–0.0029	2.42599
5	0.01781	–0.0005	0	0.00005	0	–0.00005	2.42599

Hence one root is 2.42599 correct up to five decimal places.

3.8   CHEBYSHEV METHOD

Consider the equation to be solved as $f(x) = 0$. The function $f(x)$ can be expanded by Taylor's series in the neighborhood of x_n as

$$0 = f(x) = f(x_n) + x - x_n)f'(x_n) + \ldots. \tag{3.39}$$

Equation (3.39) gives

$$x = x_n - \frac{f(x_n)}{f'(x_n)} \tag{3.40}$$

Equation (3.40) gives the $(n + 1)^{\text{th}}$ approximation to the root.

Hence
$$x_{n+1} = x_n - \frac{f(x_n)}{f'(x_n)} \tag{3.41}$$

Once again, we expand $f(x)$ by Taylor's series and retain up to the second order term, we obtain

$$0 = f(x) = f(x_n) + (x - x_n)f'(x_n) + \frac{(x - x_n)^2}{2}f''(x_n) \tag{3.42}$$

Hence $\quad f(x_{n+1}) = f(x_n) + (x_{n+1} - x_n)f'(x_n) + \frac{(x_{n+1} - x_n)^2}{2}f''(x_n) = 0 \tag{3.43}$

Substituting the value of $x_{n+1} - x_n$ from Equation (3.41) to the last term and we obtain

$$f(x_n) + (x_{n+1} - x_n)f'(x_n) + \frac{1}{2}\frac{[f(x_n)]^2}{[f'(x_n)]^2}f''(x_n) = 0 \tag{3.44}$$

Hence $\quad x_{n+1} = x_n - \dfrac{f(x_n)}{f'(x_n)} - \dfrac{1}{2}\dfrac{[f(x_n)]^2}{[f'(x_n)]^3} f''(x_n)$ (3.45)

Equation (3.45) can be recognized as the extended form of the Newton-Raphson formula and it is called the Chebyshev's formula.

The rate of convergence of this method can be shown to be cubic.

3.9 AITKEN'S Δ^2 METHOD

Suppose we have an equation

$$f(x) = 0 \tag{3.46}$$

whose roots are to be determined.

Let I be an interval containing the point $x = \alpha$.

Now, Equation (3.46) can be written as $x = \phi(x)$ such that $\phi(x)$ and $\phi'(x)$ are continuous in I and $|\phi'(x)| < 1$ for all x in I.

Denoting x_{i-1}, x_i and x_{i+1} as the three successive approximations to the desired root α, we can write

$$\alpha - x_i = \lambda(\alpha - x_{i-1}) \tag{3.47}$$

and $\qquad \alpha - x_{i+1} = \lambda(\alpha - x_i)$ (3.48)

where λ is a constant so that $|\phi'(x)| \leq \lambda \leq 1$ for all i.

Dividing Equation (3.47) with Equation (3.48), we obtain

$$\frac{\alpha - x_i}{\alpha - x_{i+1}} = \frac{\alpha - x_{i-1}}{\alpha - x_i} \tag{3.49}$$

Equation (3.49) gives

$$\alpha = x_{i+1} - \frac{(x_{i+1} - x_i)^2}{(x_{i+1} - 2x_i - x_{i-1})} \tag{3.50}$$

Now $\qquad \Delta x_i = x_{i+1} - x_i$

and $\qquad \Delta^2 x_{i-1} = (E-1)^2 x_{i-1} = (E^2 - 2E + 1)x_{i-1} = x_{i+1} - 2x_i + x_{i-1}$ (3.51)

Using Equation (3.51), Equation (3.50) can be written as

$$\alpha = x_{i+1} - \frac{(\Delta x_i)^2}{\Delta_{i-1}^2} \tag{3.52}$$

Equation (3.52) gives the successive approximation to the root α and method is known as the *Aitken's Δ^2 method*.

	Δ	Δ^2
x_{i-1}		
	Δx_{i-1}	
x_i		$\Delta^2 x_{i-1}$
	Δx_i	
x_{i+1}		

EXAMPLE 3.18

Find the root of the function $x = \left(\dfrac{1 + \cos x}{3}\right)$ correct to four decimal places using Aitken's iteration method.

Solution:

$$f(x) = \cos x - 3x + 1 \tag{E.1}$$

$$f(0) = 1$$

$$f(\pi/2) = \cos(\pi/2) - 3(\pi/2) + 1 = -3.005$$

Hence $f(0) > 0$ and $f(\pi/2) < 0$

Also $f(0) f(\pi/2) = 1(-3.005) = -3.005 < 0$

Therefore, a root exists between 0 and $\pi/2$.

Equation (E.1) can be written as

$$x = \left(\frac{1 + \cos x}{3}\right) = \phi(x)$$

Now $\phi'(x) = \dfrac{-\sin x}{3} = |\phi'(x)| = \left|\dfrac{-\sin x}{3}\right| < 1 - x \in \left(0, \dfrac{\pi}{2}\right) \tag{E.2}$

Equation (E.2) signifies that Aitken's method can be employed.

Let $x_0 = 0$ be an initial approximation to the root of (E.1).

$$x_1 = \phi(x_0) = \left(\frac{1 + \cos 0}{3}\right) = 0.66667$$

$$x_2 = \phi(x_1) = \frac{1 + \cos(0.66667)}{3} = 0.59530$$

$$x_3 = \phi(x_2) = \frac{1 + \cos(0.59530)}{3} = 0.60933$$

We can now construct the table as shown in Table 3.15.

TABLE 3.15 Aitken's method.

x	**Δx**	**Δ²**
$x_1 = 0.66667$	$\dfrac{-0.07137}{\Delta x_1}$	
$x_2 = 0.59530$		$\dfrac{0.08540}{\Delta^2 x_1}$
$x_3 = 0.60933$	$\dfrac{0.01403}{\Delta x_2}$	

Therefore

$$x_4 = x_3 - \frac{(\Delta x_2)^2}{(\Delta^2 x_1)} = 0.60933 - \frac{(0.01403)^2}{(0.08540)} = 0.60703.$$

Hence, the root is 0.6070 correct to four decimal places.

3.10 COMPARISON OF ITERATIVE METHODS

The bisection method and the method of false position always converge to an answer, provided a root is bracketed in the interval (a, b) to start with. Since the root lies in the interval (a, b), on every iteration the width of the interval is reduced until the solution is obtained. The Newton-Raphson method and the method of successive approximations require only one initial guess and on every iteration it approaches to the true solution or the exact root. The bisection method is guaranteed to converge. The bisection method may fail when the function is tangent to the axis and does not cross the x-axis at $f(x) = 0$.

The bisection method, the method of false position, and the method of successive approximations converge linearly while the Newton-Raphson method converges quadratically. The Newton-Raphson method requires a lesser number of iterations than the other three methods. One disadvantage with the Newton-Raphson method is that when the derivative $f'(x_i)$ is zero, a new starting or initial value of x must be selected to continue with the iterative procedure. The successive approximation method converges only when the condition $|\phi'(x)| < 1$ is satisfied. Table 3.16 gives a summary of the comparison of the methods presented in this chapter.

TABLE 3.16 Comparison of the methods.

No	Method	Formula	Order of convergence	Functional evaluations at each step
1.	Bisection	$x_{s1} = \dfrac{a+b}{2}$	One bit/ iteration gain	1
2.	False position	$x_s = \dfrac{af(b) - bf(a)}{f(b) - f(a)}$	1	1
3.	Newton-Raphson	$x_{i+1} = x_i - \dfrac{f(x_i)}{f'(x_i)}$	2	2
4.	Modified Newton-Raphson	$x_{i+1} = x_n - \dfrac{f_n}{f'\left[x_n - \frac{1}{2} f_n / f'_n\right]}$	3	3
5.	Successive approximation	$x_1 = \phi(x_0)$	1	1
6.	Secant	$x_{i+1} = x_i - \dfrac{f(x_i)(x_i - x_{i-1})}{f(x_i) - f(x_{i-1})}$	1.62	1
7.	Muller	$x_{n+1} = x_n + (x_n - x_{n-1})\lambda$	1.84	1
8.	Chebyshev	$x_{n+1} = x_n - \dfrac{f_n}{f'_n} - \dfrac{1}{2} \dfrac{f_n^2}{f_n'^3} f''_n$	3	3

3.11 SUMMARY

In this chapter, the techniques for the numerical solution of algebraic and transcendental equations have been presented. Numerical methods involving iterative solution of nonlinear equations are more powerful. These methods can be divided into two categories: Direct methods and indirect (or iterative) methods. The indirect or iterative methods are further divided into two categories: bracketing and open method. The bracketing methods require the limits between which the root lies, whereas the open methods require the initial estimation of the solution. The bisection and false position methods are two known examples of the bracketing methods. Among the open methods, the Newton-Raphson and the method of successive approximation are most commonly used. The most popular method for solving a nonlinear equation is the Newton-Raphson method and this method has a quadratic rate of convergence. These methods have been illustrated with examples.

EXERCISES

3.1 Use the bisection method to find a solution accurate to four decimal places for $x = \tan x$ in the interval $(4.4, 4.6)$.

3.2 Determine the solution of the equation $8 - \dfrac{9}{2}(x - \sin x) = 0$ by using the bisection method accurate to five decimal places in the interval $(2, 3)$.

3.3 Use the bisection method to compute the root of $e^x - 3x = 0$ correct to three decimal places in the interval $(1.5, 1.6)$.

3.4 Find the root of $\log x = \cos x$ and correct to two decimal places using the bisection method.

3.5 Use the bisection method to find a root of the equation $x^3 - 4x - 9 = 0$ in the interval $(2, 3)$, accurate to four decimal places.

3.6 Use the bisection method to determine a root correct to three decimal places of the equation $x \log_{10} x = 1.2$. Interval $(2, 3)$.

3.7 Use the bisection method to find a root of the equation $4.905t^2 - 15t + 5 = 0$ in the interval $(0.3, 0.4)$ with an accuracy of 4 digits.

3.8 Use the bisection method to find the root of $f(x) = x^3 - 10x^2 + 5 = 0$ that lies in the interval $(0.6, 0.8)$ correct within four decimal places.

3.9 Use the bisection method to find the root of $f(x) = x - \tan x$ in the interval $(7, 8)$ correct to four decimal places.

3.10 Use the bisection method to find the smallest positive root of $\cos x = \dfrac{1}{2} + \sin x$ in the interval $(0.41, 043)$. Use an error tolerance of $\epsilon = 0.0001$.

3.11 Use the method of false position to find solution accurate to within 10^{-4} for the function $f(x) = x - \cos x$ in the interval $(0, \pi/2)$.

3.12 Use the method of false position to find solution accurate to within 10^{-4} for the function $f(x) = x - 0.8 - 0.2 \sin x = 0$ in the interval $(0, \pi/2)$.

3.13 Repeat exercise 3.6 correct to four decimal places using the false position method.

3.14 Repeat exercise 3.7 correct to four decimal places using the false position method.

3.15 Use the method of false position to solve the equation $x \tan x + 1 = 0$ accurate to three decimal places starting with 2.5 and 3.0 as the initial approximations to the root.

3.16 Use method of false position to solve the equation $x \log x - 1 = 0$ correct to three significant figures.

3.17 Use the method of false position to solve the equation $xe^x - \cos x = 0$ correct to four decimal places in the interval $(0, 1)$.

3.18 Use the method of false position to find a root correct to three decimal places for the function $\tan x - 4x = 0$.

3.19 Use the method of false position to find a root of $f(x) = e^x - 2x^2 = 0$ with an accuracy of four digits. The root lies between 1 and 1.5.

3.20 Use the method of false position to find a root correct to three decimal places of the function $x^3 - 4x - 9 = 0$.

3.21 A root of $f(x) = x^3 - 10x^2 + 5 = 0$ lies close to $x = 0.7$. Determine this root with the Newton-Raphson method to five decimal accuracy.

3.22 A root of $f(x) = e^x - 2x^2$ lies in the interval $(1, 2)$. Determine this root with the Newton-Raphson method to five decimal accuracy.

3.23 A root of $f(x) = x^3 - x^2 - 5 = 0$ lies in the interval $(2, 3)$. Determine this root with the Newton-Raphson method for four decimal places.

3.24 Use the Newton-Raphson method to find solution accurate to within 10^{-4} for the function $f(x) = x - \cos x$ in the interval $(0, \pi/2)$.

3.25 Use the Newton-Raphson method to find solution accurate to within 10^{-4} for the function $f(x) = x - 0.8 - 0.2 \sin x = 0$ in the interval $(0, \pi/2)$.

3.26 A positive root of the equation $e^x = 1 + x + \dfrac{x^2}{2} + \dfrac{x^3}{6} e^{0.3x}$ lies in the interval $(2, 3)$. Use the Newton-Raphson method to find this root accurate to five decimal places.

3.27 Use the Newton-Raphson method to find the smallest positive root of the equation $\tan x = x$ accurate to four decimal places.

3.28 Determine the positive root of the equation $x = 2 \sin x$ accurate to three decimal places.

3.29 Use the Newton-Raphson method to estimate the root of $f(x) = e^{-x} - x$ with an initial guess of $x_0 = 0$ accurate to five decimal places.

3.30 The equation $f(x) = 0.1 - x + \dfrac{x^2}{4} - \dfrac{x^3}{36} + \dfrac{x^4}{576} + \ldots = 0$ has one root in the interval $(0, 1)$. Determine this root correct to five decimal places.

3.31 Use the successive approximation method to find correct to four significant figures a real root of $\cos x - 3x + 1 = 0$.

3.32 Use the successive approximation method to find correct to four significant figures a real root of $e^{-x} - 10x = 0$.

3.33 Use the successive approximation method to find correct to four decimal places a real root of $2x - \log_{10} x - 7 = 0$.

3.34 Use the successive approximation method to find correct to four significant figures a real root of the function $e^x \tan x - 1 = 0$.

3.35 Find the real root of the equation $x - \sin x - 0.25 = 0$ to three significant digits using the successive approximation method.

3.36 Use the method of successive approximation to find a root of the equation $e^x - 3x = 0$ in the interval $(0, 1)$ accurate to four decimal places.

3.37 Use the method of successive approximation to find a real root of $e^x - x^2 = 0$ correct to four significant figures.

3.38 Use the method of successive approximation to determine a solution accurate to within 10^{-2} for $x^4 - 3x^2 - 3 = 0$ on $[1, 2]$. Use $x_0 = 1$.

3.39 Find a root of the equation $x^3 - 3x^2 + 4 = 0$ using the modified Newton-Raphson method, starting with $x_0 = 1.8$.

3.40 Find a root of the following function with an accuracy of 4 digits using the modified Newton-Raphson method, starting with $x_0 = 1.4$. $f(x) = e^x - 2x^2 = 0$.

3.41 Find a root of the equation $x^3 - 8x - 4 = 0$ using the modified Newton-Raphson method starting with $x_0 = 2.8$ up to four significant figures.

3.42 Find a root of the equation $x^3 - 3x - 5 = 0$ using the modified Newton-Raphson method correct up to four decimal places starting with $x_0 = 2.0$.

3.43 Find a root of the equation $x^3 - x - 1 = 0$ using the modified Newton-Raphson method correct up to four decimal places starting with $x_0 = -1.5$.

3.44 Find a root of the equation $x^6 - x - 1 = 0$ using the secant method approximations: $x_0 = 2$ and $x_1 = 1.0$.

3.45 Find a root of the equation $x^3 - 75 = 0$ using the secant method with the initial approximations of $x_0 = 4$ and $x_1 = 5$.

3.46 Find a root of the equation $\tan x - \tan hx = 0$ using the secant method with initial approximations: $x_0 = 7$ and $x_1 = 7.5$.

3.47 Find a root of the equation $\cos x \cosh x - 1 = 0$ using the secant method with initial approximations: $x_0 = 4.5$ and $x_1 = 5.0$.

3.48 Find a root of the equation $\sin x - 0.1x = 0$ using the secant method with initial approximations: $x_0 = 2$ and $x_1 = 3$.

3.49 Repeat exercise 3.39 using Muller's method given that a root is near 1.0.

3.50 Repeat exercise 3.40 using Muller's method given that a root is near 4.0.

3.51 Repeat exercise 3.41 using Muller's method given that a root is near 7.0.

3.52 Repeat exercise 3.42 using Muller's method given that a root is near 4.6.

3.53 Repeat exercise 3.43 using Muller's method given that a root is near 2.8.

3.54 Find a root of the equation $\cos x - xe^x = 0$ using Aitken's Δ^2 method.

3.55 Find the root of the equation $x^3 - 5x - 11 = 0$ correct to three decimal places using Aitken's method.

3.56 Find the root of $0.5 + \sin x - x = 0$ and $x_0 = 1$ using Aitken's method.

3.57 Use Aitken's method to find a root of the equation $3x - \log_{10}x - 16 = 0$.

3.58 Use Aitken's method to find a root of the equation $e^x - 3x = 0$ lying between 0 and 1.

3.59 Use Aitken's method to find a root of the equation $x^3 + x - 1 = 0$.

3.60 Use Aitken's method to find a root of the equation $5x^3 - 20x + 3 = 0$ in the interval $(0, 1)$.

3.61 Use Aitken's method to find a root of the equation $x^3 + 2x - 2 = 0$ up to three decimal places.

3.62 Use Aitken's method to find a root of the equation $x^3 - 3x^2 + 4 = 0$.

4

NUMERICAL DIFFERENTIATION

4.1   INTRODUCTION

Numerical differentiation deals with the following problem: *given* the function $y = f(x)$ find one of its derivatives at the point $x = x_k$. Here, the term *given* implies that we either have an algorithm for computing the function, or possesses a set of discrete data points (x_i, y_i), $i = 1, 2, n$. In other words, we have a finite number of (x, y) data points or pairs from which we can compute the derivative. Numerical differentiation is a method to compute the derivatives of a function at some values of independent variable x, when the function $f(x)$ is explicitly unknown, however it is known only for a set of arguments.

Like the numerical interpolation discussed in Chapter 5, a number of formulae for differentiation are derived in this chapter. They are:

a) Derivatives based on Newton's forward interpolation formula. This formula is used to find the derivative for some given x lying near the beginning of the data table.

b) Derivatives based on Newton's backward interpolation formula. This formula is suitable to find the derivative for a point near the end of the data table.

c) Derivatives based on Stirling's interpolation formula. This formula is used to find the derivative for some point lying near the middle of the tabulated value.

A method to find the maxima and minima of a given function is also discussed in this chapter.

4.2 DERIVATIVES BASED ON NEWTON'S FORWARD INTEGRATION FORMULA

Suppose the function $y = f(x)$ is known at $(n + 1)$ equispaced points x_0, x_1, ..., x_n and they are $y_0, y_1, ..., y_n$ respectively, i.e., $y_i = f(x_i)$, $i = 0, 1, ..., n$. Let $x_i = x_0 + ih$ and $u = \dfrac{x - x_i}{h}$, where h is the spacing.

Referring to Chatper 5, Newton's forward interpolation formula is

$$y = f(x) = y_0 + u\Delta y_0 + \frac{u(u-1)}{2!}\Delta^2 y_0 + ... + \frac{u(u-1)....(u-\overline{n-1})}{n!}\Delta^n y_0$$

$$= y_0 + u\Delta y_0 + \frac{u^2 - u}{2!}\Delta^2 y_0 + \frac{u^3 - 3u^2 + 2u}{3!}\Delta^3 y_0 + \frac{u^4 - 6u^3 + 11u^2 - 6u}{4!}\Delta^4 y_0$$

$$+ \frac{u^5 - 10u^4 + 35u^3 - 50u^2 + 24u}{5!}\Delta^5 y_0 + ... \tag{4.1}$$

Differentiating Eq.(4.1) w.r.t. x, we get

$$f'(x) = \frac{1}{h}\left[\Delta y_0 + \frac{2u-1}{2!}\Delta^2 y_0 + \frac{3u^2 - 6u + 2}{3!}\Delta^3 y_0 + \frac{4u^3 - 18u^2 + 22u - 6}{4!}\Delta^4 y_0\right.$$

$$\left. + \frac{5u^4 - 40u^3 + 105u^2 - 100u + 24}{5!}\Delta^5 y_0 \right. \tag{4.2}$$

Note here that $\dfrac{du}{dx} = \dfrac{1}{h}$.

Differentiating Eq.(4.2) w.r.t. x, we obtain

$$f''(x) = \frac{1}{h^2}\left[\Delta^2 y_0 + \frac{6u-6}{3!}\Delta^3 y_0 + \frac{12u^2 - 36u + 22}{4!}\Delta^4 y_0\right.$$

$$\left. + \frac{20u^3 - 120u^2 + 210u - 100}{5!}\Delta^5 y_0 + ...\right] \tag{4.3}$$

and so on.

Equations (4.2) and (4.3) give the approximate derivatives of $f(x)$ at arbitrary point $x = x_0 + uh$.

When $x = x_0$, $u = 0$, Eqs. (4.2) and (4.3) become

$$f'(x_0) = \frac{1}{h}\left[\Delta y_0 - \frac{1}{2}\Delta^2 y_0 + \frac{1}{3}\Delta^3 y_0 - \frac{1}{4}\Delta^4 y_0 + \frac{1}{5}\Delta^5 y_0 - ....\right] \qquad (4.4)$$

and $\quad f''(x_0) = \frac{1}{h^2}\left[\Delta^2 y_0 - \Delta^3 y_0 + \frac{11}{12}\Delta^4 y_0 - \frac{5}{6}\Delta^5 y_0 + ...\right] \qquad (4.5)$

and so on.

EXAMPLE 4.1

From the following table find the value of $\dfrac{dy}{dx}$ and $\dfrac{d^2y}{dx^2}$ at the point $x = 1.0$.

x	1	1.1	1.2	1.3	1.4	1.5
y	5.4680	5.6665	5.9264	6.2551	6.6601	7.1488

Solution:

The forward difference table is

x	y	Δy	Δ²y	Δ³y
1.0	5.4680			
		0.1985		
1.1	5.6665		0.0614	
		0.2599		0.0074
1.2	5.9264		0.0688	
		0.3287		0.0074
1.3	6.2551		0.0763	
		0.4050		0.0074
1.4	6.6601		0.0837	
		0.4887		
1.5	7.1488			

Here $x_0 = 1.0$ and $h = 0.1$. Then $u = 0$ and hence

$$\frac{dy}{dx} = y'(1.0) = \frac{1}{h}\left[\Delta y_0 - \frac{1}{2}\Delta^2 y_0 + \frac{1}{3}\Delta^3 y_0 - ....\right]$$

$$= \frac{1}{0.1}\left[0.1985 - \frac{1}{2}(0.0614) + \frac{1}{3}(0.0074)\right] = 1.7020$$

$$\frac{d^2y}{dx^2} = y''(1.0) = \frac{1}{h^2}\left[\Delta y_0 - \Delta^3 y_0 + ...\right] = \frac{1}{(0.1)^2}[0.0614 - 0.0074] = 5.4040$$

EXAMPLE 4.2

Obtain the first and second derivatives of the function tabulated below at the points $x = 1.1$ and $x = 1.2$.

x	1	1.2	1.4	1.6	1.8	2.0
y	0	0.128	0.544	1.298	2.440	4.02

Solution:

We first construct the forward difference table as shown here.

x	y	Δy	Δ²y	Δ³y	Δ⁴y
1.0 → 0					
		0.128			
1.2 ↠ 0.128			0.288		
		0.416		0.05	
1.4	0.544		0.338		0
		0.754		0.05	
1.6	1.298		0.388		0
		1.142		0.05	
1.8	2.440		0.438		
		1.580			
2.0	4.02				

Since $x = 1.1$ is a nontabulated point near the beginning of the table, we take $x_0 = 1.0$ and compute

$$p = \frac{x - x_0}{h} = \frac{1.1 - 1.0}{0.2} = 0.5$$

Hence,
$$\frac{dy}{dx} = \frac{1}{h}\left[\Delta y_0 + \frac{2p-1}{2}\Delta^2 y_0 + \frac{3p^2 - 6p + 2}{6}\Delta^3 y_0 \right]$$

$$= \frac{1}{0.2}\left[0.128 + 0 + \frac{3(0.5)^2 - 6(0.5) + 2}{6}(0.05) \right] = 0.62958$$

$$\frac{d^2 y}{dx^2} = \frac{1}{h^2}\left[\Delta^2 y_0 + (p-1)\Delta^3 y_0 \right] = \frac{1}{(0.2)^2}[0.288 + (0.5 - 1)0.05] = 6.575$$

Now, $x = 1.2$ is a tabulated point near the beginning of the table. For $x = x_0 = 1.2$, $p = 0$ and

$$\frac{dy}{dx} = \frac{1}{h}\left[\Delta y_0 - \frac{1}{2}\Delta^2 y_0 + \frac{1}{3}\Delta^3 y_0\right] = \frac{1}{0.2}\left[0.416 - \frac{1}{2}(0.338) + \frac{1}{3}(0.05)\right] = 1.31833$$

$$\frac{d^2 y}{dx^2} = \frac{1}{h^2}[\Delta^2 y_0 - \Delta^3 y_0] = \frac{1}{(0.2)^2}[0.338 - 0.05] = 7.2$$

EXAMPLE 4.3

Find the first and second derivatives of the functions tabulated below at the point $x = 1.1$ and $x = 1.2$.

x	1	1.2	1.4	1.6	1.8	2.0
y	0	0.1	0.5	1.25	2.4	3.9

Solution:

First, we construct the forward difference table:

Here $x = 1.1$ is a nontabulated point near the beginning of the table. For $x_0 = 1.0$,

$$p = \frac{x - x_0}{h} = \frac{1.1 - 1.0}{0.2} = 0.5$$

Hence, $\dfrac{dy}{dx} = \dfrac{1}{h}\left[\Delta y_0 + \dfrac{2p-1}{2}\Delta^2 y_0 + \dfrac{3p^2 - 6p + 2}{6}\Delta^3 y_0\right]$

$$= \frac{1}{0.2}\left[0.1 + 0 + \frac{3(0.5)^2 - 6(0.5) + 2}{6}(0.05)\right] = 0.48958$$

$$\frac{d^2y}{dx^2} = \frac{1}{h^2}\left[\Delta^2 y_0 + (p-1)\Delta^3 y_0\right] = \frac{1}{(0.2)^2}[0.3 + (0.5-1)0.05] = 6.875$$

For $x = 1.2$, it is a tabulated point near the beginning of the table.

Let $\quad x = x_0 = 1.2, p = 0$

$$\frac{dy}{dx} = \frac{1}{h}\left[\Delta y_0 - \frac{1}{2}\Delta^2 y_0 + \frac{1}{3}\Delta^3 y_0\right] = \frac{1}{0.2}\left[0.4 - \frac{1}{2}(0.35) + \frac{1}{3}(0.05)\right] = 1.208$$

$$\frac{d^2y}{dx^2} = \frac{1}{h^2}[\Delta^2 y_0 - \Delta^3 y_0] = \frac{1}{(0.2)^2}[0.35 - 0.05] = 7.5$$

4.3 DERIVATIVES BASED ON NEWTON'S BACKWARD INTERPOLATION FORMULA

Here, we assume the function $y = f(x)$ is known at $(n+1)$ points $x_0, x_1, ..., x_n$, i.e., $y_i = f(x_i)$, $i = 0, 1, 2, ..., n$ are known. Let $x_i = x_0 + ih$, $i = 0, 1, 2, ..., n$ and $v = \frac{x - x_n}{h}$.

Then, the Newton's backward interpolation formula from Chapter 5 is given by

$$f(x) = y_n + v\nabla y_n + \frac{v(v+1)}{2!}\nabla^2 y_n + \frac{v(v+1)(v+2)}{3!}\nabla^3 y_n$$

$$+ \frac{v(v+1)(v+2)(v+3)}{4!}\nabla^4 y_n + \frac{v(v+1)(v+2)(v+3)(v+4)}{5!}\nabla^5 y_n + ... \quad (4.6)$$

When the Eq.(4.6) is differentiated w.r.t. x successively, we obtain

$$f'(x) = \frac{1}{h}\left[\nabla y_n + \frac{2v+1}{2!}\nabla^2 y_n + \frac{3v^2+6v+2}{3!}\nabla^3 y_n + \frac{4v^3+18v^2+22v+6}{4!}\nabla^4 y_n\right.$$

$$\left. + \frac{5v^4+40v^3+105v^2+100v+24}{5!}\nabla^5 y_n + ...\right] \quad (4.7)$$

$$f''(x) = \frac{1}{h^2}\left[\nabla^2 y_n + \frac{6v+6}{3!}\nabla^3 y_n + \frac{12v^2+36v+22}{4!}\nabla^4 y_n \right.$$
$$\left. + \frac{20v^3+120v^2+210v+100}{5!}\nabla^5 y_n + ...\right] \quad (4.8)$$

and so on.

Equations (4.7) and (4.8) can be used to determine the approximate differentiation of first, second, and so on, order at any point x, where $x = x_n + vh$.

If $x = x_n$, then $v = 0$.

Equations (4.7) and (4.8) become

$$f'(x_n) = \frac{1}{h}\left[\nabla y_n + \frac{1}{2}\nabla^2 y_n + \frac{1}{3}\nabla^3 y_n + \frac{1}{4}\nabla^4 y_n + \frac{1}{5}\nabla^5 y_n + ...\right] \qquad (4.9)$$

and

$$f''(x_n) = \frac{1}{h^2}\left[\nabla^2 y_n + \nabla^3 y_n + \frac{11}{12}\nabla^4 y_n + \frac{5}{6}\nabla^5 y_n + ...\right] \qquad (4.10)$$

EXAMPLE 4.4

A slider in a machine moves along a fixed straight rod. Its distance $x(m)$ along the rod is given in the following table for various values of the time t (seconds).

t(sec.)	1	2	3	4	5	6
x(m)	0.0201	0.0844	0.3444	1.0100	2.3660	4.7719

Find the velocity and acceleration of the slider at time $t = 6$ sec.

Solution:

The backward difference table is

t	x	∇x	$\nabla^2 x$	$\nabla^3 x$	$\nabla^4 x$	$\nabla^5 x$
1.0	0.0201					
2.0	0.0844	0.0643				
3.0	0.3444	0.2600	0.1957			
4.0	1.0100	0.6656	0.4056	0.2100		
5.0	2.3660	1.3560	0.6904	0.2847	0.0748	
6.0	4.7719	2.4059	1.0499	0.3595	0.0748	0.0000

Here $h = 1.0$

$$\frac{dx}{dt} = \frac{1}{h}\left[\nabla x + \frac{1}{2}\nabla^2 x + \frac{1}{3}\nabla^3 x + \frac{1}{4}\nabla^4 x + \frac{1}{5}\nabla^5 x + ...\right]$$

$$= \frac{1}{1.0}\left[2.4059 + \frac{1}{2}(1.0499) + \frac{1}{3}(0.3595) + \frac{1}{4}(0.0748) + \frac{1}{5}(0.0)\right] = 3.0694$$

$$\frac{d^2 x}{dt^2} = \frac{1}{h^2}\left[\nabla^2 x + \nabla^3 x + \frac{11}{12}\nabla^5 x + ...\right]$$

$$= \frac{1}{(1.0)^2}\left[1.0499 + 0.3595 + \frac{11}{12}(0.0748) + \frac{5}{6}(0)\right] = 1.4780$$

4.4   DERIVATIVES BASED ON STIRLING'S INTERPOLATION FORMULA

Suppose $y_{\pm i} = f(x_{\pm i})$, $i = 0, 1, ..., n$ are given for $2n + 1$ equispaced points x_0, $x_{\pm 1}, x_{\pm 2}, ..., x_{\pm n}$, where $x_{\pm i} = x_0 \pm ih$, $i = 0, 1, ..., n$.

Stirling's interpolation polynomial is given by

$$f(x) = y_0 + \frac{u}{1!}\left[\frac{\Delta y_{-1} + \Delta y_0}{2}\right] + \frac{u^2}{2!}\Delta^2 y_{-1} + \frac{u^3 - u}{3!}\left[\frac{\Delta^3 y_{-2} + \Delta^3 y_{-1}}{2}\right]$$

$$+ \frac{u^4 - u^2}{4!}\Delta^4 y_{-2} + \frac{u^5 - 5u^3 + 4u}{5!}\left[\frac{\Delta^5 y_{-3} + \Delta^5 y_{-2}}{2}\right] + ... \qquad (4.11)$$

where $u = \dfrac{x - x_0}{h}$.

When Equation (4.11) is differentiated with respect to x successively, we obtain

$$f'(x) = \frac{1}{h}\left[\frac{\Delta y_{-1} + \Delta y_0}{2} + u\Delta^2 y_{-1} + \frac{3u^2 - 1}{6}\left(\frac{\Delta^3 y_{-2} + \Delta^3 y_{-1}}{2}\right)\right.$$

$$\left. + \frac{2u^3 - u}{12}\Delta^4 y_{-2} + \frac{5u^4 - 15u^2 + 4}{120}\left(\frac{\Delta^5 y_{-3} + \Delta^5 y_{-2}}{2}\right) + ...\right] \qquad (4.12)$$

and

$$f''(x) = \frac{1}{h^2}\left[\begin{array}{l}\Delta^2 y_{-1} + u\dfrac{\Delta^3 y_{-2} + \Delta^3 - 1}{2} + \dfrac{6u^2 - 1}{12}\Delta^4 y_{-2} \\[3mm] + \dfrac{2u^3 - 3u}{12}\left(\dfrac{\Delta^5 y_{-3} + \Delta^5 y_{-2}}{2}\right) + ...\end{array}\right] \qquad (4.13)$$

At $x = x_0$, $u = 0$ and Eqs. (4.12) and (4.13) become

$$f'(x_0) = \frac{1}{h}\left[\frac{\Delta y_0 + \Delta y_{-1}}{2} - \frac{1}{6}\left(\frac{\Delta^3 y_{-1} + \Delta^3 y_{-2}}{2}\right) + \frac{1}{30}\left(\frac{\Delta^5 y_{-2} + \Delta^5 y_{-3}}{2}\right) + ...\right] \quad (4.14)$$

and

$$f''(x_0) = \frac{1}{h^2}\left[\Delta^2 y_{-1} - \frac{1}{12}\Delta^4 y_{-2} + ...\right] \quad (4.15)$$

EXAMPLE 4.5

Find $\dfrac{dy}{dx}$ and $\dfrac{d^2 y}{dx^2}$ for $x = 0.2$ for the data given in the following table.

x	0	0.1	0.2	0.3	0.4	0.5
y	0	0.10017	0.20134	0.30452	0.41076	0.52115

Solution:

Construct the following difference table.

x	y	Δy	Δ²y	Δ³y	Δ⁴y
0	0				
		0.10017			
0.1	0.10017		0.001		
		0.10017		0.00101	
0.2	0.20134		0.00201		0.00004
		0.10318		0.00105	
0.3	0.30452		0.00306		0.00004
		0.10624		0.00109	
0.4	0.41076		0.00415		
		0.11039			
0.5	0.52115				

Here, we use Stirling's formula. Hence, for $x = 0.2$, we have

$$\frac{dy}{dx} = \frac{1}{h}\left[\frac{\Delta y_{-1} + \Delta y_0}{2} - \frac{1}{6}\frac{\Delta^3 y_{-2} + \Delta^3 y_{-1}}{2}\right]$$

$$= \frac{1}{0.1}\left[\frac{0.10117 + 0.10318}{2} - \frac{1}{12}\left(0.00101 + 0.00105\right)\right] = 1.020033$$

$$\frac{d^2 y}{dx^2} = \frac{1}{h^2}\left[\Delta^2 y_{-1} - \frac{1}{12}\Delta^4 y_{-2}\right] = \frac{1}{(0.1)^2}\left[0.00201 - \frac{1}{12}\left(0.00004\right)\right] = 0.200666 \cdot$$

EXAMPLE 4.6

Compute the values of $f'(3.1)$ and $f''(3.2)$ using the following table.

x	1	2	3	4	5
f(x)	0	1.4	3.3	5.6	8.1

Solution:

The central difference table is

x	y = f(x)	Δy	Δ²y	Δ³y	Δ4y
$x_{-2} = 1$	0				
		1.4			
$x_{-1} = 2$	1.4		0.5		
		1.9		–0.1	
$x_0 = 3$	3.3		0.4		–0.1
		2.3		–0.2	
$x_1 = 4$	5.6		0.2		
		2.5			
$x_2 = 5$	8.1				

Let $\quad x_0 = 3, h = 1, \ u = \dfrac{3.1 - 3}{1} = 0.1$

$$f'(3.1) = \frac{1}{h}\left[\begin{array}{l}\dfrac{\Delta y_{-1} + \Delta y_0}{2} + u\Delta^2 y_{-1} + \dfrac{3u^2 - 1}{6}\left(\dfrac{\Delta^3 y_{-2} + \Delta^3 y_{-1}}{2}\right) \\[2mm] + \dfrac{2u^3 - u}{12}\Delta^4 y_{-2} + ...\end{array}\right]$$

$$= \frac{1}{1}\left[\frac{1.9 + 2.3}{2} + 0.1(0.4) + \frac{3(0.1)^2 - 1}{6}\left(\frac{-0.1 - 0.2}{2}\right) + \frac{2(0.1)^3 - 0.1}{12}(-0.1)\right]$$

$$= [2.1 + 0.04 + 0.02425 + 0.00082] = 2.16507.$$

$$f''(3.1) = \frac{1}{h^2}\left[\Delta^2 y_{-1} + u\left(\frac{\Delta^2 y_{-2} + \Delta^3 y_{-1}}{2}\right) + \frac{6u^2 - 1}{12}\Delta^4 y_{-2} + ...\right]$$

$$= \frac{1}{1^2}\left[0.4 + 0.1\left(\frac{-0.1 - 0.2}{2}\right) + \frac{6(0.1)^2 - 1}{12}(-0.1)\right]$$

$$= [0.4 - 0.015 + 0.00783] = 0.39283$$

4.5   MAXIMA AND MINIMA OF A TABULATED FUNCTION

From calculus, we know that if a function is differentiable, then the maximum and minimum value of that function can be determined by equating the first derivative to zero and solving for the variable. This method is extendable for the tabulated function.

Now, consider Newton's forward difference formula given in Eq.(4.1).

Differentiating Eq.(4.1) w.r.t. u, we obtain

$$\frac{dy}{du} = \Delta y_0 + \frac{2u-1}{2}\Delta^2 y_0 + \frac{3u^2 - 3u + 2}{6}\Delta^3 y_0 + \dots\dots \tag{4.16}$$

For maximum or minimum, $\frac{dy}{du} = 0$. Neglecting the term after the third difference to obtain a quadratic equation in u.

Hence $\quad \Delta y_0 + \left(u - \frac{1}{2}\right)\Delta^2 y_0 + \left(\frac{u^2}{2} - \frac{u}{2} + \frac{1}{3}\right)\Delta^3 y_0 = 0 \tag{4.17}$

or $\quad \dfrac{\Delta^3 y_0}{2}u^2 + \left[\Delta^2 y_0 - \frac{1}{2}\Delta^2 y_0\right]u + \left[\Delta y_0 - \frac{1}{2}\Delta^2 y_0 + \frac{1}{3}\Delta^3 y_0\right] = 0$

or $\quad a_0 u^2 + a_1 u + a_2 = 0 \tag{4.18}$

which gives the values of u.

Here $\quad a_0 = \frac{1}{2}\Delta^3 y_0$

$$a_1 = \Delta^2 y_0 - \frac{1}{2}\Delta^3 y_0$$

$$a_2 = \Delta y_0 - \frac{1}{2}\Delta^2 y_0 + \frac{1}{3}\Delta^3 y_0 \tag{4.19}$$

The values of x will then be obtained from $x = x_0 + uh$.

EXAMPLE 4.7

Find x correct to four decimal places for which y is maximum from the following data given in tabular form and find also the value of y.

x	1	1.2	1.4	1.6	1.8
y	0	0.128	0.544	1.298	2.44

Solution:

We first construct the forward difference table as shown here:

x	y	Δy	$\Delta^2 y$	$\Delta^3 y$
1.0	0			
		0.128		
1.2	0.128		0.288	
		0.416		0.05
1.4	0.544		0.338	
		0.754		0.05
1.6	1.298		0.388	
		1.142		
1.8	2.44			

Let $x_0 = 1.0$.

Here $a_0 = \dfrac{1}{2}(0.05) = 0.025$

$$a_1 = 0.288 - \frac{1}{2}(0.05) = 0.2630$$

$$a_2 = 0.128 - \frac{1}{2}(0.288) + \frac{1}{3}(0.05) = 0.128 - 0.144 + 0.0166 = 0.000666$$

Hence $a_0 u^2 + a_1 u + a_2 = 0$, which gives the value of u.

or $\quad 0.025u^2 + 0.263u + 0.000666 = 0$

$$u_{1,2} = \frac{-0.263 \pm \sqrt{(0.263)^2 - 4(0.025)(0.000666)}}{2(0.025)} = (0, -10.5175)$$

Hence $u = 0$ or $u = -10.5175$.

Therefore, $x = 1.0$ and $x = 1.0 - 10.5175(0.2) = -1.1035$

At $x = 1.0$, $y = 0$ and at $x - 1.1035$, we apply Newton's forward interpolation formula.

$$y = y_0 + u\Delta y_0 + \frac{u(u-1)}{2!}\Delta^2 y_0 + \frac{u(u-1)(u-2)}{3!}\Delta^3 y_0 + ...$$

$$= 0 + (-10.5175)(0.128) + \frac{(-10.5175)(-11.5175)}{2}(0.288)$$

$$+ \frac{(-10.5175)(-11.5175)(-12.5175)}{(3)(2)(1)}(0.05)$$

$$= 3.46132 \text{ (maximum value)}.$$

4.6 CUBIC SPLINE METHOD

The cubic spline method described in Section 5.7 of Chapter 5 can be used to find the first and second derivatives of a function. The method requires two steps. In the first step, the cubic splines are constructed with suitable intervals. In the second step, the first and second derivatives are determined from the appropriate cubic spline. These steps are illustrated by an example as follows:

EXAMPLE 4.8

Given $y = f(x) = \cos x$, $0 \leq x \leq \pi/2$. Determine

a) the natural cubic spline in the interval $0 \leq x \leq \pi/4$ and $\pi/4 \leq x \leq /2$

b) the first and second derivatives $f'(\pi/8)$ and $f''(\pi/8)$.

Solution:

Here $h = \dfrac{\pi}{4}$, $y_0 = \cos 0 = 1$, $y_1 = \cos \dfrac{\pi}{4} = \dfrac{1}{\sqrt{2}}$ and $y_2 = \cos \dfrac{\pi}{2} = 0$. Also, $k_0 = k_2 = 0$.

From Equation (5.85) of section 5.7,

$$k_{i-1} + 4k_i + k_{i+1} = \frac{6}{h^2}[y_{i-1} - 2y_i + y_{i+1}], \qquad i = 2, 3, \ldots, n-1 \qquad \text{(E.1)}$$

or $\qquad k_0 + 4k_1 + k_2 = \dfrac{6}{h^2}[y_0 - 2y_1 + y_2]$

or $\qquad 4k_1 = \dfrac{96}{\pi^2}\left(1 - \sqrt{2}\right)$

or $\qquad k_1 = \dfrac{24}{\pi^2}\left(1 - \sqrt{2}\right) = -1.007247$

Therefore, the cubic spline is given by Equation (5.81) of Chapter 5,

$$f(x) = \begin{cases} f_{0,\frac{\pi}{4}}(x) \\ f_{\frac{\pi}{4},\frac{\pi}{2}}(x) \end{cases} \qquad \text{(E.2)}$$

where $\quad f_{0,\frac{\pi}{4}}(x) = \dfrac{4}{\pi}\left[\dfrac{x^3}{6}k_1 - \left(1 - \dfrac{1}{\sqrt{2}} + \dfrac{\pi^2}{96}k_1\right)x + \dfrac{\pi}{4}\right] \qquad \text{(E.3)}$

and $\quad f_{\frac{\pi}{4},\frac{\pi}{2}}(x) = \dfrac{4}{\pi}\left[\dfrac{\left(\dfrac{\pi}{2} - x\right)^3}{6}k_1 - \left(\dfrac{1}{\sqrt{2}} - \dfrac{\pi^2}{96}k_1\right)\left(\dfrac{\pi}{2} - x\right)\right] \qquad \text{(E.4)}$

Hence $f'\left(\dfrac{\pi}{8}\right) = f'_{0,\frac{\pi}{4}}\left(\dfrac{\pi}{4}\right) = -0.339961$

and $f''\left(\dfrac{\pi}{8}\right) = f''_{0,\frac{\pi}{4}}\left(\dfrac{\pi}{8}\right) = -0.503623$.

4.7 SUMMARY

Numerical differentiation is not a particularly accurate process due to a conflict between round-off errors and errors inherent in interpolation. Hence, a derivative of a function can never be computed with the same precision as the function itself.

EXERCISES

4.1. From the following table of values, estimate $y'(1.0)$ and $y''(1.0)$:

(a)

x	1	2	3	4	5	6
y	-4	+3	22	59	120	211

(b)

x	1	1.5	2	2.5	3	3.5
y	5	6.125	9	14.375	23	35.625

(c)

x	1	2	3	4	5	6
y	6	25	104	309	730	1481

(d)

x	1	1.5	2	2.5	3	3.5
y	2.7	-5.5188	-27.8	-75.4688	-163.3	-309.5188

(e)

x	1	2	3	4	5	6
y	2.9	-26.2	-157.9	-523	-1307.5	-2752.6

(f)

x	1	3	5	7	9	11
y	5.2	28.4	130	367.6	798.8	1481.2

(g)

x	1	1.5	2	2.5	3	3.5
y	−48	−41.0068	−24.362	8.1098	64.084	152.7363

4.2. Find the values of $y'(3.5)$ and $y''(3.5)$ from the following table:

(a)

x	1	1.5	2	2.5	3	3.5
y	−49	−46.6250	−42	−34.3750	−23	−7.1250

(b)

x	1	1.5	2	2.5	3	3.5
y	5	6.1250	9	14.3750	23	35.6250

(c)

x	1	1.5	2	2.5	3	3.5
y	0.5470	0.4536	0.2020	−0.3284	−1.2930	−2.8814

(d)

x	1	1.5	2	2.5	3	3.5
y	0.2377	0.3108	0.4829	0.7941	1.2849	1.9953

4.3. Find the values of $y'(6.0)$ and $y''(6.0)$ from the following table:

(a)

x	1	2	3	4	5	6
y	−0.5530	−0.7740	−1.2490	−2.0380	−3.1770	−4.6780

(b)

x	1	2	3	4	5	6
y	0.0858	−0.0099	−0.3242	−0.9827	−2.1094	−3.8270

4.4. A particle is moving along a straight line. The displacement x at some time instance t are given in the following table. Find the velocity and acceleration of the particle at $t = 4$.

t	1	3	5	7	9	11
x	0.1405	0.7676	3.5135	9.9351	21.5892	40.0324

4.5. Find the values of $y'(4)$ and $y''(4)$ from the following table:

x	0	1	2	3	4
y	5	8	12	17	26

4.6. Find the values of $y'(2)$ and $y''(2)$ from the following table:

x	1.5	1.6	1.7	1.8	1.9	2.0
y	0.3328	0.5312	0.7651	1.0384	1.3552	1.7198

4.7. Compute the values of $y'(3)$ and $y''(3)$ from the following table:

(a)

x	1	2	3	4	5
y	0	1.4	1.65	5.673	8.0978

(b)

x	1	2	3	4	5
y	0.4	0.65	0.75	0.87	0.98

4.8. Compute the values of $y'(2)$ and $y''(2)$ from the following table:

(a)

x	1	1.5	2	2.5	3
y	0	0.5	1.1	3.2	5.3

(b)

x	1	1.5	2	2.5	3
y	-2	2	3.456	5.674	8.4592

4.9. Compute the values of $y'(1.2)$ and $y''(1.2)$ from the following table:

(a)

x	1	1.1	1.2	1.3	1.4
y	0.1	0.34	0.42	0.53	0.62

(b)

x	1	1.1	1.2	1.3	1.4
y	0.0254	0.0437	0.0587	0.0670	0.0780

(c)

x	1	1.1	1.2	1.3	1.4
y	0.0012	0.2342	0.5786	0.7693	0.8934

4.10. Find x for which y is maximum and also find the corresponding value of y, from the following table:

(a)

x	1	2	3	4	5	6
y	−4	+3	22	59	120	211

(b)

x	1	1.5	2	2.5	3	3.5
y	5	6.125	9	14.375	23	35.625

(c)

x	1	2	3	4	5	6
y	6	25	104	309	730	1481

(d)

x	1	1.5	2	2.5	3	3.5
y	2.7	−5.5188	−27.8	−75.4688	−163.3	−309.5188

(e)

x	1	2	3	4	5	6
y	2.9	−26.2	−157.9	−523	−1307.5	−2752.6

(f)

x	1	3	5	7	9	11
y	5.2	28.4	130	367.6	798.8	1481.2

(g)

x	1	1.5	2	2.5	3	3.5
y	−48	−41.0068	−24.362	8.1098	64.084	152.7363

4.11. Repeat exercise 5.68 (Chapter 5).

4.12. Repeat exercise 5.71 (Chapter 5).

4.13. Use the cubic spline method to find $f'(2.0)$ and $f''(2.5)$ from the following table:

x	2	3	5	6
y	13	34	136	229

4.14. Repeat exercise 4.7(a) using the cubic spline method.

4.15. Repeat exercise 4.8(a) using the cubic spline method.

4.16. Repeat exercise 4.9(a) using the cubic spline method.

4.17. Repeat exercise 4.9(b) using the cubic spline method.

5

FINITE DIFFERENCES AND INTERPOLATION

5.1   INTRODUCTION

Interpolation is the technique of estimating the value of a function for any intermediate value of the independent variable. The process of computing or finding the value of a function for any value of the independent variable outside the given range is called *extrapolation*. Here, interpolation denotes the method of computing the value of the function $y = f(x)$ for any given value of the independent variable x when a set of values of $y = f(x)$ for certain values of x are known or given.

Hence, if (x_i, y_i), $i = 0, 1, 2, ..., n$ are the set of $(n + 1)$ given data points of the function $y = f(x)$, then the process of finding the value of y corresponding to any value of $x = x_i$ between x_0 and x_n, is called *interpolation*. There are several definitions available for the term interpolation. Hiral defines interpolation as the estimation of a most likely estimate in given conditions. It is the technique of estimating a past figure. Theile's definition of interpolation is "Interpolation is the art of reading between the lines of a table" while Harper's definition is "Interpolation consists in reading a value which lies between two extreme points."

If the function $f(x)$ is known explicitly, then the value of y corresponding to any value of x can easily be obtained. On the other hand, if the function $f(x)$ is not known, then it is very hard to find the exact form of $f(x)$ with the tabulated values (x_i, y_i). In such cases, the function $f(x)$ can be replaced by a simpler, function, say, $\phi(x)$, that has the same values as $f(x)$ for $x_0, x_1, x_2, ..., x_n$. The function $\phi(x)$ is called the *interpolating* or *smoothing function* and any other value can be computed from $\phi(x)$.

If $\phi(x)$ is a polynomial, then $\phi(x)$ is called the *interpolating polynomial* and the process of computing the intermediate values of $y = f(x)$ is called the *polynomial interpolation*. In the study of interpolation, we make the following assumptions:

a) there are no sudden jumps in the values of the dependent variable for the period under consideration

b) the rate of change of figures from one period to another is uniform.

In this chapter, we present the study of interpolation based on the calculus of finite differences. The following important interpolation formulae obtained or derived based on forward, backward and central differences of a function are presented.

a) Newton's binomial expansion formula for equal intervals

b) Newton's forward interpolation formula for equal intervals

c) Newton's backward interpolation formula for equal intervals

d) Lagrange's formula for unequal intervals

e) Lagrange's formula for inverse interpolation

f) Gauss's forward interpolation formula

g) Gauss's backward interpolation formula

h) Bessel's formula

i) Stirling's formula

j) Laplace-Everett formula

5.2 FINITE DIFFERENCE OPERATORS

Consider a function $y = f(x)$ defined on (a, b). x and y are the independent and dependent variables respectively. If the points $x_0, x_1, ..., x_n$ are taken at equidistance, i.e., $x_i = x_0 + ih$, $i = 0, 1, 2, ..., n$, then the value of y, when $x = x_i$, is denoted as y_i, where $y_i = f(x_i)$. Here, the values of x are called *arguments* and the values of y are known as *entries*. The interval h is called the *difference interval.* The differences $y_1 - y_0$, $y_2 - y_1$, ..., $y_n - y_{n-1}$ are called the *first differences* of the function y. They are denoted by Δy_0, Δy_1, ..., etc., That is

$$\Delta y_0 = y_1 - y_0$$

$$\Delta y_1 = y_2 - y_1$$

$$\vdots$$

$$\Delta y_n = y_n - y_{n-1} \tag{5.1}$$

The symbol Δ in Eq. (5.1) is called the *difference operator.*

5.2.1 Forward Differences

The *forward difference* or simply *difference* operator is denoted by Δ and may be defined as

$$\Delta f(x) = f(x + h) - f(x) \tag{5.2}$$

or writing in terms of y, at $x = x_i$, Eq.(5.2) becomes

$$\Delta f(x_i) = f(x_i + h) - f(x_i) \tag{5.3}$$

or $\qquad\qquad \Delta y_i = y_{i+1} - y_i \qquad\qquad i = 0, 1, 2, ..., n - 1$

The differences of the first differences are called the *second differences* and they are denoted by $\Delta^2 y_0$, $\Delta^2 y_1$, ..., $\Delta^2 y_n$.

Hence $\quad \Delta^2 y_0 = \Delta y_1 - \Delta y_0 = (y_2 - y_1) - (y_1 - y_0) = y_2 - 2y_1 + y_0$

$$\Delta^2 y_1 = \Delta y_2 - \Delta y_1 = (y_3 - y_2) - (y_2 - y_1) = y_3 - 2y_2 + y_1$$

$$\Delta^3 y_0 = \Delta^2 y_1 - \Delta^2 y_0 = (y_3 - 2y_2 + y_1) - (y_2 - 2y_1 + y_0) = y_3 - 3y_2 + 3y_1 - y_0$$

$$\Delta^3 y_1 = y_4 - 3y_3 + 3 y_2 - y_1, \text{ etc.}$$

Generalizing, we have

$$\Delta^{n+1} f(x) = \Delta[\Delta^n f(x)], \text{ i.e., } \Delta^{n+1} y_i = \Delta[\Delta^n y_i], n = 0, 1, 2, ... \tag{5.4}$$

Also, $\quad \Delta^{n+1} f(x) = \Delta^n[f(x + h) - f(x)] = \Delta^n f(x + h) - \Delta^n f(x)$

and $\quad \Delta^{n+1} y_i = \Delta^n y_{i+1} - \Delta^n y_i, n = 0, 1, 2, ...$ \tag{5.5} where $\Delta^0 \equiv$ identity operator, i.e., $\Delta^0 f(x) = f(x)$ and $\Delta^1 = \Delta$.

TABLE 5.1 Forward difference table.

x	y	Δy	$\Delta^2 y$	$\Delta^3 y$	$\Delta^4 y$	$\Delta^5 y$
x_0	y_0					
		Δy_0				
x_1	y_1		$\Delta^2 y_0$			
		Δy_1		$\Delta^3 y_0$		
x_2	y_2		$\Delta^2 y_1$		$\Delta^4 y_0$	
		Δy_2		$\Delta^3 y_1$		$\Delta^5 y_0$
x_3	y_3		$\Delta^2 y_2$		$\Delta^4 y_1$	
		Δy_3		$\Delta^3 y_2$		
x_4	y_4		$\Delta^2 y_3$			
		Δy_4				
x_5	y_5					

The forward differences for the arguments $x_0, x_1, ..., x_5$ are shown in Table 5.1. Table 5.1 is called a *diagonal difference table* or *forward difference table*. The first term in Table 5.1 is y_0 and is called the *leading term*. The differences $\Delta y_0, \Delta^2 y_0, \Delta^3 y_0, ...,$ are called the *leading differences*. Similarly, the differences with fixed subscript are called *forward differences*.

5.2.2 Backward Differences

The *backward difference operator* is denoted by ∇ and it is defined as

$$\nabla f(x) = f(x) - f(x - h) \tag{5.6}$$

Equation (5.6) can be written as

$$\nabla y_i = y_i - y_{i-1}, \qquad i = n, n-1, ..., 1. \tag{5.7}$$

or $\qquad \nabla y_1 = y_1 - y_0, \nabla y_2 = y_2 - y_1, ..., \nabla y_n = y_n - y_{n-1}$ $\qquad$ (5.8)

The differences in Eq.(5.8) are called *first differences*. The *second differences* are denoted by

$$\nabla^2 y_2, \nabla^2 y_3, ..., \nabla^2 y_n.$$

Hence $\qquad \nabla^2 y_2 = \nabla(\nabla y_2) = \nabla(y_2 - y_1) = \nabla y_2 - \nabla y_1 = (y_2 - y_1) - (y_1 - y_0)$
$$= y_2 - 2y_1 + y_0.$$

Similarly, $\quad \nabla^2 y_3 = y_3 - 2y_2 + y_1, \nabla^2 y_4 = y_4 - 2y_3 + y_2$, and so on.

Generalizing, we have

$$\nabla^k y_i = \nabla^{k-1} y_i - \nabla^{k-1} y_{i-1}, \qquad i = n, n - 1, ..., k \qquad (5.9)$$

where $\nabla^0 y_i = y_i, \nabla^1 y_i = \nabla y_i.$

The backward differences written in a tabular form is shown in Table 5.2. In Table 5.2, the differences $\nabla^n y$ with a fixed subscript "i" lie along the diagonal upward sloping.

TABLE 5.2 Backward difference table.

x	y	∇y	$\nabla^2 y$	$\nabla^3 y$	$\nabla^4 y$
x_0	y_0				
		∇y_1			
x_1	y_1		$\nabla^2 y_2$		
		∇y_2		$\nabla^3 y_3$	
x_2	y_2		$\nabla^2 y_3$		$\nabla^4 y_4$
		∇y_3		$\nabla^3 y_4$	
x_3	y_3		$\nabla^2 y_4$		
		∇y_4			
x_4	y_4				

Table 5.2 is called the *backward difference* or *horizontal table*.

5.2.3 Central Differences

The central difference operator is denoted by the symbol δ and is defined by

$$\delta f(x) = f(x + h/2) - f(x - h/2)$$

where h is the interval of differencing.

In terms of y, the first central difference is written as

$$\delta y_1 = y_{i+1/2} - y_{i-1/2} \qquad (5.10)$$

where $y_{i+1/2} = f(x_i + h/2)$ and $y_{i-1/2} = f(x_i - h/2)$.

Hence $\delta y_{1/2} = y_1 - y_0, \delta y_{3/2} = y_2 - y_1, ..., \delta y_{n-1/2} = y_n - y_{n-1}.$

The second central differences are given by

$$\delta^2 y_i = \delta y_{i+1/2} - \delta y_{i-1/2} = (y_{i+1} - y_i) - (y_i - y_{i-1}) = y_{i+1} - 2y_i + y_{i-1}.$$

Generalizing

$$\delta^n y_i = \delta^{n-1} y_{i+1/2} - \delta^{n-1} y_{i-1/2}. \tag{5.11}$$

The central difference table for the seven arguments $x_0, x_1, ..., x_4$ is shown in Table 5.3.

TABLE 5.3 Central difference table.

x	y	δ	δ^2	δ^3	δ^4	δ^5	δ^6
x_0	y_0						
		$\delta y_{1/2}$					
x_1	y_1		$\delta^2 y_1$				
		$\delta y_{3/2}$		$\delta^3 y_{3/2}$			
x_2	y_2		$\delta^2 y_2$		$\delta^4 y_2$		
		$\delta y_{5/2}$		$\delta^3 y_{5/2}$		$\delta^5 y_{5/2}$	
x_3	y_3		$\delta^2 y_3$		$\delta^4 y_3$		$\delta^6 y_3$
		$\delta y_{7/2}$		$\delta^3 y_{7/2}$		$\delta^5 y_{7/2}$	
x_4	y_4		$\delta^2 y_4$		$\delta^4 y_4$		
		$\delta y_{9/2}$		$\delta^3 y_{9/2}$			
x_5	y_5		$\delta^2 y_5$				
		$\delta y_{11/2}$					
x_6	y_6						

It is noted in Table 5.3 that all odd differences have fraction suffices and all the even differences are with integral suffices.

EXAMPLE 5.1

a) Construct the forward difference table and the horizontal table for the following data:

x	1	2	3	4	5
y = f(x)	4	6	9	12	17

b) Construct a forward difference table for the following data

x	0	10	20	30
y	0	0.174	0.347	0.518

c) Construct a difference table for $y = f(x) = x^3 + 2x + 1$ for $x = 1, 2, 3, 4, 5$.

d) Obtain the backward differences for the function $f(x) = x^3$ from $x = 1$ to 1.05 to two decimals chopped.

Solution:

a) The forward and the horizontal or backward difference tables are shown in Tables 5.4 and 5.5.

TABLE 5.4 Forward difference table.

x	f(x)	Δf(x)	Δ²f(x)	Δ³f(x)	Δ⁴f(x)
1	4				
		2			
2	6		1		
		3		1	
3	9		0		3
		3		2	
4	12		2		
		5			
5	17				

TABLE 5.5 Horizontal or backward difference table.

x	f(x)	Δf(x)	Δ²f(x)	Δ³f(x)	Δ⁴f(x)
1	4				
2	6	2			
3	9	3	1		
4	12	3	0	1	
5	17	5	2	2	3

b) Table 5.6 shows the forward difference operations.

TABLE 5.6

x	y	Δy	Δ²y	Δ³y
0	0			
		0.174		
10	0.174		−0.001	
		0.173		−0.001
20	0.347		−0.002	
		0.171		
30	0.518			

c) Table 5.7 shows the forward difference table.

TABLE 5.7

x	x = f(x)	Δy	Δ²y	Δ³y
1	4			
		9		
2	13		12	
		21		6
3	34		18	
		39		6
4	73		24	
		63		
5	136			

d) The following Table 5.8 shows the backward differences.

TABLE 5.8

x	y = f(x)	∇y	∇²y	∇³y	∇⁴y
1.00	1				
		0.030			
1.01	1.030		0.001		
		0.031		−0.001	
1.02	1.061		0.000		0.002
		0.031		0.001	
1.03	1.092		0.001		−0.001
		0.032		0.000	
1.04	1.124		0.001		
		0.033			
1.05	1.157				

Note the typical oscillations in sign and growth of the entries.

5.2.4 Error Propagation in a Difference Table

Let $y_0, y_1, y_2, ..., y_n$ be the true values of a function and suppose the value y_4 to be affected with an error $\in$, so that its erroneous value is $y_4 + \in$. Then the successive differences of the y are as shown in Table 5.9.

TABLE 5.9 Error propagation in a difference table.

y	Δy	$\Delta^2 y$	$\Delta^3 y$
y_0			
	Δy_0		
y_1		$\Delta^2 y_0$	
	Δy_1		$\Delta^3 y_0$
y_2		$\Delta^2 y_0$	
	Δy_2		$\Delta^3 y_1$
y_3		$\Delta^2 y_1$	
	Δy_3		$\Delta^3 y_2 + \in$
y_4		$\Delta^2 y_3 + \in$	
	$\Delta y_4 + \in$		$\Delta^3 y_3 - 3\in$
$y_5 + \in$		$\Delta^2 y_4 - 2\in$	
	$\Delta y_5 - \in$		$\Delta^3 y_4 + 3\in$
y_6		$\Delta^2 y_5 + \in$	
	Δy_6		$\Delta^3 y_5 - \in$
y_7		$\Delta^2 y_6$	
	Δy_7		$\Delta^3 y_6$
y_8		$\Delta^2 y_7$	
	Δy_8		
y_9			

Table 5.9 shows that the effect of an error increases with the successive differences, that the coefficients of the $\in$'s are the binomial coefficients with alternating signs, and that the algebraic sum of the errors in any difference column is zero. The same effect is also true for the horizontal difference in Table 5.2.

EXAMPLE 5.2

Table 5.10 gives the values of a polynomial of degree five. It is given that $f(4)$ is in error. Correct the value of $f(4)$.

TABLE 5.10

x	1	2	3	4	5	6	7
$y = f(x)$	0.975	−0.6083	−3.5250	−5.5250	−6.3583	4.2250	36.4750

Solution:

It is given that $y = f(x)$ is a polynomial of degree five. Hence $\Delta^5 y$ must be a constant and $f(4)$ is in error.

Let $-5.5250 + \in$ be the true or correct value. The difference table is shown in Table 5.11.

TABLE 5.11

x	y	Δy	$\Delta^2 y$	$\Delta^3 y$	$\Delta^4 y$	$\Delta^5 y$
1	0.975					
		−1.5833				
2	−0.6083		−1.3333			
		−2.9167		$2.25 + \in$		
3	−3.5250		$0.9167 + \in$		$-2 - 4\in$	
		$-2 + \in$		$0.25 - 3\in$		$12 + 10\in$
4	$-5.5250 + \in$		$1.1667 - 2\in$		$10 + 6\in$	
		$-0.8333 - \in$		$10.25 + 3\in$		$-10 - 10\in$
5	−6.3583		$11.4667 + \in$		$0 - 4\in$	
		10.5833		$10.25 - \in$		
6	4.2250		21.6667			
		32.2500				
7	36.4750					

Since the fifth differences of y are constant, we have

$$12 + 10\in = -10 - 10\in$$

or $\qquad 20\in = -22$

or $\qquad \in = -1.1$

Hence $\quad f(4) = -5.5250 + \in = -5.5250 - 1.1$

or $\qquad f(4) = -6.6250$

EXAMPLE 5.3

The following is a table of values of a polynomial of degree 5. It is given that $f(3)$ is in error. Correct the error.

TABLE 5.12

x	0	1	2	3	4	5	6
y	1	2	33	254	1054	3126	7777

Solution:

It is given that $y = f(x)$ is a polynomial of degree 5.

Hence $\Delta^5 y$ must be constant; $f(3)$ is in error.

Let $254 + \epsilon$ be the true value, now we form the difference table as shown in Table 5.13.

TABLE 5.13

x	y	Δy	$\Delta^2 y$	$\Delta^3 y$	$\Delta^4 y$	$\Delta^5 y$
0	1					
		1				
1	2		30			
		31		$160 + \epsilon$		
2	33		$190 + \epsilon$		$200 - 4\epsilon$	
		$221 + \epsilon$		$360 - 3\epsilon$		$220 + 10\epsilon$
3	$254 + \epsilon$		$550 - 2\epsilon$		$420 + 6\epsilon$	
		$1771 - \epsilon$		$1780 + 3\epsilon$		$20 - 10\epsilon$
4	1054		$1330 + \epsilon$		$440 - 4\epsilon$	
		2101		$1220 - \epsilon$		
5	3126		12550			
		4651				
6	7777					

Since the fifth differences of y are constant

$$220 + 10\epsilon = 20 - 10\epsilon$$

$\Rightarrow \qquad 20\epsilon = -220$

$\Rightarrow \qquad \epsilon = -10$

Hence $\qquad f(3) = 254 + \epsilon$

$\Rightarrow \qquad f(3) = 244$

EXAMPLE 5.4

Table 5.14 shows a difference table. Find the location of error.

TABLE 5.14

x	$y = x^3$	Δ	Δ^2	Δ^3	Δ^4
5	125	7.651	0.306	0.006	0
5.1	132.651	7.957	0.312	0.006	0
5.2	140.608	8.269	0.318	0.006	−0.027
5.3	148.877	8.587	0.324	−0.021	0.081
5.4	157.464	8.911	0.303	0.060	−0.081
5.5	166.375	9.214	0.363	−0.021	0.027
5.6	175.616	9.577	0.342	0.006	0
5.7	185.193	9.919	0.348	0.006	
5.8	195.112	10.267	0.354		
5.9	205.379	10.621			
6	216				

Solution:

The location of an error is obviously centered on 0.060 in the third difference. Instead of 9.214, one has wrongly entered it as 9.241.

5.2.5 Properties of the Operator Δ

1. If c is a constant then $\Delta c = 0$.
   Proof:

   Let $\qquad f(x) = c$

   Hence $\qquad f(x + h) = c$, where h is the interval of differencing.

   Hence $\qquad \Delta f(x) = f(x + h) - f(x) = c - c = 0$

   or $\qquad \Delta c = 0$

2. Δ is distributive, i.e.,

$$\Delta[f(x) \pm g(x)] = \Delta f(x) \pm \Delta g(x).$$

   Proof:
$$\Delta[f(x) + g(x)] = [f(x + h) + g(x + h)] - [f(x) + g(x)]$$
$$= f(x + h) - f(x) + g(x + h) - g(x)$$
$$= \Delta f(x) + \Delta g(x).$$

Similarly, we have

$$\Delta[f(x) - g(x)] = \Delta f(x) - \Delta g(x)$$

3. If c is a constant then

$$\Delta[cf(x)] = c\Delta f(x).$$

From properties 2 and 3 above, it is observed that Δ is a linear operator.

Proof:

$$\Delta[cf(x)] = cf(x + h) - cf(x) = c[f(x + h) - f(x)] = c\Delta f(x)$$

Hence $\Delta[cf(x)] = c\Delta f(x).$

4. If m and n are positive integers, then $\Delta^m \Delta^n f(x) = \Delta^{m+n} f(x).$

Proof:

$$\Delta^m \Delta^n f(x) = (\Delta \times \Delta \times \Delta \ \dots \ m \text{ times}) \, (\Delta \times \Delta \ \dots \ n \text{ times}) \, f(x)$$

$$= (\Delta\Delta\Delta \ \dots \ (m + n) \text{ times}) \, f(x)$$

$$= \Delta^{m+n} f(x).$$

In a similar manner, we can prove the following properties:

5. $\Delta[f_1(x) + f_2(x) + \ \dots \ + f_n(x)] = \Delta f_1(x) + \Delta f_2(x) + \ \dots \ + \Delta f_n(x).$

6. $\Delta[f(x)g(x)] = f(x) \, \Delta g(x) + g(x) \, \Delta f(x).$

7. $\Delta \left[\dfrac{f(x)}{g(x)} \right] = \dfrac{g(x)\Delta f(x) - f(x)\Delta g(x)}{g(x)g(x + h)}$

5.2.6 Difference Operators

a) Shift operator, E:

The shift operator is defined as

$$Ef(x) = f(x + h) \tag{5.12}$$

or $$Ey_i = y_{i+1} \tag{5.13}$$

Hence, shift operator sifts the function value y_i to the next higher value y_{i+1}. The second shift operator gives

$$E^2 f(x) = E[Ef(x)] = E[f(x + h)] = f(x + 2h) \tag{5.14}$$

E is linear and obeys the law of indices. Generalizing,

$$E^n f(x) = f(x + nh) \text{ or } E^n y_i = y_{i+nh}$$ (5.15)

The inverse shift operator E^{-1} is defined as

$$E^{-1} f(x) = f(x - h)$$ (5.16)

In a similar manner, second and higher inverse operators are given by

$$E^{-2} f(x) = f(x - 2h) \text{ and } E^{-n} f(x) = f(x - nh)$$

The more general form of E operator is given by

$$E^r f(x) = f(x + rh)$$ (5.17)

where r is positive as well as negative rationals.

b) **Average operator, μ:**

The average operator μ is defined as

$$\mu f(x) = \frac{1}{2} [f(x + h/2) + f(x - h/2)]$$

i.e., $\qquad \mu y_i = \frac{1}{2} [y_{i+1/2} + y_{i-1/2}]$ (5.18)

c) **Differential operator, D**

The differential operator is usually denoted by D, where

$$Df(x) = \frac{d}{dx} f(x) = f'(x)$$

$$D^2 f(x) = \frac{d^2}{dx^2} f(x) = f''(x)$$ (5.19)

5.2.7 Relation Between the Operators

In order to develop approximations to differential equations, following summary of operators is useful.

TABLE 5.15

Operator	Definition
Forward difference operator Δ	$\Delta f(x) = f(x+h) - f(x)$
Backward difference operator ∇	$\nabla f(x) = f(x) - f(x-h)$
Central difference operator δ	$\delta f(x) = f(x+h/2) - f(x-h/2)$
Shift operator E	$Ef(x) = f(x+h)$
Average operator μ	$\mu f(x) = 0.5[f(x+h/2) - f(x-h/2)]$
Differential operator D	$Df(x) = f'(x)$

Here h is the difference interval. For linking different operators with differential operator D we consider Taylor's formula:

$$f(x + h) = f(x) + hf'(x) + \frac{1}{2!}h^2 f''(x) + \ldots\ldots$$

In operator notation, we can write it as:

$$Ef(x) = \left[1 + hD + \frac{1}{2!}(hD)^2 + \ldots\right] f(x)$$

This series in brackets is the expression for the exponential and hence we can write

$$E = e^{hD}$$

This relation can be used by symbolic programs such as Maple or Mathematica to analyze the accuracy of finite difference scheme.

From the definition of Δ, we know that

$$\Delta f(x) = f(x + h) - f(x)$$

where h is the interval of differencing. Using the operator E we can write

$$\Delta f(x) = Ef(x) - f(x)$$

$$\Rightarrow \qquad \Delta f(x) = (E - 1) f(x).$$

The above relation can be expressed as an identity

$$\Delta = E - 1$$

i.e., $$E = 1 + \Delta$$

Proof: $E\Delta f(x) = E(f(x+h) - f(x))$

$$= Ef(x+h) - Ef(x)$$

$$= f(x+2h) - f(x+h)$$

$$= \Delta f(x+h)$$

$$= \Delta Ef(x)$$

Hence $E\Delta = \Delta E$.

EXAMPLE 5.5

Show that $\Delta \log f(x) = \log\left[1 + \dfrac{\Delta f(x)}{f(x)}\right]$.

Solution:

Let h be the interval of differencing

$$f(x+h) = Ef(x) = (\Delta + 1)f(x) = \Delta f(x) + f(x)$$

$$\Rightarrow \frac{f(x+h)}{f(x)} = \frac{\Delta f(x)}{f(x)} + 1,$$

Taking logarithms on both sides we get

$$\log\left[\frac{f(x+h)}{f(x)}\right] = \log\left[1 + \frac{\Delta f(x)}{f(x)}\right]$$

$$\Rightarrow \log f(x+h) - \log f(x) = \log\left[1 + \frac{\Delta f(x)}{f(x)}\right]$$

$$\Rightarrow \Delta \log f(x) = \log\left[1 + \frac{\Delta f(x)}{f(x)}\right].$$

EXAMPLE 5.6

Evaluate $\left(\dfrac{\Delta^2}{E}\right)x^3$.

Solution:

Let h be the interval of differencing

$$\left(\frac{\Delta^2}{E}\right)x^3 = (\Delta^2 E^{-1})\,x^3$$

$$= (E-1)^2\,E^{-1}x^3$$

$$= (E^2 - 2E + 1)\,E^{-1}x^3$$

$$= (E - 2 + E^{-1})x^3$$

$$= Ex^3 - 2x^3 + E^{-1}\,x^3$$

$$= (x+h)^3 - 2x^3 + (x-h)^3$$

$$= 6xh.$$

Note: If $h = 1$, then $\left(\dfrac{\Delta^2}{E}\right)x^3 = 6x$.

EXAMPLE 5.7

Prove that $e^x = \dfrac{\Delta^2}{E}\,e^x \cdot \dfrac{Ee^x}{\Delta^2 e^x}$, the interval of differencing being h.

Solution:

We know that

$$Ef(x) = f(x+h)$$

Hence $E\,e^x = e^{x+h}$,

Again $\Delta e^x = e^{x+h} - e^x = e^x(e^h - 1)$

$$\Rightarrow \Delta^2 e^x = e^x \cdot (e^h - 1)^2$$

Hence, $\left(\dfrac{\Delta^2}{E}\right)e^x = (\Delta^2 E^{-1})\,e^x = \Delta^2 e^{x-h}$

$$= e^{-h}(\Delta^2 e^x) = e^{-h}e^x(e^h - 1)^2$$

Therefore, the right-hand side $= e^{-h}e^x\,(e^h - 1)\dfrac{e^{x+h}}{e^x(e^h - 1)} = e^x$.

Relation between E and ∇:

$$\nabla f(x) = f(x) - f(x - h) = f(x) - E^{-1}f(x)$$

$$\Rightarrow \nabla = 1 - E^{-1}$$

$$\nabla = \frac{E-1}{E}.$$

EXAMPLE 5.8

Prove the following (a) $(1 + \Delta)(1 - \Delta) = 1$ (b) $\Delta\nabla = \Delta - \nabla$.

Solution:

a) $(1 + \Delta)(1 - \nabla) f(x) = EE^{-1}f(x) = E f(x - h) = f(x) = 1 \cdot f(x)$.

   $\therefore (1 + \Delta)(1 - \nabla) = 1$

b) $\nabla\Delta f(x) = (E - 1)(1 - E^{-1})f(x)$

   $= (E - 1)[f(x) - f(x - h)]$

Proofs for the relations among the operators:

1. $\qquad\qquad \Delta = E - 1$

   Since $\qquad \Delta f(x) = f(x + h) - f(x)$

   or $\qquad \Delta f(x) = E[f(x)] - f(x) = (E - 1)f(x)$

   Since $f(x)$ is arbitrary, so ignoring it, we have

   $\Delta = E - 1$ or $E = 1 + \Delta$

2. $\qquad\qquad \nabla = 1 - E^{-1}$

   We have $\quad \nabla f(x) = f(x) - f(x - h) = f(x) - E^{-1}[f(x)] = (1 - E^{-1})f(x)$

   Hence $\quad \nabla = 1 - E^{-1}$

3. $\qquad\qquad \delta = E^{1/2} - E^{-1/2}$

   We have $\quad \delta[f(x)] = f(x + h/2) - f(x - h/2) = E^{1/2} \cdot [f(x)] - E^{-1/2} \cdot [f(x)] =$
   $\qquad\qquad (E^{1/2} - E^{-1/2})f(x)$

   Hence $\quad \delta = E^{1/2} - E^{-1/2}$

4. $\qquad\qquad \Delta = E\nabla = \nabla E = \delta E^{1/2}$

   We have $\quad E\nabla[f(x)] = E[f(x) - f(x - h)] = E[f(x)] - E[f(x - h)] = f(x + h) -$
   $\qquad\qquad f(x) = \Delta f(x)$

   Hence $\quad E\nabla = \Delta$

   Again, $\quad \nabla E[f(x)] = \nabla f(x + h) = f(x + h) - f(x) = \Delta f(x)$

   Hence $\quad \nabla E = \Delta$

Also, $\delta E^{1/2} \cdot [f(x)] = \delta[f(x + h/2)] = f(x + h) - f(x) = \Delta f(x)$

Hence $\delta E^{1/2} = \Delta$

5. $$E = e^{hD}$$

where $$D = \frac{d}{dx}$$

We know $E[f(x)] = f(x + h) = f(x) + hf'(x) + \dfrac{h^2}{2!} f''(x) + ...,$ by Taylor's series

$$= f(x) + hDf(x) + \frac{h^2}{2!} D^2 f(x) + ...$$

$$= \left(1 + hD + \frac{h^2 D^2}{2!} + ... \right) f(x) = e^{hD} \cdot f(x)$$

Hence $E = e^{hD}$.

6. $u = \dfrac{1}{2}(E^{1/2} + E^{-1/2})$

Since $u[f(x)] = \dfrac{1}{2}[f(x + h/2) + f(x - h/2)$

$$= \frac{1}{2}[E^{1/2} f(x) + E^{-1/2} f(x)] = \frac{1}{2}[E^{1/2} + E^{-1/2}]f(x)$$

Hence $u = \dfrac{1}{2}(E^{1/2} + E^{-1/2})$.

7. $\Delta \nabla = \Delta \nabla = \delta^2$

Since $\Delta \nabla f(x) = \Delta(f(x) - f(x - h)) = \Delta f(x) - \Delta f(x - h)$

$$= [f(x + h) - f(x)] - [f(x) - f(x - h)]$$

$$= \delta \cdot f(x + h/2) - \delta f(x - h/2) = \delta^2 f(x)$$

Hence $\Delta \nabla = \delta^2$

Also $\nabla \Delta f(x) = \nabla (f(x + h) - f(x)) = \nabla f(x + h) - \nabla f(x)$

$$= [f(x + h) - f(x)] - [f(x) - f(x - h)]$$

$$= \delta \cdot f(x + h/2) - \delta f(x - h/2) = \delta^2 f(x)$$

Hence $\nabla \Delta = \delta^2$

8. $(1 + \Delta)(1 - \Delta) = 1$

L.H.S. $= E \cdot E^{-1} = E^{1-1} = E^0 = 1 = $ R.H.S.

Hence the result. The relationships among the various operators are shown in Table 5.16.

TABLE 5.16 Relationship among the operators.

	E	Δ	∇	δ
E	E	$\Delta + 1$	$(1 - \nabla)^{-1}$	$1 + \dfrac{1}{2}\delta^2 + \delta\sqrt{\left(1 + \dfrac{1}{4}\delta^2\right)}$
Δ	$E - 1$	Δ	$(1 - \nabla)^{-1} - 1$	$\dfrac{1}{2}\delta^2 + \delta\sqrt{\left(1 + \dfrac{1}{4}\delta^2\right)}$
∇	$1 - E^{-1}$	$1 - (1 + \Delta)^{-1}$	∇	$-\dfrac{1}{2}\delta^2 + \delta\sqrt{\left(1 + \dfrac{1}{4}\delta^2\right)}$
δ	$E^{1/2} - E^{-1/2}$	$\Delta(1 + \Delta)^{-1/2}$	$\nabla(1 - \nabla)^{-1/2}$	δ
u	$\dfrac{1}{2}(E^{1/2} + E^{-1/2})$	$\left(1 + \dfrac{1}{2}\Delta\right)(1 + \Delta)^{1/2}$	$\left(1 - \dfrac{1}{2}\Delta\right)(1 - \Delta)^{-1/2}$	$\sqrt{\left(1 + \dfrac{1}{4}\delta^2\right)}$

5.2.8 Representation of a Polynomial Using Factorial Notation

A polynomial of degree n can be expressed as a fractional polynomial of the same degree. Let $f(x)$ be a polynomial of degree which is to be expressed in factorial notation and let

$$f(x) = a_0 + a_1 x^1 + a_2 x^2 + \ldots + a_n x^n \tag{5.20}$$

where $a_0, a_1, \ldots, a_n$ are constants and $a_0 \neq 0$ then

$$\Delta f(x) = \Delta[a_0 + a_1 x^1 + \ldots + a_n x^n]$$

$$\Rightarrow \Delta f(x) = a_1 + 2a_2 x^1 + \ldots + ra_n x^{(n-1)}$$

Hence $\Delta^2 f(x) = \Delta[a_1 + 2a_2 x^1 + \ldots + ra_n x^{(n-1)}]$

or $\Delta^2 f(x) = 2a_2 + 2 \times 3a_3 x^1 + \ldots + n(n-1)x^{(n-2)}$

$$\ldots$$

$$\Delta^r f(x) = a_n^r (n-1) \ldots 2 \times 1 x^{(0)} = a_n^{r!}.$$

Substituting $x = 0$ in the above, we obtain

$$f(0) = a_0, \frac{\Delta f(0)}{1!} = a_1, \frac{\Delta^2 f(0)}{2!} = a_2, \ldots, \frac{\Delta^n f(0)}{n!} = a_n$$

Putting the values of $a_0, a_1, a_2, \ldots, a_n$ in Eq. (5.20), we get

$$f(x) = f(0) + \frac{\Delta f(0)}{1!} x^1 + \frac{\Delta^2 f(0)}{2!} x^2 + \ldots + \frac{\Delta^n f(0)}{n!} x^n.$$

EXAMPLE 5.9

Evaluate (a) $\left(\dfrac{\Delta^2}{E}\right) x^2$ (b) $\Delta \sin x$

(c) $\Delta \log x$ (d) $\Delta \tan^{-1} x$.

Solution:

a) $\left(\dfrac{\Delta^2}{E}\right) x^2 = \left[\dfrac{(E-1)^2}{E}\right] x^2 = \left(\dfrac{E^2 - 2E + 1}{E}\right) x^2$

$= (E - 2 + E^{-1}) x^2 = Ex^2 - 2x^2 + E^{-1} x^2 = (x+1)^2 - 2x^2$
$+ (x+1)^2 = 2.$

b) $\Delta \sin x = \sin(x+h) - \sin x$

$= 2 \cos\left(\dfrac{x+h+x}{2}\right) \sin\left(\dfrac{x+h-x}{2}\right)$

$= 2 \cos\left(x + \dfrac{h}{2}\right) \sin\dfrac{h}{2}$

Hence $\Delta \sin x = 2 \cos\left(x + \dfrac{h}{2}\right) \sin\dfrac{h}{2}$.

c) $\Delta \log x = \log(x+h) - \log x$

$= \log\dfrac{x+h}{x} = \log\left[1 + \dfrac{h}{x}\right]$

Hence $\Delta \log x = \log\left[1 + \dfrac{h}{x}\right]$

$\Delta \tan^{-1} = \tan^{-1}(x+h) - \tan^{-1}$

$= \tan^{-1}\left[\dfrac{x+h-x}{1+(x+h)x}\right] = \tan^{-1}\left[\dfrac{h}{1+hx+x^2}\right].$

EXAMPLE 5.10

Find   a) $\Delta^2 e^x$ b) $\Delta \log x$

Solution:

a) $\Delta^2 e^x = \Delta(\Delta e^x) = \Delta[e^{x+h} - e^x]$

$\quad\quad = \Delta[e^x(e^h - 1)] = (e^h - 1)\, \Delta e^x$

$\quad\quad = (e^h - 1)(e^{x+h} - e^x) = (e^h - 1)e^x$

Hence $\Delta^2 e^x = (e^h - 1)^2 e^x$

b)   See sol. E5.9 (c).

EXAMPLE 5.11

Evaluate $\left(\dfrac{\Delta^2}{E}\right)x^3$.

Solution:

Let h = interval of differencing.

$$\left(\frac{\Delta^2}{E}\right)x^3 = (\Delta^2 E^{-1})x^3 = (E-1)^2 E^{-1} x^3$$

$$= (E^2 - 2E + 1)E^{-1}x^3 = (E - 2 + E^{-1})x^3 = Ex^3 - 2x^3 + E^{-1}x^3$$

$$= (x + h)^3 - 2x^3 + (x - h)^3 = 6xh$$

EXAMPLE 5.12

Given $u_0 = 1,\, u_1 = 11,\, u_2 = 21,\, u_3 = 28,\, u_4 = 30$, find $\Delta^4 u_0$.

Solution:

$$\Delta^4 u_0 = (E - 1)^4 u_0 = (E^4 - 4c_1 E^3 + 4c_2 E^2 - 4c_3 E + 1)u_0$$

$$= E^4 y_0 - 4E^3 u_0 + 6E^2 u_0 - 4Eu_0 + u_0 = u_4 - 4u_3 + 6u_2 - 4u_1 + u_0$$

$$= 30 - 112 + 126 - 44 + 1 = 1.$$

EXAMPLE 5.13

Estimate the missing term in the following table.

x	0	1	2	3	4
$y = f(x)$	4	3	4	?	12

Solution:

We are given four values, so the third differences are constant and the fourth differences are zero.

Hence $\Delta^4 f(x) = 0$ for all values of x.

That is $(E - 1)^4 f(x) = 0$

$$(E^4 - 4E^3 + 6E^2 - 4E + 1)f(x) = 0$$

$$E^4 f(x) - 4E^3 f(x) + 6E^2 f(x) - 4Ef(x) + f(x) = 0$$

$$f(x + 4) - 4f(x + 3) + 6f(x + 2) - 4f(x + 1) + f(x) = 0$$

where the interval of differencing is 1.

Now substituting $x = 0$, we obtain

$$f(4) + 4f(3) + 6f(2) - 4f(1) + f(0) = 0$$

$$12 + 4f(3) + 6(4) - 4(3) + 4 = 0$$

or $\quad f(3) = 7$.

EXAMPLE 5.14

Find $\quad \Delta^3 (1 - 3x)(1 - 2x)(1 - x)$.

Solution:

Let $\quad f(x) = (1 - 3x)(1 - 2x)(1 - x) = -6x^3 + 11x^2 - 6x + 1$

Here, $f(x)$ is a polynomial of degree three and the coefficient of x^3 is (-6).

Hence $\quad \Delta^3 f(x) = (-6)3! = -36$.

EXAMPLE 5.15

Evaluate $\quad \Delta(e^{ax} \log bx)$.

Solution:

Let $\quad f(x) = e^{ax}$ and $g(x) = \log bx$.

Hence $\quad \Delta f(x) = e^{a(x+h)} - e^{ax} = e^{ax}(e^{ah} - 1)$

$$\Delta g(x) = \log b(x+h) - \log bx = \log\left(1 + \frac{h}{x}\right)$$

Also $\quad \Delta(f(x).g(x)) = f(x+h)\, \Delta g(x) + g(x).\, \Delta f(x)$

$$= e^{a(x+h)} \log(1 + h/x) + \log bx.e^{ax}(e^{ah} - 1)$$

$$= e^{ax}.[e^{ah} \log(1 + h/x) + (e^{ah} - 1)\log bx].$$

EXAMPLE 5.16

If m is a positive integer and the interval of differencing is 1, show that

$$\Delta^2 x^{(-m)} = m(m+1)x^{(-m-2)}$$

Solution:

$$x^{(-m)} = \frac{1}{(x+1)(x+2)....(x+m)},$$

$$\Delta[x^{(-m)}] = \frac{1}{(x+2)(x+1)....(x+m+1)} - \frac{1}{(x+1)....(x+m)}$$

$$= \frac{1}{(x+2)....(x+m)}\left[\frac{1}{(x+m+1)} - \frac{1}{(x+1)}\right]$$

$$= m\frac{(-1)}{(x+1)(x+2)....(x+m+1)} = (-m)x^{(-m-1)}$$

$$\Delta^2(x^{(-m)}) = (-m)(-m-1)x^{(-m-2)} = m(m+1)x^{(-m-2)}.$$

EXAMPLE 5.17

Express $f(x) = 3x^3 + x^2 + x + 1$, in the factorial notation, interval of differencing being unity.

Solution:

Here $f(x)$ is a polynomial of degree 3.

∴ we can write

$$f(x) = f(0) + \frac{\Delta f(0)}{1!}x^1 + \frac{\Delta^2 f(0)}{2!}x^2 + \frac{\Delta^3 f(0)}{3!}x^3.$$

The interval of differencing is unit and finding the values of the function at $x = 0, 1, 2, 3$, we get

$$f(0) = 1, f(1) = 6, f(2) = 31, f(3) = 94.$$

The difference table (Table 5.17) for the above values is given below:

TABLE 5.17

x	f(x)	$\Delta f(x)$	$\Delta^2 f(x)$	$\Delta^3 f(x)$
0	1			
		5		
1	6		20	
		25		16
2	31		38	
		63		
3	94			

From the table we have $f(0) = 1$, $\Delta f(0) = 5$, $\Delta^2 f(0) = 20$, $\Delta^3 f(0) = 18$.

Substituting the above values in $f(x)$, we get

$$f(x) = 1 + 5x^1 + \frac{20}{2!}x^2 + \frac{18}{3!}x^3,$$

Hence $$f(x) = 3x^3 + 10x^2 + 5x + 1.$$

5.3   INTERPOLATION WITH EQUAL INTERVALS

Here, we assume that for function $y = f(x)$, the set of $(n + 1)$ functional values $y_0, y_1, ..., y_n$ are given corresponding to the set of $(n + 1)$ equally spaced values of the independent variable, $x_i = x_0 + ih$, $i = 0, 1, ..., n$, where h is the spacing.

5.3.1 Missing Values

Let a function $y = f(x)$ is given for equally spaced values $x_0, x_1, x_2, ..., x_n$ of the argument and $y_0, y_1, y_2, ..., y_n$ denote the corresponding values of the function. If one or more values of $y = f(x)$ are missing, we can determine the missing values by employing the relationship between the operators E and Δ.

5.3.2 Newton's Binomial Expansion Formula

Suppose y_0, y_1, y_2, ..., y_n denote the values of the function $y = f(x)$ corresponding to the values $x_0, x_0 + h, x_0 + 2h, ..., x_0 + nh$ of x. Let one of the values of y is missing since n values of the functions are known.

Therefore, we have

$$\Delta^n y_0 = 0$$

or $$(E - 1)^n y_0 = 0 \qquad (5.21)$$

Expanding Eq. (5.21), we have

$$[E^n - {}^nC_1 E^{n-1} + {}^nC_2 E^{n-2} + ... + (-1)^n]y_0 = 0 \qquad (5.22)$$

or $$E^n y_0 - nE^{n-1}y_0 + \frac{n(n-1)}{2!} E^{n-2}y_0 + ... + (-1)^n y_0 = 0$$

or $$y_n - ny_{n-1} + \frac{n(n-1)}{2} y_{n-2} + ... + (-1)^n y_0 = 0 \qquad (5.23)$$

Equation (5.23) is quite useful in determining the missing values without actually constructing the difference table.

EXAMPLE 5.18

Determine the missing entry in the following table.

x	0	1	2	3	4
$y = f(x)$	1	4	17	–	97

Solution:

Let $y_0 = 1$, $y_1 = 4$, $y_2 = 17$ and $y_4 = 97$. We are given four values of y. Let y be a polynomial of degree 3.

Hence $\Delta^4 y_0 = 0$

or $$(E - 1)^4 y_0 = 0$$

$$(E^4 - 4E^3 + 6E^2 - 4E + 1)y_0 = 0$$

$$E^4 y_0 - 4E^3 y_0 + 6E^2 y_0 - 4E y_0 + y_0 = 0$$

or $\quad y_4 - 4y_3 + 6y_2 - 4y_1 + y_0 = 0$

That is $\quad 97 - 4(y_3) + 6(17) - 4(4) + 1 = 0$

or $\quad y_3 = 46.$

EXAMPLE 5.19

Find the missing entry in the following table.

x	0	1	2	3	4	5
y = f(x)	1	3	11	–	189	491

Solution:

Here, we are given $y_0 = 1$, $y_1 = 3$, $y_2 = 11$, $y_4 = 189$ and $y_5 = 491$. Since five values are given, we assume that y is a polynomial of degree 4.

Hence $\quad \Delta^5 y_0 = 0$

or $\quad (E - 1)^5 y_0 = 0$ $\hspace{4cm}$ (E.1)

$\quad (E^5 - 5E^4 + 10E^3 - 10E^2 + 5E - 1) y_0 = 0$

or $\quad y_5 - 5y_4 + 10y_3 - 10y_2 + 5y_1 - y_0 = 0$ $\hspace{2cm}$ (E.2)

Substituting the given values for $y_0, y_1, \ldots, y_5$ in Eq. (E.2), we get

$$491 - 5(189) + 10y_3 - 10(11) + 5(3) - 1 = 0$$

or $\quad 10y_3 = 550$

or $\quad y_3 = 55.$

EXAMPLE 5.20

Find the missing entries in the following table.

x	0	1	2	3	4	5
y = f(x)	1	–	11	28	–	116

Solution:

Here, we are given $y_0 = 1$, $y_2 = 11$, $y_3 = 28$, and $y_5 = 116$. Since three values are known, we assume $y = f(x)$ as a polynomial of degree three.

Hence $\Delta^4 y_0 = 0$

or $(E - 1)^4 y_0 = 0$

That is $(E^4 - 4E^3 + 6E^2 - 4E + 1)y_0 = 0$

or $y_4 - 4y_3 + 6y_2 - 4y_1 + y_0 = 0$

    $y_4 - 4(28) + 6(11) - 4y_1 + 1 = 0$

    $y_4 - 4y_1 = 45$ (E.1)

and $\Delta^5 y_0 = 0$

or $(E - 1)^5 y_0 = 0$

or $(E^5 - 5E^4 + 10E^3 - 10E^2 + 5E - 1)y_0 = 0$

    $y_5 - 5y_4 + 10y_3 - 10y_2 + 5y_1 - y_0 = 0$

    $116 - 5y_4 + 10(28) - 10(11) + 5y_1 - 1 = 0$

or $-5y_4 + 5y_1 = -285$ (E.2)

Solving Equations (E.1) and (E.2), we obtain

$$y_1 = 4 \text{ and } y_4 = 61.$$

5.3.3 Newton's Forward Interpolation Formula

Let $y = f(x)$, which takes the values $y_0, y_1, y_2, ..., y_n$, that is the set of $(n + 1)$ functional values $y_0, y_1, y_2, ..., y_n$ are given corresponding to the set of $(n + 1)$ equally spaced values of the independent variable, $x_i = x_0 + ih$, $i = 0, 1, 2, ...,$ n where h is the spacing. Let $\phi(x)$ be a polynomial of the nth degree in x taking the same values as y corresponding to $x = x_0, x_1, ..., x_n$. Then, $\phi(x)$ represents the continuous function $y = f(x)$ such that $f(x_i) = \phi(x_i)$ for $i = 0, 1, 2, ..., n$ and at all other points $f(x) = \phi(x) + R(x)$, where $R(x)$ is called the *error term* (reminder term) of the interpolation formula.

Let $\phi(x) = a_0 + a_1(x - x_0) + a_2(x - x_0)(x - x_1) + a_3(x - x_0)(x - x_1)(x - x_2) + ...$

$$+ a_n(x - x_0)(x - x_1)(x - x_2) .... (x - x_{n-1})$$ (5.24)

and $\phi(x_i) = y_i$ $i = 0, 1, 2, ..., n$ (5.25)

The constants $a_0, a_1, a_2, ...,$ can be determined as follows:

Substituting $x = x_0, x_1, x_2, ..., x_n$ successively in Eq. (5.24), we get

$$a_0 = y_0 \tag{5.26}$$

$$y_1 = a_0 + a_1(x_1 - x_0)$$

or $\quad y_1 = y_0 + a_1(x_1 - x_0) \qquad$ [using Eq. (5.26)]

$$a_1 = \frac{y_1 - y_0}{x_1 - x_0} = \frac{\Delta y_0}{h} \tag{5.27}$$

$$y_2 = a_0 + a_1(x_2 - x_0) + a_2(x_2 - x_0)(x_2 - x_1)$$

or $\quad y_2 - y_0 - a_1(x_2 - x_0) = a_2(x_2 - x_0)(x_2 - x_1)$

or $\quad (y_2 - y_0) - \dfrac{(y_1 - y_0)}{(x_1 - x_0)}(x_2 - x_0) = a_2(x_2 - x_0)(x_2 - x_1)$

or $\quad (y_2 - y_0) - \dfrac{(y_1 - y_0)2h}{h} = a_2 2hh$

or $\quad a_2 = \dfrac{y_2 - 2y_1 + y_0}{2h^2} = \dfrac{\Delta^2 y_0}{2!h^2} \tag{5.28}$

Similarly, we obtain

$$a_3 = \frac{\Delta^3 y_0}{3!h^3}, ..., a_n = \frac{\Delta^n y_0}{n!h^n}$$

Hence, from Eq. (5.24), we have

$$\phi(x) = y_0 + \frac{\Delta y_0}{h}(x - x_0) + \frac{\Delta^2 y_0}{2!h^2}(x - x_0)(x - x_1) + \frac{\Delta^n y_0}{n!h^n}(x - x_0)(x - x_1)....(x - x_{n-1}) \tag{5.29}$$

Let $\qquad x = x_0 + uh$

or $\qquad x - x_0 = uh$

and $\qquad x - x_1 = (x - x_0) - (x_1 - x_0) = uh - h = (u - 1)h \tag{5.30}$

$$x - x_2 = (x - x_1) - (x_2 - x_1) = (u - 1)h - h = (u - 2)h, \text{ etc.}$$

Using the values from Eq. (5.30), Eq. (5.29) reduces to

$$\phi(x) = y_0 + u\Delta y_0 + \frac{u(u-1)}{2!}\Delta^2 y_0 + \frac{u(u-1)(u-2)}{3!}\Delta^3 y_0 + ...$$
$$+ \frac{u(u-1)....(u-(n-1))}{n!}\Delta^n y_0 \qquad (5.31)$$

The formula given in Eq. (5.31) is called *Newton's forward interpolation formula*. This formula is used to interpolate the values of y near the beginning of a set of equally spaced tabular values. This formula can also be used for extrapolating the values of y a little backward of y_0.

EXAMPLE 5.21

Given that $\sqrt{15500} = 124.4990$, $\sqrt{15510} = 124.5392$, $\sqrt{15520} = 124.5793$, and $\sqrt{15530} = 124.6194$, find the value of $\sqrt{15516}$.

Solution:

The difference table is given below:

x	$y = \sqrt{x}$	Δy	$\Delta^2 y$
15500 x_0	124.4990 y_0		
		0.0402	
15510	124.5392		0 $\Delta^2 y_0$
		0.0401	
15520	124.5793		0
		0.0401	
15530	124.6194		

Here $x_0 = 15500$, $h = 10$ and $x = 15516$

$$u = \frac{x - x_0}{h} = \frac{15516 - 15500}{10} = 1.6$$

Newton's forward difference formula is

$$f(x) = y_0 + u\Delta y_0 + \frac{u(u-1)}{2!}\Delta^2 y_0 + ...$$

or $f(15516) = 124.4990 + 1.6(0.0402) + 0 = 124.56323$

EXAMPLE 5.22

A second degree polynomial passes through the points $(1, -1)$, $(2, -2)$, $(3, -1)$, and $(4, 2)$. Find the polynomial.

Solution:

The difference table is constructed with the given values of x and y as shown here:

x	y	Δy	$\Delta^2 y$	$\Delta^3 y$
1	−1			
		−1		
2	−2		2	
		1		0
3	−1		2	
		3		
4	2			

Here $x_0 = 1, h = 1, y_0 = -1, \Delta y_0 = -1$ and $\Delta^2 y_0 = 2$

$$u = \frac{x - x_0}{h} = (x - 1)$$

From Newton's forward interpolation formula, we have

$$y = f(x) = y_0 + u\Delta y_0 + \frac{u(u-1)}{2!}\Delta^2 y_0 + ...$$

or $f(x) = -1 + (x-1)(-1) + \dfrac{(x-1)(x-1-1)}{2}.2 = x^2 - 4x + 2$

EXAMPLE 5.23

Find $y = e^{3x}$ for $x = 0.05$ using the following table.

x	0	0.1	0.2	0.3	0.4
e^{3x}	1	1.3499	1.8221	2.4596	3.3201

Solution:

The difference table is shown here:

x	$y = e^{3x}$	Δy	$\Delta^2 y$	$\Delta^3 y$	$\Delta^4 y$
0.00	1.0000				
		0.3499			
0.10	1.3409		0.1224		
		0.4723		0.0428	
0.20	1.8221		0.1652		0.0150
		0.6375		0.0578	
0.30	2.4596		0.2230		
		0.8605			
0.40	3.3201				

We have $x_0 = 0.00$, $x = 0.05$, $h = 0.1$

Hence $u = \dfrac{x - x_0}{h} = \dfrac{0.05 - 0.00}{0.1} = 0.5$

Using Newton's forward formula

$$f(x) = y_0 + u\Delta y_0 + \frac{u(u-1)}{2!}\Delta^2 y_0 + \frac{u(u-1)(u-2)}{3!}\Delta^3 y_0$$
$$+ \frac{u(u-1)(u-2)(u-3)}{4!}\Delta^4 y_0 + ...$$

$$f(0.05) = 1.0 + 0.5(0.3499) + \frac{0.5(0.5-1)}{2}(0.1224) + \frac{(0.5)(0.5-1)(0.5-2)}{6}(0.0428)$$
$$+ \frac{0.5(0.5-1)(0.5-2)(0.5-3)}{24}(0.0150)$$
$$f(0.05) = 1.16172.$$

EXAMPLE 5.24

The values of sin x are given below for different values of x. Find the value of sin 42°.

x	40	45	50	55	60
$y = f(x) = \sin x$	0.6428	0.7071	0.7660	0.8192	0.8660

Solution:

$x = 42°$ is near the starting value $x_0 = 40°$. Hence, we use Newton's forward interpolation formula.

x	$y = \sin x$	Δy	$\Delta^2 y$	$\Delta^3 y$	$\Delta^4 y$
40°	0.6428				
		0.0643			
45°	0.7071		−0.0054		
		0.0589		−0.0004	
50°	0.7660		−0.0058		0
		0.0531		−0.0004	
55°	0.8192		−0.0062		
		0.0469			
60°	0.8660				

$$u = \frac{x - x_0}{h} = \frac{42° - 40°}{5} = 0.4$$

We have $y_0 = 0.6428$, $\Delta y_0 = 0.0643$, $\Delta^2 y_0 = -0.0054$, $\Delta^3 y_0 = -0.0004$

Putting these values in Newton's forward interpolation formula we get

$$f(x) = y_0 + u\Delta y_0 + \frac{u(u-1)}{2!}\Delta^2 y_0 + \frac{u(u-1)(u-2)}{3!}\Delta^3 y_0 + \dots$$

$$f(42°) = 0.6428 + 0.4(0.0643) + \frac{0.4(0.4-1)}{2}(-0.0054)$$

$$+ \frac{0.4 - (0.4-1)(0.4-2)}{6}(-0.0004) = 0.66913.$$

EXAMPLE 5.25

The profits of a company (in thousands of rupees) are given below:

Year (x)	1990	1993	1996	1999	2002
Profit y = f(x)	120	100	111	108	99

Calculate the total profits between 1990–2002.

Solution:

The forward difference table is constructed as shown further:

x	y	Δy_0	$\Delta^2 y_0$	$\Delta^3 y_0$	$\Delta^4 y_0$
1990	120				
		−20			
1993	100		31		
		11		−45	
1996	111		−14		53
		−3		8	
1999	108		−6		
		−9			
2002	99				

To calculate profits at 1991:

Let $x_0 = 1990, x = 1991, h = 3, p = \dfrac{x - x_0}{h} = 0.33$

Using Newton's forward interpolation formula, we obtain

$$y(1991) = y_0 + u\Delta y_0 + \frac{u(u-1)}{2!}\Delta^2 y_0 + \frac{u(u-1)(u-2)}{3!}\Delta^3 y_0$$

$$+ \frac{u(u-1)(u-2)(u-3)}{4!}\Delta^4 y_0$$

$$= 120 + 0.33(-20) + \frac{0.33(0.33-1)}{2}(31) + \frac{0.33(0.33-1)(0.33-2)}{6}(-45)$$

$$+ \frac{0.33(0.33-1)(0.33-2)(0.33-3)}{24}(53) = 104.93$$

or   104.93 thousand rupees.

As an example, consider the difference table (Table 5.18) of $f(x) = \sin x$ for $x = 0°$ to $50°$:

TABLE 5.18

x(deg)	f(x) = sin x	Δ	Δ^2	Δ^3	Δ^4	Δ^5
0	0	0.1736				
10	0.1736	0.1684	−0.0052	−0.0052		
20	0.3420	0.1580	−0.0104	−0.0048	0.0004	0
30	0.5000	0.1428	−0.0152	−0.0044	0.0004	
40	0.6425	0.1232	−0.0196			
50	0.766					

Since the fourth-order differences are constant, we conclude that a quartic approximation is appropriate. In order to determine sin 5° from the table, we use Newton's forward difference formula (to fourth order); thus, taking $x_j = 0$,

taking $x_j = 0$, we find $a = \dfrac{5-0}{10} = \dfrac{1}{2}$.

Hence $\quad \sin 5° = \sin 0° + \frac{1}{2}(0.1736) + (\frac{1}{2})(\frac{1}{2})(-\frac{1}{2})(-0.0052)$

$\qquad + (1/6)(\frac{1}{2})(-\frac{1}{2})(-3/2)(-0.0052) + (1/24)(\frac{1}{2})(-\frac{1}{2})(-3/2)(-5/2)(0.0004)$

$\qquad = 0 + 0.0868 + 0.0006(5) - 0.0003(3) - 0.0000(2) = 0.0871.$

In order to determine sin 45° from the table, we use Newton's backward difference formula (to fourth order); thus, taking $x_j = 40$, we find $b = \dfrac{45-40}{10} = \dfrac{1}{2}$

and $\quad \sin 45° = \sin 40° + \dfrac{1}{2}(0.1428) + \dfrac{1}{2}\dfrac{1}{2}\dfrac{3}{2}(-0.0152) + \dfrac{1}{6}\dfrac{1}{2}\dfrac{3}{2}\dfrac{5}{2}(-0.0048)$

$\qquad + \dfrac{1}{24}\dfrac{1}{2}\dfrac{3}{2}\dfrac{5}{2}\dfrac{7}{2}(0.0004)$

$\qquad = 0.6428 + 0.0714 - 0.0057 - 0.0015 + 0.00001 = 0.7071$

EXAMPLE 5.26

If $f(x)$ is known at the following data points

x_i	0	1	2	3	4
f_i	1	7	23	55	109

Find $f(0.5)$ using Newton's forward difference formula.

Solution:

Forward difference table is prepared as shown in Table 5.19.

TABLE 5.19

x	f	Δf	$\Delta^2 f$	$\Delta^3 f$	$\Delta^4 f$
0	1				
		6			
1	7		10		
		16		6	
2	23		16		0
		32		6	
3	55		22		
		54			
4	109				

By Newton's forward difference formula

$$f(x_0 + a\,h) = \left[f_0 + a\Delta f_0 + \frac{a(a-1)}{2!}\Delta^2 f_0 + \frac{a(a-1)(a-2)}{3!}\Delta^3 f_0 \right]$$

To find $f(0.5)$:

At $x = 0.5$, $a = (x - x_0)\,/\,h = (0.5 - 0)/1 = 0.5$

$$\text{Hence } f(0.5) = \left[1 + 0.5 \times 6 + \frac{0.5(0.5-1)}{2!}10 + \frac{0.5(0.5-1)(0.5-2)}{3!}6 \right]$$

$$= 1 + 3 + 2.5 \times (-0.5) + (-0.25)(-1.5)$$

$$= 3.125$$

EXAMPLE 5.27

Find $f(0.15)$ using Newton backward difference formula from Table 5.20.

TABLE 5.20

x	f(x)	∇f	$\nabla^2 f$	$\nabla^3 f$	$\nabla^4 f$
0.1	0.09983				
		0.09884			
0.2	0.19867		−0.00199		
		0.09685		−0.00156	
0.3	0.29552		−0.00355		0.00121
		0.0939		−0.00035	
0.4	0.38942		−0.0039		
		0.09			
0.5	0.97943				

Solution:

Using Newton-Gregory's formula:

$$f(x) = \left[f_n + n\nabla f_n + \frac{b(b+1)}{2!}\nabla^2 f_n + \frac{b(b+1)(b+2)}{3!}\nabla^3 f_n + \frac{b(b+1)(b+2)(b+3)}{4!}\nabla^4 f_n \right]$$

where in present case: $h = 0.1$, $n = 5$

$b = (x - x_n)\,/\,h = (0.15 - 0.5)/0.1 = -3.5$

Hence, $f(0.15) = 0.97943 +$

$$-3.5 \times 0.09 + \frac{-3.5(-3.5+1)}{2!}(-0.0039) + \frac{-3.5(-3.5+1)(-3.5+2)}{3!}$$

$$(-0.00035) + \frac{-3.5(-3.5+1)(-3.5+2)(-3.5+3)}{4!}(0.00121)$$

$= 0.97943 - 0.315 - 0.01706 + 0.000765625 + 0.00033086$

$= 0.14847$

5.3.4 Newton's Backward Interpolation Formula

Newton's forward interpolation formula is not suitable for interpolation values of y near the end of a table of values.

Let $y = f(x)$ be a function that takes the values $y_0, y_1, y_2, \ldots, y_n$ corresponding to the values $x_0, x_1, x_2, \ldots, x_n$ of the independent variable x. Let the values of x be equally spaced with h as the interval of differencing.

That is $\quad x_i = x_0 + ih, \qquad\qquad i = 0, 1, 2, \ldots, n$

Let $\phi(x)$ be a polynomial of the nth degree in x taking the same values of y corresponding to $x = x_0, x_1, \ldots, x_n$. That is, $\phi(x)$ represents $y = f(x)$ such that $f(x_i) = \phi(x_i)$, $i = 0, 1, 2, \ldots$. Hence, we can write $\phi(x)$ as

$$\phi(x_i) = y_i, \qquad\qquad i = n, n-1, \ldots, 1, 0$$

and $\quad x_{n-i} = x_{n-ih}, \qquad\qquad i = 1, 2, \ldots, n$

Let $\quad \phi(x) = a_0 + a_1(x - x_n) + a_2(x - x_n)(x - x_{n-1}) + \ldots +$
$$a_n(x - x_n)(x - x_{n-1}) \ldots (x - x_0) \qquad\qquad (5.32)$$

Substituting $x = x_n, x_{n-1}, \ldots, x_1, x_0$ successively, we obtain

$$a_0 = y_n \qquad\qquad (5.33)$$

$$y_{n-1} = a_0 + a_1(x_{n-1} - x_n)$$

or $\qquad a_1 = \dfrac{y_{n-1} - y_n}{x_{n-1} - x_n} = \dfrac{\nabla y_n}{h} \qquad\qquad (5.34)$

Similarly, we obtain

$$a_2 = \frac{\nabla^2 y_n}{2! h^2}, \ldots, a_n = \frac{\nabla^n y_n}{n! h^n} \qquad\qquad (5.35)$$

Substituting the values from Equations (5.33), (5.34), and (5.35) in Eq. (5.32), we get

$$\phi(x) = y_n + \frac{\nabla y_n}{h}(x - x_n) + \frac{\nabla^2 y_n}{2!h^2}(x - x_n)(x - x_{n-1}) + ...$$

$$+ \frac{\nabla^n y_n}{n!h^n}(x - x_n)(x - x_{n-1})....(x - x_0) \qquad (5.36)$$

Now, setting $x = x_n + vh$, we obtain

$$x - x_n = vh$$
$$x - x_{n-1} = (v + 1)h$$
$$.....$$
$$x - x_0 = (v + n - 1)h$$

Hence, Eq. (5.36) reduces to

$$\phi(x) = y_n + \frac{v(v+1)}{2!}\nabla^2 y_n + ... + v(v+1)....\frac{(v+n-1)}{n!}\nabla^n y_n \qquad (5.37)$$

where $v = \dfrac{x - x_n}{h}$.

The formula given in Eq. (5.37) is called *Newton's backward interpolation formula*. This formula is used for interpolating values of y near the end of the tabulated values and also is used for extrapolating values of y a little backward of y_n.

EXAMPLE 5.28

Calculate the value of $f(84)$ for the data given in the table below:

x	40	50	60	70	80	90
f(x)	204	224	246	270	296	324

Solution:

The value of 84 is near the end of Table 5.21. Hence, we use Newton's backward interpolation formula. The difference table is shown below.

TABLE 5.21

x	f(x)	∇	∇^2	∇^3	∇^4	∇^5
40	204					
		20				
50	224		2			
		22		0		
60	246		2		0	
		24		0		0
70	270		2		0	
		26		0		
80	296		2			
		28				
90	324					

We have $x_n = 90$, $x = 84$, $h = 10$, $t_n = y_n = 324$, $\nabla t_n = \nabla y_n = 28$, $\nabla^2 y_n = 2$ and $fh = fh$.

$\nabla^3 y_n = \nabla^4 y_n = \nabla^5 y_n = 0$,

$$u = \frac{x - x_n}{h} = \frac{84 - 90}{10} = -0.6$$

From Newton's backward formula

$$f(84) = t_n + u\nabla t_n + \frac{u(u+1)}{2}\nabla^2 t_n + \ldots$$

$$f(84) = 324 - 0.6 \times 28 + \frac{(-0.6)(-0.6+1)}{2}2 = 324 - 16.8 - 0.24 = 306.96.$$

EXAMPLE 5.29

Use the Gauss forward formula to find y at $x = 30$ given the following table of values:

x	21	25	29	33	37
y	18.4708	17.8144	17.1070	16.3432	15.5154

Solution: We construct the following difference Table 5.22:

TABLE 5.22

x	y	Δy	$\Delta^2 y$	$\Delta^3 y$	$\Delta^4 y$
$x_0 - 2h = 21$	18.4708				
		−0.6564			
$x_0 - h = 25$	17.8144		−0.0510		
		−0.7074		−0.0054	
$x_0 = 29$	17.1070		−0.0564		−0.002
		−0.7638		−0.0076	
$x + h + 33$	16.3432		−0.0640		
		−0.8278			
$x_0 + 2h = 37$	15.5154				

Here $h = 4, u = \dfrac{30 - 29}{4} = \dfrac{1}{4} = 0.25.$

$u = 0.25$ lies between 0 and 1.

Hence, Gauss's forward formula is suitable. Substituting in Gauss's interpolation formula

$$y = y_0 + u\Delta y_0 + \frac{u(u-1)}{2!}\Delta^2 y_{-1} + \frac{(u+1)u(u-1)}{3!}\Delta^3 y_{-1}$$

$$+ \frac{(u+1)u(u-1)(u-2)}{4!}\Delta^4 y_{-2} + \dots.$$

We get

$$y_{0.25} = f(0.25) = 17.1070 + (0.25)(-0.7638) + \frac{(0.25)(-0.75)}{2} \times (-0.0564)$$

$$+ \frac{(1.25)(0.25)(-0.75)}{6} \times (-0.0076) + \frac{(1.25)(0.25)(-0.75)(-1.75)}{24}(-0.0022)$$

$$= 16.9216.$$

EXAMPLE 5.30

From the following table estimate the number of students who obtained marks in computer programming between 75 and 80.

Marks:	35–45	45–55	55–65	65–75	75–85
No. of students:	20	40	60	60	20

Solution:

The cumulative frequency table is shown in Table 5.30.

TABLE 5.30

Marks less than (x)	No. of students (y)	∇y	$\nabla^2 y$	$\nabla^3 y$	$\nabla^4 y$
45	20				
55	60	40			
65	120	60	20		
75	180	60	0	−20	
85	200	20	−40	−40	−20

To find the number of students with marks less than 80

Let $x_n = 85, x = 80, h = 10, p = \dfrac{x - x_n}{h} = -0.5$

Then using Newton's backward interpolation formula, we obtain

$$y = y_n + p\nabla y_n + \frac{v(v+1)}{2!}\nabla^2 y_n + \frac{v(v+1)(v+2)}{3!}\nabla^3 y_n + \frac{v(v+1)(v+2)(v+3)}{4!}\nabla^4 y_n$$

$$= 200 + (-0.5)(20) + \frac{-0.5(-0.5+1)}{2}(-40) + \frac{-0.5(-0.5+1)(-0.5+2)}{6}(-40)$$

$$+ \frac{-0.5(-0.5+1)(-0.5+2)(-0.5+3)}{24}(-20) = 198.2813$$

So, the number of students getting marks in computer programming between 75 and 80

$$= 198 - 180 = 18.$$

5.3.5 Error in the Interpolation Formula

Let $\phi(x)$ denote the *interpolating polynomial*. Also, let the function $f(x)$ be continuous and possess continuous derivatives within the interval (x_0, x_n). Now, defining the auxiliary function $F(t)$ as

$$F(t) = f(t) - \phi(t) - \{f(x) - \phi(x)\}\frac{(t-x_0)(t-x_1)....(t-x_n)}{(x-x_0)(x-x_1)....(x-x_n)} \tag{5.38}$$

The expression $(t - x_0)(t - x_1)....(t - x_n)$ is a polynomial of degree $(n + 1)$ in t and the coefficient of $t = 1$.

Hence, the $(n + 1)^{th}$ derivative f polynomial is $(n + 1)!$ That is

$$F^{n+1}(\xi) = f^{n+1}(\xi) - \{f(x) - \phi(x)\}\frac{(n+1)!}{(x - x_0)(x - x_1)....(x - x_n)} = 0 \quad (5.39)$$

or $\quad f(x) - \phi(x) = \frac{f^{n+1}(\xi)}{(n+1)!}(x - x_0)(x - x_1)....(x - x_n) \quad (5.40)$

Let $R(x)$ denote the error in the formula. Then

$$R(x) = f(x) - \phi(x)$$

Hence $\quad R(x) = \frac{f^{n+1}(\xi)}{(n+1)!}(x - x_0)(x - x_1)....(x - x_n)$

Since $x - x_0 = uh$ or $x - x_1 = (u - 1)h$, ... $(x - x_h) = (u - n)h$ where h is the interval of differencing, we have

Error $\quad R(x) = \frac{h^{n+1}f^{n+1}(\xi)}{(n+1)!}u(u-1)(u-2)....(u-n)$

Now, employing the relation

$$D = \frac{1}{h}\Delta$$

we have $\quad D^{n+1} = \frac{1}{h^{n+1}}\Delta^{n+1}$

or $\quad f^{n+1}(\xi) = \frac{\Delta^{n+1}f(x_0)}{n+1} \quad (5.41)$

The error in the forward interpolation formula is given by

$$R(x) = \frac{\Delta^{n+1}y_0}{(n+1)!}u(u-1)(u-2)....(u-n) \quad (5.42)$$

In a similar manner, by taking the auxiliary function $F(t)$ in the form

$$F(t) = f(t) - \phi(t) - \{f(x) - \phi(x)\} \frac{(t - x_n)(t - x_{n-1})....(t - x_0)}{(x - x_n)(x - x_{n-1})....(x - x_0)}$$

and proceeding as above, we obtain the error in Newton's backward interpolation formula as

$$R(x) = \frac{\nabla^{n+1} y_n}{(n+1)!} u(u+1)....(u+n) \qquad (5.43)$$

where $\quad u = \dfrac{x - x_n}{h}$.

EXAMPLE 5.31

Using Newton's forward interpolation formula find the value of cos 52° from the following data and estimate the error.

x	45°	50°	55°	60°
y = cos x	0.7071	0.6428	0.5736	0.5

Solution:

The difference table is given below:

x	y = cos x	Δy	$\Delta^2 y$	$\Delta^3 y$
45°	0.7071			
		−0.0643		
50°	0.6428		−0.0049	
		−0.0692		0.0005
55°	0.5736		−0.0044	
		−0.0736		
60°	0.5			

Here $\quad x_0 = 45°$, $x_1 = 52°$, $y_0 = 0.7071$, $\Delta y_0 = -0.0643$, $\Delta^2 y_0 = -0.0049$ and $\Delta^3 y_0 = 0.0005$.

$$u = \frac{x - x_0}{h} = \frac{52° - 45°}{5°} = 1.4$$

From Newton's forward interpolation formula

$$y = u_0 + u\Delta y_0 + \frac{u(u-1)}{2!}\Delta^2 y_0 + \frac{u(u-1)(u-2)}{3!}\Delta^3 y_0 \ldots$$

Hence $\quad y = f(52) = 0.7071 + 1.4(-0.0643) + \dfrac{1.4(1.4-1)}{2}(-0.0049)$

$$+ \frac{(1.4)(1.4-1)(1.4-2)}{6}(0.0005) = 0.615680$$

Error $\quad = \dfrac{u(u-1)(u-2)\ldots(u-n)}{n+1}\Delta^{n+1} y_0 \qquad$ where $n = 2$.

$$= \frac{u(u-1)(u-2)}{3!}\Delta^3 y_0 = \frac{1.4(1.4-1)(1.4-2)}{6}(0.0005)$$

Error $\quad = -0.000028$.

5.4 INTERPOLATION WITH UNEQUAL INTERVALS

Newton's forward and backward interpolation formulae are applicable only when the values of n are given at equal intervals. In this section, we present Lagrange's formula for unequal intervals.

5.4.1 Lagrange's Formula for Unequal Intervals

Let $y = f(x)$ be a real valued continuous function defined in an interval $[a, b]$. Let $x_0, x_1, x_2, \ldots, x_n$ be $(n+1)$ distinct points that are not necessarily equally spaced and the corresponding values of the function are $y_0, y_1, \ldots, y_n$. Since $(n+1)$ values of the function are given corresponding to the $(n+1)$ values of the independent variable x, we can represent the function $y = f(x)$ is a polynomial in x of degree n.

Let the polynomial is represented by

$$f(x) = a_0(x-x_1)(x-x_2)\ldots(x-x_n) + a_1(x-x_0)(x-x_2)\ldots(x-x_n)$$
$$+ a_2(x-x_0)(x-x_1)(x-x_3)\ldots(x-x_n) + \ldots$$
$$+ a_n(x-x_0)(x-x_1)\ldots(x-x_{n-1}) \tag{5.44}$$

Each term in Eq. (5.44) being a product of n factors in x of degree n, putting $x = x_0$ in Eq. (5.44) we obtain

$$f(x) = a_0(x_0-x_1)(x_0-x_2)\ldots(x_0-x_n)$$

or $\qquad a_0 = \dfrac{f(x_0)}{(x_0 - x_1)(x_0 - x_2)....(x_0 - x_n)}$

Putting $x = x_2$ in Eq. (5.44) we obtain

$$f(x_1) = a_1(x_1 - x_0)(x_1 - x_2)....(x_1 - x_n)$$

or $\qquad a_1 = \dfrac{f(x_1)}{(x_1 - x_0)(x_1 - x_2)....(x_1 - x_n)}$

Similarly putting $x = x_2$, $x = x_3$, $x = x_n$ in Eq. (5.44) we obtain

$$a_2 = \dfrac{f(x_2)}{(x_2 - x_0)(x_2 - x_1)....(x_2 - x_n)}$$

$$\vdots \qquad \qquad \vdots$$

and $\qquad a_n = \dfrac{f(x_n)}{(x_n - x_0)(x_n - x_1)....(x_n - x_{n-1})}$

Substituting the values of a_0, a_1, ..., a_n in Eq. (5.44) we get

$$y = f(x) = \dfrac{(x - x_1)(x - x_2)....(x - x_n)}{(x_0 - x_1)(x_0 - x_2)....(x_0 - x_n)} f(x_0) + \dfrac{(x - x_0)(x - x_2)....(x - x_n)}{(x_1 - x_0)(x_1 - x_2)....(x_1 - x_n)} f(x_1) + ...$$

$$+ \dfrac{(x - x_0)(x - x_1)....(x - x_{n-1})}{(x_n - x_0)(x_n - x_1)....(x_n - x_{n-1})} f(x_n) \qquad \qquad (5.45)$$

The formula given by Eq. (5.45) is known as the *Lagrange's interpolation formula*.

EXAMPLE 5.32

Apply Lagrange's interpolation formula to find a polynomial which passes through the points (0, −20), (1, −12), (3, −20), and (4, −24).

Solution:

We have $x_0 = 0, x_1 = 1, x_2 = 3, x_3 = 4, y_0 = f(x_0) = -20, y_1 = f(x_1) = -12, y_2 = f(x_2) = -20$ and $y_3 = f(x_3) = -24$.

The Lagrange's interpolation formula is

$$f(x) = \frac{(x-x_1)(x-x_2)(x-x_3)}{(x_0-x_1)(x_0-x_2)(x_0-x_3)} f(x_0) + \frac{(x-x_0)(x-x_2)(x-x_3)}{(x_1-x_0)(x_1-x_2)(x_1-x_3)} f(x_1)$$

$$+ \frac{(x-x_0)(x-x_1)(x-x_3)}{(x_2-x_0)(x_2-x_1)....(x_2-x_3)} f(x_2) + \frac{(x-x_0)(x-x_1)(x-x_2)}{(x_3-x_0)(x_3-x_1)(x_3-x_2)} f(x_3).$$

Hence $\quad f(x) = \dfrac{(x-1)(x-3)(x-4)}{(0-1)(0-3)(0-4)}(-20) + \dfrac{(x-0)(x-3)(x-4)}{(1-0)(1-3)(1-4)}(-12)$

$$+ \frac{(x-0)(x-1)(x-4)}{(3-0)(3-1)(3-4)}(-20) + \frac{(x-0)(x-1)(x-3)}{(4-0)(4-1)(4-3)}(-24)$$

or $\quad f(x) = x^3 - 8x^2 + 15x + 20$ is the required polynomial.

EXAMPLE 5.33

Using Lagrange's interpolation formula find a polynomial that passes the points $(0, -12)$, $(1,0)$, $(3,6)$, $(4,12)$.

Solution:

We have $x_0 = 0$, $x_1 = 1$, $x_2 = 3$, $x_3 = 4$, $y_0 = f(x_0) = -12$, $y_1 = f(x_1) = 0$, $y_2 = f(x_2) = 6$, $y_3 = f(x_3) = 12$.

Using Lagrange's interpolation formula, we can write

$$f(x) = \frac{(x-x_1)(x-x_2)(x-x_3)}{(x_0-x_1)(x_0-x_2)(x_0-x_3)} f(x_0) + \frac{(x-x_0)(x-x_2)(x-x_3)}{(x_1-x_0)(x_1-x_2)(x_1-x_3)} f(x_1)$$

$$+ \frac{(x-x_0)(x-x_1)(x-x_3)}{(x_2-x_0)(x_2-x_1)(x_2-x_3)} f(x_2) + \frac{(x-x_0)(x-x_1)(x-x_2)}{(x_3-x_0)(x_3-x_1)(x_3-x_2)} f(x_3)$$

Substituting the values, we get:

$$f(x) = -\frac{(x-1)(x-3)(x-4)}{12} \times 12 - \frac{(x-0)(x-3)(x-4)}{6} \times 0$$

$$+ \frac{(x-0)(x-1)(x-4)}{-6} \times 6 + \frac{(x-0)(x-1)(x-3)}{12} \times 12$$

$$= -(x-1)(x-3)(x-4) + -(x-0)(x-1)(x-4) + (x-0)(x-1)(x-3)$$

EXAMPLE 5.34

Using Lagrange's interpolation formula, find the value of y corresponding to $x = 10$ from the following data.

x	5	6	9	11
$y = f(x)$	380	−2	196	508

Solution:

Lagrange's interpolation formula is

$$y = f(x) = \frac{(x - x_1)(x - x_2)(x - x_3)}{(x_0 - x_1)(x_0 - x_2)(x_0 - x_3)} y_0 + \frac{(x - x_0)(x - x_2)(x - x_3)}{(x_1 - x_0)(x_1 - x_2)(x_1 - x_3)} y_1$$

$$+ \frac{(x - x_0)(x - x_1)(x - x_3)}{(x_2 - x_0)(x_2 - x_1)(x_2 - x_3)} y_2 + \frac{(x - x_0)(x - x_1)(x - x_2)}{(x_3 - x_0)(x_3 - x_1)(x_3 - x_2)} y_3 \qquad \text{(E.1)}$$

Here, we have $x_0 = 5$, $x_1 = 6$, $x_2 = 9$, $x_3 = 11$, $y_0 = 380$, $y_1 = -2$, $y_2 = 196$ and $y_3 = 508$. Substituting these values in Eq. (E.1), we get

$$f(10) = \frac{(10-6)(10-9)(10-11)}{(5-6)(5-9)(5-11)} \times (380) + \frac{(10-5)(10-9)(10-11)}{(6-5)(6-9)(6-11)} \times (-2)$$

$$+ \frac{(10-5)(10-6)(10-11)}{(9-5)(9-6)(9-11)} \times (196) + \frac{(10-5)(10-6)(10-9)}{(11-5)(11-6)(11-9)} \times (508)$$

or $\qquad f(10) = 330.$

5.4.2 Hermite's Interpolation Formula

Hermite's interpolation formula provides an expression for a polynomial passing through given points with given slopes. The Hermite interpolation accounts for the derivatives of a given function. Let x_i, f_i, f_i', (for $i = 0,1,2,...n$) be given.

The polynomial $f(x)$ of degree $(2n + 1)$ for which $f(x_i) = f_i$ and $f'(x_i) = f'_i$ is given by:

$$f(x) = \sum_{j=0}^{n} h_j(x) f_i + \sum_{j=0}^{n} \bar{h}_j(x) f_i'$$

where $\quad h_j(x) = 1 - \dfrac{q_n''(x_j)}{q_n'(x_j)}(x - x_j)[L_j(x)]^2$

$$\bar{h}_j(x) = (x - x_j)[L_j(x)]^2$$

$$q_n(x) = (x - x_0)\,(x - x_1)....(x - x_n)$$

$$L_j(x) = \dfrac{q_n(x)}{(x - x_j)q_n'(x_j)}$$

It is used to write the interpolation formulae in finite element analysis. Famous cubic polynomials are derived from two points with their slopes. It is used to represent bending motion of a beam. For example, in the case of a beam finite element, suppose we need to obtain cubic polynomials that satisfy the following cases:

1. Consider: $y = ax^3 + bx^2 + cx + d$ in [0, 1].
2. Apply conditions

	@ $x = 0$	@ $x = 1$
Case 1:	$y = 1, y' = 0$	$y = y' = 0$
Case 2:	$y = 0, y' = 1$	$y = y' = 0$
Case 3:	$y = 0, y' = 0$	$y = 1, y' = 0$
Case 4:	$y = 0, y' = 0$	$y = 0, y' = 1$

3. Solve each case for a, b, c, d.

Then we obtain:

$$y(x) = 1 + 0x - 1x^2 + 2x^2(x - 1) = 2x^3 - 3x^2 + 1$$

$$y(x) = 0 + 1x - 1x^2 + 1x^2(x - 1) = x^3 - 2x^2 + x$$

$$y(x) = 0 + 0x + 1x^2 - 2x^2(x - 1) = -2x^3 + 3x^2$$

$$y(x) = 0 + 0x + 0x^2 + 1x^2(x - 1) = x^3 - x^2$$

These polynomials are plotted in Figure 5.1.

For cases involved with higher order derivatives, the principle is same. When $y^{(n)}(x_i)$ is used, all lower derivatives and $y(x_i)$ itself must be included in the constraints. For example, you can not have $y'(x_i)$ as a constraint but not $y(x_i)$, nor $y^{(2)}(x_i)$ but not $y'(x_i)$ and $y(x_i)$.

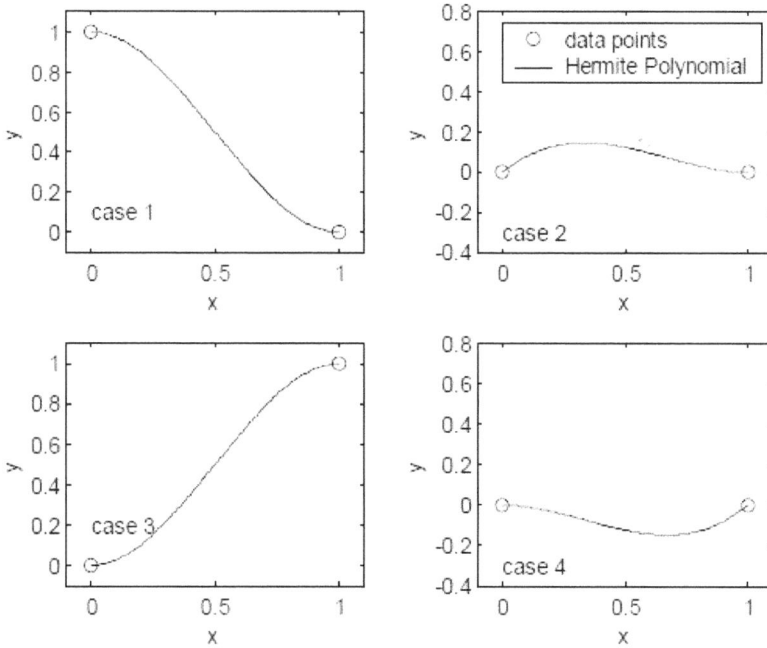

FIGURE 5.1 Hermite interpolation.

EXAMPLE 5.35

Construct displacements in a beam element from Hermite polynomials.

Solution:

Consider the beam of length L. The Hermite polynomials are:

$$N_1(x) = 2\left(\frac{x}{L}\right)^3 - 3\left(\frac{x}{L}\right)^2 + 1$$

$$N_2(x) = \frac{x^3}{L^2} - 2\frac{x^2}{L} + x$$

$$N_3(x) = -2\left(\frac{x}{L}\right)^3 + 3\left(\frac{x}{L}\right)^2$$

$$N_4(x) = \frac{x^3}{L^2} - \frac{x^2}{L}$$

These polynomial interpolation functions may be thought of as the fundamental modes of deflection. The deflection w(x) of any statically loaded beam can be written in terms of these modes as

$$w(x) = N_1 W_1 + N_2 \theta_1 + N_3 W_2 + N_4 \theta_2$$

where the subscripts associate quantities with positions (or nodes) 1 and 2 on the beam and W_i, θ_i, $i = 1,2$, are the deflection and slope, respectively, at each node.

5.4.3 Inverse Interpolation

In interpolation, we estimate the missing value of the function $y = f(x)$ corresponding to a value x intermediate between two given values. In inverse interpolation, we interpolate the argument x corresponding to an intermediate value y of the entry.

5.4.4 Lagrange's Formula for Inverse Interpolation

In the Lagrange interpolation formula y is expressed as a function of x as

$$y = f(x) = \frac{(x-x_1)(x-x_2)....(x-x_n)}{(x_0-x_1)(x_0-x_2)....(x_0-x_n)} y_0 + \frac{(x-x_0)(x-x_2)....(x-x_n)}{(x_1-x_0)(x_1-x_2)....(x_1-x_n)} y_1 + ...$$

$$+ \frac{(x-x_0)(x-x_1)....(x-x_{n-1})}{(x_n-x_0)(x_n-x_1)....(x_n-x_{n-1})} y_n \qquad (5.46)$$

By interchanging x and y in Eq. (5.46) we can express x as a function of y as follows:

$$x = \frac{(y-y_1)(y-y_2)....(y-y_n)}{(y_0-y_1)(y_0-y_2)....(y_0-y_n)} x_0 + \frac{(y-y_0)(y-y_2)....(y-y_n)}{(y_1-y_0)(y_1-y_2)....(y_1-y_n)} x_1 + ...$$

$$+ \frac{(y-y_0)(y-y_1)....(y-y_{n-1})}{(y_n-y_0)(y_n-y_1)....(y_n-y_{n-1})} x_n \qquad (5.47)$$

Equation (5.47) can be used for inverse interpolation.

EXAMPLE 5.36

The following table gives the values of y corresponding to certain values of x. Find the value of x when $y = 167.59789$ by applying Lagrange's inverse interpolation formula.

x	1	2	5	7
y = f(x)	1	12	117	317

Solution:

Here $x_0 = 1, x_1 = 2, x_2 = 5, x_3 = 7, y_0 = 1, y_1 = 12, y_2 = 117, y_3 = 317$ and $y = 167.59789$.

The Lagrange's inverse interpolation formula is given by

$$x = \frac{(y-y_1)(y-y_2)(y-y_3)}{(y_0-y_1)(y_0-y_2)(y_0-y_3)}x_0 + \frac{(y-y_0)(y-y_2)(y-y_3)}{(y_1-y_0)(y_1-y_2)(y_1-y_3)}x_1$$

$$+ \frac{(y-y_0)(y-y_1)(y-y_3)}{(y_2-y_0)(y_2-y_1)(y_2-y_3)}x_2 + \frac{(y-y_0)(y-y_1)(y-y_2)}{(y_3-y_0)(y_3-y_1)(y_3-y_2)}x_3$$

Hence, $x = \dfrac{(167.59789-12)(167.59789-117)(167.59789-317)}{(1-12)(1-117)(1-317)}(1)$

$$+ \frac{(167.59789-1)(167.59789-117)(167.59789-317)}{(12-1)(12-117)(12-317)}(12)$$

$$+ \frac{(167.59789-1)(167.59789-12)(167.59789-317)}{(117-1)(117-117)(117-317)}(117)$$

$$+ \frac{(167.59789-1)(167.59789-12)(167.59789-117)}{(317-1)(317-12)(317-117)}(317)$$

or $x = 5.65238$.

5.5   CENTRAL DIFFERENCE INTERPOLATION FORMULAE

In this section, we derive some important interpolation formulae by means of central differences of a function that are quite frequently employed in engineering and scientific computations.

In particular, we develop central difference formulae which are best suited for interpolation near the middle of a tabulated data set. The following central difference formulae are presented:

1. Gauss's forward interpolation formula
2. Gauss's backward interpolation formula
3. Bessel's formula
4. Stirling's formula
5. Laplace-Everett formula

Let the function $y = y_x = f(x)$ be given for $(2n + 1)$ equispaced values of argument x_0, $x_0 \pm h$, $x_0 \pm 2h$, ..., x_0, x_h. The corresponding values of y be y_i ($i = 0$, ± 1, ± 2, ..., $\pm n$). Also, let $y = y_0$ denote the central ordinate corresponding to $x = x_0$. We can then form the difference table as shown in Table 5.23. Table 5.24 shows the same Table 5.23 written using the Sheppard's operator δ, in which the relation $\delta = \Delta E^{-1/2}$ was used. Tables 5.22 and 5.24 are known as *central difference tables*.

TABLE 5.23 Central difference table.

x	y	Δy	$\Delta^2 y$	$\Delta^3 y$	$\Delta^4 y$	$\Delta^5 y$	$\Delta^6 y$
$x_0 - 3h$	y_{-3}						
		Δy_{-3}					
$x_0 - 2h$	y_{-2}		$\Delta^2 y_{-3}$				
		Δy_{-2}		$\Delta^3 y_{-3}$			
$x_0 - h$	y_{-1}		$\Delta^2 y_{-2}$		$\Delta^4 y_{-3}$		
		Δy_{-1}		$\Delta^3 y_{-2}$		$\Delta^5 y_{-3}$	
x_0	y_0		$\Delta^2 y_{-1}$		$\Delta^4 y_{-2}$		$\Delta^6 y_{-3}$
		Δy_0		$\Delta^3 y_{-1}$		$\Delta^5 y_{-2}$	
$x_0 + h$	y_1		$\Delta^2 y_0$		$\Delta^4 y_{-1}$		
		Δy_1		$\Delta^3 y_0$			
$x_0 + 2h$	y_2		$\Delta^2 y_1$				
		Δy_2					
$x_0 + 3h$	y_3						

TABLE 5.24 Central differences written in terms of Sheppard's operator δ.

x	y	δy	$\delta^2 y$	$\delta^3 y$	$\delta^4 y$	$\delta^5 y$	$\delta^6 y$
$x_0 - 3h$	y_{-3}						
		$\delta y_{-5/2}$					
$x_0 - 2h$	y_{-2}		$\delta^2 y_{-2}$				
		$\delta y_{-3/2}$		$\delta^3 y_{-3/2}$			
$x_0 - h$	y_{-1}		$\delta^2 y_{-1}$		$\delta^4 y_{-1}$		
		$\delta y_{-1/2}$		$\delta^3 y_{-1/2}$		$\delta^5 y_{-1/2}$	
x_0	y_0		$\delta^2 y_0$		$\delta^4 y_0$		$\delta^6 y_0$
		$\delta y_{1/2}$		$\delta^3 y_{1/2}$		$\delta^5 y_{1/2}$	
$x_0 + h$	y_1		$\delta^2 y_1$		$\delta^4 y_1$		
		$\delta y_{3/2}$		$\delta^3 y_{3/2}$			
$x_0 + 2h$	y_2		$\delta^2 y_2$				
		$\delta y_{5/2}$					
$x_0 + 3h$	y_3						

5.5.1 Gauss's Forward Interpolation Formula

Newton's forward interpolation formula is

$$y = f(x) = y_0 + u\Delta y_0 + \frac{u(u-1)}{2!}\Delta^2 y_0 + \frac{u(u-1)(u-2)}{3!}\Delta^3 y_0 \ldots, \quad (5.48)$$

where $u = \dfrac{x - x_0}{h}$ and $x = x_0$ is the origin.

In deriving the Gauss's forward interpolation formula, we assume the differences lie on the bottom solid lines in Table 5.25 and they are of the form

$$y_p = y_0 + G_1 \Delta y_0 + G_2 \Delta^2 y_{-1} + G_3 \Delta^3 y_{-1} + G_4 \Delta^4 y_{-2} + \ldots. \quad (5.49)$$

TABLE 5.25 Gauss's forward and backward interpolation formulae.

x	y	Δy	$\Delta^2 y$	$\Delta^3 y$	$\Delta^4 y$	$\Delta^5 y$	$\Delta^6 y$
x_{-4}	y_{-4}						
		Δy_{-4}					
x_{-3}	y_{-3}		$\Delta^2 y_{-4}$				
		Δy_{-3}		$\Delta^3 y_{-4}$			
x_{-2}	y_{-2}		$\Delta^2 y_{-3}$		$\Delta^4 y_{-4}$		
		Δy_{-2}		$\Delta^3 y_{-3}$		$\Delta^5 y_{-4}$	
x_{-1}	y_{-1}		$\Delta^2 y_{-2}$		$\Delta^4 y_{-3}$		$\Delta^6 y_{-4}$
		Δy_{-1}		$\Delta^3 y_{-2}$		$\Delta^5 y_{-3}$	
x_0	y_0		$\Delta^2 y_{-1}$		$\Delta^4 y_{-2}$		$\Delta^6 y_{-3}$
		Δy_0		$\Delta^3 y_{-1}$		$\Delta^5 y_{-2}$	
x_1	y_1		$\Delta^2 y_0$		$\Delta^4 y_{-1}$		$\Delta^6 y_{-2}$
		Δy_1		$\Delta^3 y_0$		$\Delta^5 y_{-1}$	
x_2	y_2		$\Delta^2 y_1$		$\Delta^4 y_0$		
		Δy_2					
x_3	y_3		$\Delta^2 y_2$				
		Δy_3					
x_4	y_4						

where G_1, G_2, ..., G_n are coefficients to be determined. From Newton's forward interpolation formula [Eq. (5.48)], we have

$$y_p = E^p y_0 = (1+\Delta)^p y_0 = y_0 + u\Delta y_0 + \frac{u(u-1)}{2!}\Delta^2 y_2 + \frac{u(u-1)(u-2)}{3!}\Delta^3 y_0 + ...$$
(5.50)

Now $\quad \Delta^2 y_{-1} = \Delta^2 E^{-1} y_0 = \Delta^2 (1+\Delta)^{-1} y_0 = \Delta^2 (1 - \Delta + \Delta^2 - \Delta^3 + ...)y_0 = \Delta^2 y_0$
$\quad\quad\quad - \Delta^3 y_0 + \Delta^4 y_0 - \Delta^5 y_0 + ...$

$\Delta^3 y_{-1} = \Delta^3 y_0 - \Delta^4 y_0 + \Delta^5 y_0 - \Delta^6 y_0 + ...$

$\Delta^4 y_{-2} = \Delta^4 E^{-2} y_0 = \Delta^4 (1-\Delta)^{-2} y_0 = \Delta^4 (1 - 2\Delta + 3\Delta^2 - 4\Delta^3 + ...)y_0 = \Delta^4 y_0$
$\quad\quad\quad - 2\Delta^5 y_0 + 3\Delta^6 y_0 - 4\Delta^7 y_0 + ...$

and so on.

Hence Eq. (5.49) becomes

$$y_p = y_0 + G_1 \Delta y_0 + G_2(\Delta^2 y_0 - \Delta^3 y_0 + \Delta^4 y_0 - \Delta^5 y_0 + ...) + G_3(\Delta^3 y_0 - \Delta^4 y_0$$
$$+ \Delta^5 y_0 - \Delta^6 y_0 + ...) + G_4(\Delta^4 y_0 - 2\Delta^5 y_0 + 3\Delta^6 y_0 - 4\Delta^7 y_0) + ... \quad (5.51)$$

Comparing Equations (5.50) and (5.51), we have

$$G_1 = u$$

$$G_2 = \frac{u(u-1)}{2!}$$

$$G_3 = \frac{(u+1)u(u-1)}{3!}$$

$$G_4 = \frac{(u+1)(u)(u-1)(u-2)}{4!}, \text{ etc.} \tag{5.52}$$

Hence, the Gauss's forward interpolation formula can be written as

$$y_p = y_0 + u\Delta y_0 + \frac{u(u-1)}{2!}\Delta^2 y_0 + \frac{(u+1)u(u-1)}{3!}\Delta^3 y_0$$

$$+ \frac{(u+1)u(u-1)(u-2)}{4!}\Delta^4 y_0 + ... \tag{5.53}$$

This formula given by Eq. (5.53) can be used to interpolate the values of y for u ($0 < u < 1$) measured forwardly from the origin.

Equation (5.53) can also be written as

$$y = f(x) = y_0 + u\Delta y_0 + \frac{u(u-1)}{2!}(\Delta^2 y_{-1} + \Delta^3 y_{-1}) + \frac{u(u-1)(u-2)}{3!}(\Delta^3 y_{-1} + \Delta^4 y_{-1})$$

$$+ \frac{u(u-1)(u-2)(u-3)}{4!}(\Delta^4 y_{-1} + \Delta^5 y_{-1}) + ... \tag{5.54}$$

or $y_4 = f(x) = y_0 + u\Delta y_0 + \frac{u(u-1)}{2!}\Delta^2 y_{-1} + \frac{(u+1)u(u-1)}{3!}\Delta^3 y_{-1}$

$$+ \frac{(u+1)u(u-1)(u-2)}{4!}\Delta^4 y_{-2} + ... \tag{5.55}$$

Equation (5.55) is known as *Gauss's forward interpolation formula*. Gauss's forward interpolation formula employs odd differences above the central line through y_0 and even differences on the central line. Gauss's forward formula is used to interpolate the values of the function for the value of u such that $0 < u < 1$.

5.5.2 Gauss Backward Interpolation Formula

Gauss's backward interpolation formula uses the differences that lie on the upper dashed line in Table 5.8 and can be assumed of the form

$$y_p = y_0 + G_1' \Delta y_{-1} + G_2' \Delta^2 y_{-1} + G_3' \Delta^3 y_{-2} + G_4' \Delta^4 y_{-2} + ... \qquad (5.56)$$

where $G_1', G_2', G_3', ..., G_n'$ are coefficients to be determined.

Now following the procedure described in Section 5.5.1 and comparing with Newton's backward interpolation formula, we find

$$G_1' = u$$

$$G_2' = \frac{u(u+1)}{2!}$$

$$G_3' = \frac{(u+2)(u+1)(u-1)u}{3!}$$

$$G_4' = \frac{(u+1)(u)(u-1)(u-2)}{4!}, \text{ etc.}$$

Hence $\quad y = f(x) = y_0 + \dfrac{u}{1!}(\Delta y_{-1} + \Delta^2 y_{-1}) + \dfrac{u(u-1)}{2!}(\Delta^2 y_{-1} + \Delta^3 y_{-1}) + ... \quad (5.57)$

or
$$y_p = y_0 + \frac{u}{1!}\Delta y_{-1} + \frac{u(u+1)}{2!}\Delta^2 y_{-1} + \frac{(u+1)u(u-1)}{3!}\Delta^3 y_{-2}$$
$$+ \frac{(u+2)(u+1)u(u-1)}{4!}\Delta^4 y_{-2} + ... \qquad (5.58)$$

Equation (5.58) is called *Gauss's backward interpolation formula*. Gauss's backward interpolation formula employs odd differences below the central line through y_0 and even differences on the central line as shown in Table 5.8. Gauss's backward interpolation formula is used to interpolate line value of the function for a negative value of u which lies between -1 and 0 $(-1 < u < 0)$.

EXAMPLE 5.37

Use Gauss's forward interpolation formula to find y for $x = 20$ given that

x	11	15	19	23	27
y	19.5673	18.8243	18.2173	17.1236	16.6162

Solution:

The difference table constructed is shown here:

x	y	Δy	$\Delta^2 y$	$\Delta^3 y$	$\Delta^4 y$
11	19.5673				
		−0.743			
15	18.8243		0.1360		
		−0.607		−0.6227	
19	18.2173		−0.4867		1.69570
		−1.0937		1.0730	
23	17.1236		0.5863		
		−0.5074			
27	16.6162				

Here $\quad h = 4, \; u = \dfrac{x - x_0}{h} = \dfrac{20 - 19}{4} = 0.25$

Gauss's forward interpolation formula is

$$y = y_0 + u\Delta y_0 + \frac{u(u-1)}{2!}\Delta^2 y_{-1} + \frac{(u+1)u(u-1)}{3!}\Delta^3 y_{-1}$$

$$+ \frac{(u+1)(u)(u-1)(u-2)}{4!}\Delta^4 y_{-2}$$

$$= 18.21730 + 0.25(-1.09370) + \frac{0.25(0.25-1)}{2}(-0.48670)$$

$$+ \frac{(0.25+1)(0.25)(0.25-1)}{6}(1.07300)$$

$$+ \frac{(0.25+1)(0.25)(0.25-1)(0.25-2)}{24}(1.69570)$$

$$y_{20} = 17.97657.$$

EXAMPLE 5.38

Use Gauss's backward interpolation formula to find the sales for the year 1986 from the following data:

Year	1951	1961	1971	1981	1991	2001
Sales (in thousands)	13	17	22	28	41	53

Solution:

Here $h = 10$, $x = 1986$ and $x_0 = 1991$.

$$u = \frac{x - x_0}{h} = \frac{1986 - 1991}{10} = -0.5$$

x	y	Δy	$\Delta^2 y$	$\Delta^3 y$	$\Delta^4 y$	$\Delta^5 y$
−4	13					
		4				
−3	17		1			
		5		0		
−2	22		1		6	
		6		6		−20
−1	28		7		−14	
		13		−8		
0	41		−1			
		12				
1	53					

Gauss's backward interpolation formula is

$$y = y_0 + u\Delta y_{-1} + \frac{u(u+1)}{2!}\Delta^2 y_{-1} + \frac{(u+1)(u)(u-1)}{3!}\Delta^3 y_{-2}$$
$$+ \frac{(u-1)(u)(u+1)(u+2)}{4!}\Delta^4 y_{-2} + ...$$

or $y = 41 + (-0.5)(13) + \dfrac{(-0.5)(-0.5+1)}{2}(-1.0) + \dfrac{(-0.5+1)(-0.5)(-0.5-1)}{6}(-8)$

$$\frac{(-0.5-1)(-0.5)(-0.5+1)(-0.5+2)}{24}(-14) = 33.79688.$$

5.5.3 Bessel's Formula

Bessel's formula uses the differences as shown in Table 5.26 in which brackets mean that the average has to be taken.

TABLE 5.26

$\vdots$	$\vdots$	$\vdots$	$\vdots$	$\vdots$	$\vdots$	$\vdots$	$\vdots$
x_{-1}	y_{-1}						
x_0 x_1	$\begin{pmatrix} y_0 \\ y_1 \end{pmatrix}$	Δy_0	$\begin{pmatrix} \Delta^2 y_{-1} \\ \Delta^2 y_0 \end{pmatrix}$	$\Delta^3 y_{-1}$	$\begin{pmatrix} \Delta^4 y_{-2} \\ \Delta^4 y_{-1} \end{pmatrix}$	$\Delta^5 y_{-2}$	$\begin{pmatrix} \Delta^6 y_{-3} \\ \Delta^6 y_{-2} \end{pmatrix}$
$\vdots$	$\vdots$	$\vdots$	$\vdots$	$\vdots$	$\vdots$	$\vdots$	$\vdots$

Hence, Bessel's formula assumes the form

$$y_p = \frac{y_0 + y_1}{2} + A_1 \Delta y_0 + A_2 \Delta y_0 \left[\frac{\Delta^2 y_{-1} + \Delta^2 y_0}{2} \right] + A_3 \Delta^3 y_{-1} + A_4 \left[\frac{\Delta^4 y_{-2} + \Delta^4 y_{-1}}{2} \right] + ...$$

$$= y_0 + \left(A_1 + \frac{1}{2} \right) \Delta y_0 + A_2 \left[\frac{\Delta^2 y_{-1} + \Delta^2 y_0}{2} \right] + A_3 \Delta^3 y_{-1} + A_4 \left[\frac{\Delta^4 y_{-2} + \Delta^4 y_{-1}}{2} \right] + ...$$

$$(5.59)$$

Newton's forward difference interpolation formula is given by

$$y_p = y_0 + u\Delta y_0 + \frac{u(u-1)}{2!}\Delta^2 y_0 + \frac{u(u-1)(u-2)}{3!}\Delta^3 y_0$$
$$+ \frac{u(u-1)(u-2)(u-3)}{4!}\Delta^4 y_0 + ...$$

$$(5.60)$$

Now, comparing Equations (5.59) and (5.60) and after simplifying the differences, we get

$$\left(A_1 + \frac{1}{2} \right) = u$$

$$A_2 = \frac{u(u-1)}{2!}$$

$$A_3 = \frac{u(u-1)\left(u - \frac{1}{2} \right)}{3!}$$

$$A_4 = \frac{(u+1)(u)(u-1)(u-2)}{4!} \text{ etc.} \qquad (5.61)$$

Hence, Bessel's formula in Eq. (5.59) becomes

$$y_p = y_0 + u\Delta y_0 + \frac{u(u-1)}{2!}\left[\frac{\Delta^2 y_{-1} + \Delta^2 y_0}{2}\right] + \frac{u(u-1)\left(u-\frac{1}{2}\right)}{3!}\Delta^3 y_{-1}$$

$$+ \frac{(u+1)(u)(u-1)(u-2)}{4!}\left[\frac{\Delta^4 y_{-2} + \Delta^4 y_{-1}}{2}\right] + ... \tag{5.62}$$

Using the central differences notation, Eq. (5.62) can be written as

$$y_p = y_0 + u\delta y_{1/2} + \frac{u(u-1)}{2!}\mu\delta^2 y_{1/2} + \frac{\left(u-\frac{1}{2}\right)u(u-1)}{3!}\delta^3 y_{1/2}$$

$$+ \frac{(u+1)u(u-1)(u-2)}{4!}\mu\delta^4 y_{1/2} + ... \tag{5.63}$$

where $\quad \mu\delta^2 y_{1/2} = \frac{1}{2}\left[\Delta^2 y_{-1} + \Delta^2 y_0\right]$

$$\mu\delta^4 y_{1/2} = \frac{1}{2}\left[\Delta^4 y_{-2} + \Delta^4 y_{-1}\right], \text{etc.} \tag{5.64}$$

EXAMPLE 5.39

Apply Bessel's interpolation formula to obtain y_{25}, given that $y_{20} = 2860$, $y_{24} = 3167$, $y_{28} = 3555$, and $y_{32} = 4112$.

Solution:

The difference table is shown here:

	x	y_x	Δy	$\Delta^2 y$	$\Delta^3 y$
x_{-1}	20	2860			
			307		
x_0	24	3167		81	
			388		88
x_1	28	3555		169	
			557		
x_2	32	4112			

Here $\quad x_0 = 24, h = 4$ and $u = \dfrac{x - x_0}{h} = \dfrac{25 - 24}{4} = 0.25$

The Bessel's formula is

$$y = y_0 + u\Delta y_0 + \frac{u(u-1)}{2}\left[\frac{\Delta^2 y_{-1} + \Delta^2 y_0}{2}\right] + \frac{u(u-1)(u-0.5)}{6}\Delta^3 y_{-1}$$

$$= 3167 + 0.25(388) + \frac{0.25(0.25 - 1)}{2}\left[\frac{81 + 169}{2}\right]$$

$$+ \frac{0.25(0.25 - 1)(0.25 - 0.5)}{6}(88) = 3252.96875$$

5.5.4 Stirling's Formula

Consider the mean of Gauss's forward and backward interpolation formula given by Equations (5.54) and (5.57), we get

$$y_p = y_0 + u\left[\frac{\Delta y_{-1} + \Delta y_0}{2}\right] + \frac{u^2}{2}\Delta^2 y_{-1} + \frac{u(u^2 - 1)}{3!}\left[\frac{\Delta^3 y_{-1} + \Delta^3 y_{-2}}{2}\right]$$

$$+ \frac{u^2(u^2 - 1)}{4!}\Delta^4 y_{-2} + \frac{u(u^2 - 1)(u^2 - 4)}{5!}\left[\Delta^5 y_{-2} + \Delta^5 y_{-3}\right] \qquad (5.65)$$

Equation (5.65) is known as *Stirling's formula*. In the central differences notation, Stirling's formula given by Eq. (5.65) becomes

$$y_p = y_0 + u\mu\delta y_0 + \frac{u^2}{2!}\delta^2 y_0 \left[\frac{u(u^2 - 1)}{3!}\right]\mu\delta^3 y_0 + \frac{u^2(u^2 - 1^2)}{4!}\delta^4 y_0 + ... \qquad (5.66)$$

where $\quad \mu\delta y_0 = \dfrac{1}{2}\left[\Delta y_0 + \Delta y_{-1}\right] = \dfrac{1}{2}\left[\delta y_{1/2} + \delta y_{-1/2}\right]$

and $\quad \mu\delta^3 y_0 = \dfrac{1}{2}\left[\Delta^3 y_{-1} + \Delta^3 y_{-2}\right] = \dfrac{1}{2}\left[\delta^3 y_{1/2} + \delta^3 y_{-1/2}\right] \qquad (5.67)$

Stirling's formula gives the most accurate result for $-0.25 \le u \le 0.25$. Hence, x_0 should be selected such that u satisfies this inequality.

EXAMPLE 5.40

Use Stirling's interpolation formula to find y_{28}, given that $y_{20} = 48234$, $y_{25} = 47354$, $y_{30} = 46267$, $y_{35} = 44978$, and $y_{40} = 43389$.

Solution:

Here $x = 30$ as origin and $h = 5$. Therefore $u = \dfrac{28 - 30}{5} = -0.4$. The difference table is shown here:

x	$u = \dfrac{x - 30}{5}$	y_u	Δy_u	$\Delta^2 y_u$	$\Delta^3 y_u$	$\Delta^4 y_u$
20	−2	48234				
			−880			
25	−1	47354		−207		
			−1087		5	
30	0	46267		−202		−103
			−1289		−98	
35	1	44978		−300		
			−1589			
40	2	43389				

Stirling's interpolation formula is

$$y_u = y_0 + u\left[\frac{\Delta^2 y_0 + \Delta^2 y_{-1}}{2}\right] + \frac{u^2 \Delta^2 y_{-1}}{2} + \frac{u(u^2 - 1)}{6}\left[\frac{\Delta^3 y_{-1} + \Delta^3 y_{-2}}{2}\right]$$
$$+ \frac{u^2(u^2 - 1)}{24}\Delta^4 y_{-2} + \dots.$$
$$= 46267 + (-0.4)\left[\frac{-1087 - 1289}{2}\right] + \frac{(-0.4)^2}{2}(202) + \frac{(-0.4)(-0.4^2 - 1)}{6}\left[\frac{5 - 98}{2}\right]$$
$$+ \frac{(-0.4)^2(-0.4^2 - 1)}{24}(-103) = 46724.0128.$$

5.5.5 Laplace-Everett Formula

Eliminating odd differences in Gauss's forward formula (Eq. (5.54)) by using the relation

$$\Delta y_0 = y_1 - y_0$$

We have $\Delta^3 y_{-1} = \Delta^2 y_0 - \Delta^2 y_{-1}$

$$\Delta^5 y_{-2} = \Delta^4 y_{-1} - \Delta^4 y_{-2} \ldots,$$

Hence

$$y = f(x)y_0 + \frac{u}{1!}(y_1 - y_0) + \frac{u(u-1)}{2!}\Delta^2 y_{-1} + \frac{(u+1)u(u-1)}{3!}(\Delta^2 y_0 - \Delta^2 y_{-1})$$

$$+ \frac{(u+1)u(u-1)(u-2)}{4!}\Delta^4 y_{-2} + \frac{(u+2)(u+1)u(u-1)(u-2)}{5!}(\Delta^4 y_{-1} - \Delta^4 y_{-2}) + \ldots$$

$$= (1-u)y_0 + uy_1 + u(u-1)\left[\frac{1}{1\times 2} - \frac{u+1}{1\times 2\times 3}\right]\Delta^2 y_{-1} + \frac{(u+1)u(u-1)}{3!}\Delta^2 y_0$$

$$+ (u+1)u(u-1)(u-2)\left[\frac{1}{1\times 2\times 3\times 4} - \frac{u+2}{5}\right]\Delta^2 y_{-2}$$

$$+ \frac{(u+2)(u+1)u(u-1)(u-2)}{5!}\Delta^4 y_{-1} + \ldots$$

$$= (1-u)y_0 + \frac{uy_1}{1!} - \frac{u(u-1)(u-2)}{3!}\Delta^2 y_{-1} + \frac{(u+1)u(u-1)}{3!}\Delta^2 y_0$$

$$- \frac{(u+1)u(u-1)(u-2)(u-3)}{3!}\Delta^4 y_{-2} + \frac{(u+2)(u+1)u(u-1)(u-2)}{5!}\Delta^4 y_{-1} + \ldots$$

$$(5.68)$$

Writing $v = 1 - u$, i.e., $u = 1 - v$ and changing the terms (5.68) with a negative sign we get

$$y = vy_0 + \frac{u}{1!}y_1 + \frac{(v+1)v(v-1)}{3!}\Delta^2 y_{-1} + \frac{(u+1)u(u-1)}{3!}\Delta^2 y_0$$

$$+ \frac{(v+2)(v+1)v(v-1)(v-2)}{5!}\Delta^2 y_{-2} + \frac{(u+2)(u+1)u(u-1)(u-2)}{5!}\Delta^4 y_{-1} + \ldots$$

$$(5.69)$$

Equation (5.69) can be written as

$$y_4 = f(x) = vy_0 + \frac{v(v^2 - 1^2)}{3!}\Delta^2 y_{-1} + \frac{v(v^2 - 1^2)(u^2 - 2^2)}{5!}\Delta^4 y_{-2} + \ldots + uy_1$$

$$+ \frac{u(u^2 - 1^2)}{3!}\Delta^2 y_0 + \frac{u(u^2 - 1^2)(u^2 - 2^2)}{5!}\Delta^2 y_{-1} + \ldots \qquad (5.70)$$

Equation (5.70) is known as *Laplace-Everett formula*. Eq. (5.71) uses only even differences of the function.

EXAMPLE 5.41

Use Everett's interpolation formula to find the value of y when $x = 1.60$ from the following table.

x	1.0	1.25	1.50	1.75	2.0	2.25
$y = f(x)$	1.0543	1.1281	1.2247	1.3219	1.4243	1.4987

Solution:

The difference table is shown here:

i	x_i	y_i	Δy_i	$\Delta^2 y_i$	$\Delta^3 y_i$	$\Delta^4 y_i$
−2	1.00	1.0543				
			0.0738			
−1	1.25	1.1281		0.0228		
			0.0966		−0.0222	
0	1.50	1.2247		0.006		0.0268
			0.0972		0.0046	
1	1.75	1.3219		0.00520		−0.0378
			0.1024		−0.0332	
2	2.0	1.4243		−0.0280		
			0.0744			
3	2.25	1.4987				

Here $\qquad x_0 = 1.50$ and $h = 0.25$.

Therefore $\qquad v = \dfrac{x - x_0}{h} = \dfrac{1.60 - 1.50}{0.25} = 0.4$

and $\qquad u = 1 - v = 1 - 0.4 = 0.6$

Everett's interpolation formula is

$$y = \left[vy_1 + \frac{v(v^2 - 1^2)}{3!} \Delta^2 y_0 + \frac{v(v^2 - 1^2)(v^2 - 2^2)}{5!} \Delta^4 y_{-1} \right]$$
$$+ \left[uy_0 + \frac{u(u^2 - 1^2)}{3!} \Delta^2 y_{-1} + \frac{u(u^2 - 1^2)(u^2 - 2^2)}{5!} \Delta^4 y_{-2} \right]$$

$$= \left[0.4(1.3219) + \frac{0.4(0.16-1)}{6}(0.00520) + \frac{0.4(0.16-1)(0.16-4)}{120}(-0.03780) \right]$$

$$+ \left[0.6(1.2247) + \frac{0.6(0.36-1)}{6}(0.0006) + \frac{0.6(0.36-1)(0.36-4)}{120}(0.02680) \right]$$

$$= 1.26316.$$

5.5.6 Selection of an Interpolation Formula

In general, the selection of an interpolation formula depends to a great extent on the position of the interpolated value in the given data.

a) Use Newton's forward interpolation formula to find a tabulated value near the beginning of the table.

b) Use Newton's backward interpolation formula to find a value near the end of the table.

c) Use either Stirling's, or Bessel's, or the Laplace-Everett formula to find an interpolated value near the center of the table.

The coefficients in the central difference formulae are smaller and converge faster than those in Newton's forward or Newton's backward interpolation formulae. Also, after a few terms, the coefficients in Stirling's formula decrease more rapidly than those of Bessel's formula. Similarly, the coefficients of Bessel's formula decrease more rapidly than those of Newton's forward or backward formula. Hence, wherever possible, central difference formula are preferred over Newton's formulae. However, as described in (a), (b), and (c) above, the right selection of an interpolation formula greatly depends on the position of the interpolated value in the given tabular data set.

5.6 DIVIDED DIFFERENCES

Let the function $y = f(x)$ be given at the point $x_0, x_1, x_2, ..., x_n$ (that need not be equally spaced) $f(x_0), f(x_1), f(x_2), ..., f(x_n)$, denote the $(n+1)$ values the function at the points $x_0, x_1, x_2, ..., x_n$.

Then the first divided differences of $f(x)$ for the arguments x_0, x_1 is defined as

$$\frac{f(x_0) - f(x_1)}{x_0 - x_1}.$$

It is denoted by $f(x_0, x_1)$ or by $[x_0, x_1]$

Likewise, $f(x_1, x_2) = \dfrac{f(x_1) - f(x_2)}{x_1 - x_2}$,

$$f(x_2, x_3) = \dfrac{f(x_2) - f(x_3)}{x_2 - x_3}, \text{ etc.}$$

The second divided difference for the arguments x_0, x_1, x_2 is defined as

$$f(x_0, x_1, x_2) = \dfrac{f(x_0, x_1) - f(x_1, x_2)}{x_0 - x_2},$$

similarly, the third differences for the arguments x_0, x_1, x_2, x_3 is defined as

$$f(x_0, x_1, x_2, x_3) = \dfrac{f(x_0, x_1, x_2) - f(x_1, x_2, x_3)}{x_0 - x_3}$$

The first divided differences are called the divided differences of order one, the second divided differences are called the divided differences of order two and so on.

The divided difference table (Table 5.11) is given below:

TABLE 5.11

Argument, x	Entry	$\nabla f(x)$	$\nabla^2 f(x)$	$\nabla^3 f(x)$
x_0	$f(x_0)$			
		$f(x_0, x_1)$		
x_1	$f(x_1)$		$f(x_0, x_1, x_2)$	
		$f(x_1, x_2)$		$f(x_0, x_1, x_2, x_3)$
x_2	$f(x_2)$		$f(x_1, x_2, x_3)$	
		$f(x_2, x_3)$		
x_3	$f(x_3)$			

EXAMPLE 5.42

If $f(x) = \dfrac{1}{x}$, then find the divided differences $f(a, b)$ and $f(a, b, c)$

Solution:

Given $f(x) = f''_{i,i+1}(x)$,

$$\Rightarrow f(a, b) = \frac{f(a) - f(b)}{a - b} = \frac{\dfrac{1}{a} - \dfrac{1}{b}}{(a - b)} = \frac{b - a}{ab(a - b)} = -\frac{1}{ab}$$

and $\quad f(a, b, c) = \dfrac{f(a, b) - f(b, c)}{a - c}$

$$= \frac{\dfrac{-1}{ab} - \left(\dfrac{-1}{bc}\right)}{a - c} = \frac{1}{b}\left(\frac{-c + a}{ac}\right)\frac{1}{a - c} = \frac{1}{abc}$$

Hence $\quad f(a, b, c) = \dfrac{1}{abc}$.

EXAMPLE 5.43

Prepare the divided difference table for the following data

TABLE 5.43

x	1	3	4	6	10
f(x)	0	18	58	190	920

Solution:

Table 5.43(a) shows the divided differences.

TABLE 5.43(a)

x	f(x)	$\nabla f(x)$	$\nabla^2 f(x)$	$\nabla^3 f(x)$	$\nabla^4 f(x)$
1	0	9	10.33333	−0.33333	0.207672
3	18	40	8.666667	1.535714	
4	58	66	19.41667		
6	190	182.5			
10	920				

5.6.1 Newton's Divided Difference Interpolation Formula

A function $f(x)$ is written in terms of divided differences as follows:

$$f(x) = f(x_0) + (x - x_0)f(x_0, x_1) + (x - x_0), (x - x_1)f(x_0, x_1, x_2)$$
$$+ (x - x_0)(x - x_1)(x - x_2)f(x_0, x_1, x_2, x_3)$$
$$+ (x - x_0)(x - x_1)(x - x_2)(x - x_3)f(x_0, x_1, x_2, x_3, x_4)$$
$$+ \dots$$

EXAMPLE 5.44

Find the form of the function $f(x)$ under suitable assumption from the following data.

x	0	1	2	5
$f(x)$	2	3	12	147

Solution:

The divided difference table (Table 5.28) is given as:

TABLE 5.28

x	$f(x)$	∇f	$\nabla^2 f$	$\nabla^3 f$
0	2			
		1		
1	3		4	
		9		1
2	12		9	
		45		
5	147			

We have $x_0 = 0, f(x_0) = 2, f(x_0, x_1) = 1, f(x_0, x_1, x_2) = 4, f(x_0, x_1, x_2, x_3) = 1$.

Newton's divided difference interpolation formula for this case is:

$$f(x) = f(x_0) + (x - x_0)f(x_0, x_1) + (x - x_0)(x - x_1)f(x_0, x_1, x_2) + (x - x_0)(x - x_1)(x - x_2)$$
$$f(x_0, x_1, x_2, x_3).$$

Substituting all constants we get:

$$f(x) = 2 + (x - 0)1 + (x - 0)(x - 1)4 + (x - 0)(x - 1)(x - 2)1$$

Hence $f(x) = x^3 + x^2 - x + 2$.

EXAMPLE 5.45

Derive the equation of the interpolating polynomial for the data given in Table 5.29 below:

TABLE 5.29

x(deg)	f(x)
0	3
1	2
2	7
3	24
4	59
5	118

Solution:

First form the divided difference table as shown in Table 5.30:

TABLE 5.30

x	f(x)	∇f	$\nabla^2 f$	$\nabla^3 f$	$\nabla^4 f$
0	3	−1	3	1	0
1	2	5	6	1	0
2	7	17	9	1	
3	24	35	12		
4	59	59			
5	118				

Using Newton's divided difference formula, the interpolating polynomial is:

$$f(x) = f(x_0) + (x - x_0) f(x_0, x_1) + (x - x_0)(x - x_1) f(x_0, x_1, x_2) + (x - x_0)(x - x_1)(x - x_2) f(x_0, x_1, x_2, x_3).$$

$$= 3 - x + 3x(x - 1) + x(x - 1)(x - 2)$$

Figure 5.2 shows the variation of the function with actual values and those obtained from polynomial.

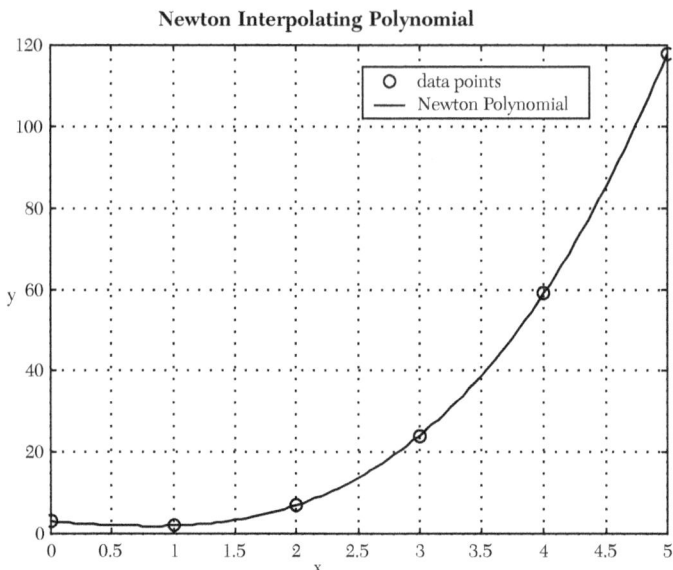

FIGURE 5.2 Newton's polynomial.

Main advantage of divided difference table approach is that it has less computational operations. We do not need to write the polynomial and then use the C^0 condition to calculate the constants. Secondly, it is much easier to incorporate in a computer code. It is important to realize that both the Lagrange and Newton polynomials are C^0 continuous and each would generate the same result.

5.7 CUBIC SPLINE INTERPOLATION

Generally, we use only one polynomial to describe the data over the entire range. Here, we will use different continuous polynomials to describe the function in each interval of known points. This type of approximation is called the *piecewise polynomial approximation*. Therefore, for $n + 1$ set of data, there will be n piecewise polynomials formed. Splines of different degree are available in the literature. However, cubic splines are most widely used.

Cubic spline interpolation method interpolates a function between a given set of data points by means of piecewise smooth polynomials. Here, the curve passes through the given set of data points. The slope and its curvature are continuous at each point. The advantage of cubic spline interpolation method is that these polynomials are of a lower degree and less oscillatory, therefore describing the given data more accurately. Cubic spline interpolation method is very powerful and widely used. It has several applications in numerical differentiation, integration, solution of boundary value problems and plotting two and three-dimensional graphs.

With a cubic spline, an expression for the second derivative can be obtained which will describe the behavior of the data most accurately within each interval.

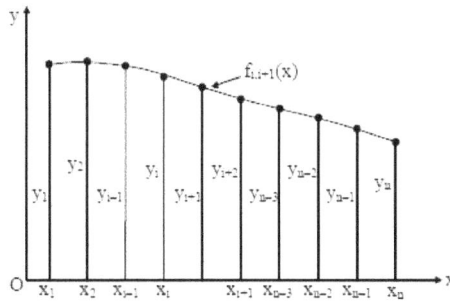

FIGURE 5.3 Cubic spline.

The second derivatives of the spline is zero at the end points. Since these end conditions occur naturally in a beam model (in strength of materials), the resulting curve is known as the *natural cubic spline*. The pins, i.e., the data points, are called the *knots* of the spline in a beam model.

Figure 5.3 shows a cubic spline that spans n knots. Let us denote $f_{i,i+1}(x)$ be the cubic polynomial that spans the segment between knots i and $i + 1$. In Figure 5.3, we note that the spline is a *piecewise cubic curve*, assembled together form the $n - 1$ cubics $f_{1,2}(x), f_{2,3}(x), ..., f_{n-1,n}(x)$, all of which have different coefficients.

Denoting the second derivative of the spline at knot i by k_i, the continuity of second derivatives requires that

$$f''_{i-1,i}(x_i) = f''_{i,i+1}(x_i) = k_i \tag{5.71}$$

In Eq. (5.71), k_i is unknown, except for

$$k_1 = k_n = 0 \tag{5.72}$$

We know that the expression for $f''_{i,i+1}(x)$ is linear and the starting point for obtaining the coefficients of $f_{i,i+1}(x)$ is $f''_{i,i+1}(x)$.

Hence, we can write using Lagrange's two-point interpolation,

$$f''_{i,i+1}(x) = k_i \ell_i(x) + k_{i+1} \ell_{i+1}(x) \tag{5.73}$$

where $\quad \ell_i(x) = \dfrac{x - x_{i+1}}{x_i - x_{i+1}}$

and $\quad \ell_{i+1}(x) = \dfrac{x - x_i}{x_{i+1} - x_i} \tag{5.74}$

Hence $\quad f''_{i,i+1}(x) = \dfrac{k_i(x - x_{i-1}) - k_{i+1}(x - x_i)}{x_i - x_{i+1}} \tag{5.75}$

Integrating Eq. (5.75) twice with respect to x, we get

$$f''_{i,i+1}(x) = \dfrac{k_i(x - x_{i+1})^3 - k_{i+1}(x - x_i)^3}{6(x_i - x_{i+1})} + A(x - x_{i+1}) - B(x - x_i) \tag{5.76}$$

or $\quad = \dfrac{k_i(x - x_{i+1})^3 - x_{i+1}(x - x_i)^3}{6(x_i - x_{i+1})} + Cx + D \tag{5.77}$

where A and B are constants of integration, $C = A - B$ and $D = -Ax_{i+1} + Bx_i$.

Now applying the condition $f_{i,i+1}(x_i) = y_i$, Eq. (5.76) becomes

$$\dfrac{k_i(x_i - x_{i+1})^3}{6(x_i - x_{i-1})} + A(x_i - x_{i+1}) = y_i \tag{5.78}$$

Hence $\quad A = \dfrac{y_i}{x_i - x_{i+1}} - \dfrac{k_i}{6}(x_i - x_{i+1}) \tag{5.79}$

Similarly, applying the condition $f_{i,i+1}(x_{i+1}) = y_{i+1}$, gives

$$B = \dfrac{y_{i+1}}{x_i - x_{i+1}} - \dfrac{k_{i+1}}{6}(x_i - x_{i+1}) \tag{5.80}$$

From Equations (5.79) and (5.80), we obtain

$$f_{i,i+1}(x) = \dfrac{k_i}{6}\left[\dfrac{(x - x_{i+1})^3}{x_i - x_{i+1}} - (x - x_{i+1})(x_i - x_{i+1})\right]$$
$$-\dfrac{k_{i+1}}{6}\left[\dfrac{(x - x_i)^3}{x_i - x_{i+1}} - (x - x_i)(x_i - x_{i+1})\right] + \dfrac{y_i(x - x_{i+1}) - y_{i+1}(x - x_i)}{x_i - x_{i+1}} \tag{5.81}$$

We note here that the second derivatives k_i of the spline at the interior knots are found from the slope continuity conditions

$$f'_{i-1}(x_i) = f'_{i,i+1}(x_i), \qquad i = 1, 2, 3, ..., n-1 \tag{5.82}$$

Applying the conditions given by Eq. (5.82) in Eq. (5.81) and after some mathematical manipulations or simplifications, we obtain the simultaneous equations:

$$k_{i-1}(x_{i-1} - x_i) + 2k_i(x_{i-1} - x_{i+1}) + k_{i+1}(x_i - x_{i+1})$$

$$= 6\left[\frac{y_{i-1} - y_i}{x_{i-1} - x_i} - \frac{y_i - y_{i+1}}{x_i - x_{i+1}}\right], \qquad i = 2, 3, ..., n-1 \tag{5.83}$$

If the data points are equally spaced at intervals h, then, we have

$$h = x_{i-1} - x_i = x_i - x_{i+1} \tag{5.84}$$

and Eq. (5.83) becomes

$$k_{i-1} + 4k_i + k_{i+1} = \frac{6}{h^2}\left[y_{i-1} - 2y_i + y_{i+1}\right], \quad i = 2, 3, ..., n-1 \tag{5.85}$$

There are two boundary conditions normally used. They are:

1. *Natural boundary condition:*

   The second derivatives of the data at the end points x_0 and x_n are arbitrarily assumed to be zero. This condition is known as the *free or natural boundary condition*. The polynomials resulting from this condition are referred to as *natural or free cubic splines*. They may not provide very accurate values close to the boundaries, but they are accurate enough in the interior region.

2. *Clamped boundary condition:*

   When the first derivative of the data is known at the end point x_0 and x_n, the corresponding boundary conditions are known. This condition is known as the *clamped boundary condition*.

EXAMPLE 5.46

Given the data points:

x	1	2	3	4	5
y	13	15	12	9	13

Find the natural cubic spline interpolation at $x = 3.4$.

Solution:

For equally spaced knots, the equations for the curvatures are written as

$$k_{i-1} + 4k_i + k_{i+1} = \frac{6}{h^2}(y_{i-1} - 2y_i + y_{i+1}), \quad i = 2, 3, 4$$

Here we have $k_1 = k_5$ and $h = 1$.

Hence

$$4k_2 + k_3 = 6[13 - 2(15) + 12] = -30$$

$$k_2 + 4k_3 + k_4 = 6[15 - 2(12) + 9] = 0$$

$$k_3 + 4k_4 = 6[12 - 2(9) + 13] = 42 \tag{E.1}$$

Solving Eq. (E.1), we obtain

$$k_2 = -7.286 \qquad k_3 = -0.857 \qquad k_4 = 10.714$$

The interpolant between knots 2 and 3 is given by

$$f_{3,4}(x) = \frac{k_3}{6}\left[\frac{(x-x_4)^3}{x_3-x_4} - (x-x_4)(x_3-x_4)\right] - \frac{k_4}{6}\left[\frac{(x-x_3)^3}{x_3-x_4} - (x-x_3)(x_3-x_4)\right]$$

$$+ \frac{y_3(x-x_4) - y_4(x-x_3)}{x_3-x_4}$$

Hence, the natural cubic spline interpolation at $x = 3.4$ is

$$f_{3,4}(3.4) = \frac{-0.857}{6}\left[\frac{(3.4-4)^3}{3-4} - (3.4-4)(3-4)\right] - \frac{10.714}{6}\left[\frac{(3.4-3)^3}{3-4} - (3.4-3)(3-4)\right]$$

$$+ \frac{12(3.4-4) - 9(3.4-3)}{3-4} = 0.054848 - 0.599984 + 10.8 = 10.2552.$$

EXAMPLE 5.47

Find the natural spline that passes through the points given below:

i	1	2	3
x_i	0	1	2
y_i	0	2	1

Find the first and second derivatives at $x = 1$ noting that the interpolant consists of two cubics, one valid in $0 \le x \le 1$, the other in $1 \le x \le 2$.

Solution:

For natural spline, we have $k_1 = k_3 = 0$. The equation for k_1 is given by

$$k_1 + 4k_2 + k_3 = \frac{6}{h^2}(y_1 - 2y_2 + y_3)$$

Hence, $\quad 0 + 4k_2 + 0 = \dfrac{6}{1^2}[0 - 2(2) + 1]$

or $\quad k_2 = -4.5$

The interpolant in $0 \le x \le 1$ is given by

$$f_{1,2}(x) = -\frac{k_2}{6}\left[\frac{(x-x_1)^3}{x_1 - x_2} - (x - x_1)(x_1 - x_2)\right] + \frac{y_1(x-x_2) - y_2(x-x_1)}{x_1 - x_2}$$

$$= \frac{4.5}{6}\left(\frac{(x-0)^3}{0-1} - (x-0)(0-1)\right) + \frac{0 - 2(x-0)}{0-1} = 0.75x^3 + 2.75x$$

The interpolant in $1 \le x \le 2$ is given by

$$f_{2,3}(x) = -\frac{k_2}{6}\left[\frac{(x-x_3)^3}{x_2 - x_3} - (x - x_3)(x_2 - x_3)\right] + \frac{y_2(x-x_3) - y_3(x-x_2)}{x_2 - x_3}$$

$$-\frac{4.5}{6}\left(\frac{(x-2)^3}{1-2} - (x-2)(1-2)\right) + \frac{2(x-2) - (x-1)}{1-2}$$

$$= -0.75(x-2)^3 - 1.75x + 4.5$$

Now $\quad f'_{1,2}(x) = -3(0.75)x^2 + 2.75 = -2.25x^2 + 2.75$

$\quad\quad\quad f'_{2,3}(x) = 3(0.75)(x-2)^2 - 1.75 = 2.25(x-2)^2 - 1.75$

$\quad\quad\quad f'_{1,2}(1) = -2.25(1)^2 + 2.75 = 0.5$

$\quad\quad\quad f'_{2,3}(1) = 2.25(1-2)^2 - 1.75 = 0.5$

$\quad\quad\quad f''_{1,2}(1) = -2.25(2) = -4.5$

$\quad\quad\quad f''_{2,3}(1) = 2.25(2)(1-2) = -4.5$

Hence $\quad f'_{1,2}(1) = f'_{2,3}(1) = 0.5$

and $\quad f''_{1,2}(1) = f''_{2,3}(1) = -4.5$

EXAMPLE 5.48

Use the end conditions with a cubic spline that has constant second derivatives within its first and last segments (noting the end segments are parabolic). The end conditions for this spline are given as $k_1 = k_2$ and $k_{n-1} = k_n$. The data points are given below:

i	1	2	3	4
x	0	1	2	3
y	1	1	0.5	0

Solution:

With evenly spaced knots, the equations for the curvatures are given by

$$k_{i-1} + 4k_i + k_{i+1} = \frac{6}{h^2}(y_{i-1} - 2y_i + y_{i+1}), \quad i = 2, 3 \tag{E.1}$$

With $k_1 = k_2$, $k_4 = k_3$ and $h = 1$, Eq. (E.1) become

$$5k_2 + k_3 = 6(1 - 2(1) + 0.5) = -3 \tag{E.2}$$

$$k_2 + 5k_3 = 6[1 - 2(0.5) + 0] = 0$$

Solving Eq. (E.2), we get $k_2 = -5/8$, $k_3 = 1/8$. The interpolant can now be evaluated from

$$f_{i,i+1}(x) = \frac{k_i}{6}\left[\frac{(x - x_{i+1})^3}{x_i - x_{i+1}} - (x - x_{i+1})(x_i - x_{i+1})\right] - \frac{k_{i+1}}{6}\left[\frac{(x - x_i)^3}{x_i - x_{i+1}} - (x - x_i)(x_i - x_{i+1})\right]$$

$$+ \frac{y_i(x - x_{i+1}) - y_{i+1}(x - x_i)}{x_i - x_{i+1}} \tag{E.3}$$

Substituting $x_i - x_{i+1} = -1$ and $i = 3$, Eq. (E.3) becomes

$$f_{3,4}(x) = \frac{k_3}{6}[-(x - x_4)^3 + (x - x_4)] - \frac{k_4}{6}[-(x - x_3)^3 + (x - x_3) - y_3(x - x_4) + y_4(x - x_3)]$$

Hence, $f_{3,4}(2.6) = \dfrac{1/8}{6}[-(2.6 - 3)^3 + (2.6 - 3)] - \dfrac{1/8}{6}[-(2.6 - 2)^3 + (2.6 - 2)]$

$$-0.5(2.6 - 3) + 0 = 0.1853$$

5.8 SUMMARY

Interpolation is the method of computing the value of the function $y = f(x)$ for any given value of the independent variable x when a set of values of $y = f(x)$ for certain values of x are given. The study of interpolation is based on the assumption that there are no sudden jump in the values of the dependent variable for the period under consideration. In this chapter, the study of interpolation was presented based on the calculus of finite differences. Some important interpolation formulae by means of forward, backward, and central differences of a function, which are frequently used in scientific and engineering calculations were also presented.

EXERCISES

5.1 Show that

(a) $\Delta \nabla = \Delta - \nabla$

(b) $\nabla = \Delta E^{-1}$

(c) $E^n = (1 + \Delta)^n$

(d) $e^x = \left(\dfrac{\Delta^2}{E} \right) e^x \dfrac{Ee^x}{\Delta^2 e^x}$ $(h = \text{interval of differencing})$

(e) $\Delta^n \sin(ax + b) = \left(2\sin\dfrac{a}{2} \right)^n \sin\left[ax + b + n\left(\dfrac{a + \pi}{2} \right) \right]$
(interval of differencing $= 1$)

(f) $\Delta^2 = (1 + \Delta)\delta^2$

(g) $\Delta^3 y_2 = \nabla^3 y_5$

(h) $\delta = \Delta(1 + \Delta)^{-1/2}$

(i) $\nabla = 1 - (1 + \nabla)^{-1}$

(j) $\dfrac{\Delta}{\nabla} - \dfrac{\nabla}{\Delta} = \Delta + \nabla$

5.2 Find the following:

(a) Δe^{ax}

(b) $\Delta \sin x$

(c) $\Delta \tan^{-1}x$

(d) $\Delta\left[\dfrac{5x+12}{x^2+5x+6}\right]$

(e) $\Delta^n\left(\dfrac{1}{x}\right)$

(f) $\left(\dfrac{\Delta^2}{E}\right)x^3$ (with interval of differencing = 1)

(g) $\dfrac{\Delta}{(1+x^2)}$

(h) $\Delta \sin(ax+b)$

(i) $\Delta^2(3e^x)$

5.3 Construct a forward difference table for the following data:

(a)

x	45	55	65	75
y = f(x)	20	60	120	180

(b)

x	40	50	60	70	80	90
y = f(x)	204	224	246	270	296	324

5.4 Construct a difference table for $y = x^3 + 2x + 3$ for $x = 1, 2, 3, 4, 5$.

5.5 Given $u_0 = 1$, $u_1 = 5$, $u_2 = 10$, $u_3 = 30$, $u_4 = 30$, find $\Delta^4 u_0$.

5.6 Given $u_0 = 5$, $u_1 = 24$, $u_2 = 81$, $u_3 = 200$, $u_4 = 100$ and $u_5 = 8$, find $\Delta^5 u_0$.

5.7 Estimate the missing term in the following tables:

(a)

x	1	2	3	4	5
y = f(x)	5	14	?	74	137

(b)

x	1	2	3	4	5
y = f(x)	8	17	38	?	140

(c)

x	0	1	2	3	4
y = f(x)	3	2	3	?	11

5.8 If m is a positive integer and the interval of differencing is 1, show that $x^m = x(x-1) \ldots [x-(x-1)]$.

5.9 Express the following in the factorial notation. Take interval of differencing as equal to 1.

(a) $y = f(x) = 3x^3 + x^2 + x + 1$

(b) $y = f(x) = x^4 - 5x^3 + 3x + 4$

5.10 Find the missing entry in the following tables:

(a)

x	0	1	2	3	4
y = f(x)	1	3	13	—	81

(b)

x	0	1	2	3	4
y = f(x)	1	0	—	28	69

(c)

x	0	1	2	3	4
y = f(x)	1	-2	-1	—	37

(d)

x	0	1	2	3	4
y = f(x)	1	4	—	28	61

(e)

x	0	1	2	3	4
y = f(x)	-6	-3	4	—	54

5.11 Find the missing entry in the following tables:

(a)

x	0	1	2	3	4
y = f(x)	1	3	—	55	189

(b)

x	0	1	2	3	4
y = f(x)	1	−3	−1	—	165

(c)

x	0	1	2	3	4
y = f(x)	−31	−35	—	5	133

(d)

x	0	1	2	3	4
y = f(x)	−23	−27	—	13	141

(e)

x	0	1	2	3	4
y = f(x)	2	−2	0	—	166

5.12 Interpolate the missing entries in the following tables:

(a)

x	0	1	2	3	4
y = f(x)	1	—	13	—	81

(b)

x	0	1	2	3	4
y = f(x)	1	−2	—	—	37

(c)

x	0	1	2	3	4
y = f(x)	1	—	11	—	61

(d)

x	0	1	2	3	4
y = f(x)	−6	—	4	—	54

(e)

x	0	1	2	3	4
y = f(x)	−6	—	12	—	118

5.13 Given that $\sqrt{12600} = 112.24972$, $\sqrt{12610} = 112.29426$, $\sqrt{12620} = 112.33877$, $\sqrt{12630} = 112.38327$. Find the value of $\sqrt{12616}$.

5.14 Evaluate $y = e^{2x}$ for $x = 0.25$ from the data in the following table.

x	0.2	0.3	0.4	0.5	0.6
e^{2x}	1.49182	1.82212	2.22554	2.71828	3.32012

5.15 The values of sin x are given below for different values of x. Find the value of sin 42°.

x	40°	45°	50°	55°	60°
y = sin x	0.64279	0.70711	0.76604	0.81915	0.86603

5.16 In an examination the number of students who obtained marks between certain limits was as follows:

Marks	30-40	40-50	50-60	60-70	70-80
No. of students	18	40	64	50	28

Find the number of students whose scores lie between 70 and 75.

5.17 From the following table estimate the number of students who obtained marks in the examination between 50 and 55.

Marks	35-45	45-5	55-65	65-75	75-85
No. of students	31	42	51	35	31

5.18 A second degree polynomial passes through the points $(2, -1)$, $(3, 1)$, $(4, 5)$, and $(5, 11)$. Find the polynomial.

5.19 A second degree polynomial passes through the points $(1, 5)$, $(2, 11)$, $(3, 19)$, and $(4, 29)$. Find the polynomial.

5.20 Find a cubic polynomial which takes the following values.

x	0	1	2	3
f(x)	1	4	17	46

5.21 Refer to Exercise 5.11. Find f (1.5).

5.22 Refer to Exercise 5.10. Find f (3.5).

5.23 The table below gives the values of $f(x)$ for $0.10 \leq x \leq 0.30$. Find f (0.12) and f (0.26).

x	0.1	0.15	0.2	0.25	0.30
f(x)	0.0998	0.1494	0.1987	0.2474	0.2955

5.24 The population (in thousands) of a small town is given in the following table. Estimate the population in the years 1965 and 1995.

Year, x	1961	1971	1981	1991	2001
Population y = f(x) (in thousands)	46	66	81	93	101

5.25 Using Newton's forward interpolation formula, find the value of sin 52° from the following data. Estimate the error.

x	40°	45°	50°	55°	60°
y = sin x	0.64279	0.70711	0.76604	0.81915	0.86603

5.26 Find the polynomial of degree three relevant to the following data using Lagrange's interpolation formula.

x	1	2	3	5
f(x)	−12	−14	−20	−20

5.27 Find the polynomial of the least degree that attains the prescribed values at the given point using Lagrange's interpolation formula.

x	1	2	4	5
y = f(x)	−27	−44	−84	−95

5.28 Find the polynomial of degree three relevant to the following data using Lagrange's interpolation formula.

x	1	3	5	6
y = f(x)	71	115	295	466

5.29 Find the polynomial of degree three relevant to the following data using Lagrange's interpolation formula.

x	0	1	2	4
y = f(x)	2	5	12	62

5.30 Using Lagrange's interpolation formula, find the value of *y* corresponding to *x* = 8 from the following table:

x	1	3	6	9
y = f(x)	71	115	466	1447

5.31 Using Lagrange's interpolation formula, find the value of *y* corresponding to *x* = 6 from the following table:

x	0	3	5	7
y = f(x)	2	29	117	317

5.32 Using Lagrange's interpolation formula, find the value of *y* corresponding to *x* = 4 from the following table:

x	0	1	3	5
y = f(x)	−20	−12	−20	−20

5.33 Using Lagrange's interpolation formula, find the value of *y* corresponding to *x* = 9 from the following table:

x	5	6	11	13
y = f(x)	380	−2	508	1020

5.34 The following table gives the values of *y* corresponding to certain values of *x*. Find the value of *x* when *y* = 420.61175 by applying Lagrange's inverse interpolation formula.

x	1	2	5	6
y = f(x)	71	82	295	466

5.35 The following table gives the values of *y* corresponding to certain values of *x*. Find the value of *x* when *y* = −76.0188 by applying Lagrange's inverse interpolation formula.

x	1	2	4	5
y = f(x)	−27	−65	−84	−95

5.36 The following table gives the values of y corresponding to certain values of x. Find the value of x when $y = 89.64656$ by applying Lagrange's inverse interpolation formula.

x	1	2	5	6
$y = f(x)$	71	82	295	466

5.37 The following table gives the values of y corresponding to certain values of x. Find the value of x when $y = -16.875$ by applying Lagrange's inverse interpolation formula.

x	0	1	3	5
$y = f(x)$	-20	-12	-20	-20

5.38 Apply Gauss's forward interpolation formula to find the value of $f(x)$ at $x = 11$ from the following table:

x	1	5	9	13	17
$f(x)$	13	16	18	21	26

5.39 Find the value of $f(x)$ at $x = 10$ by applying Gauss's forward interpolation formula from the following data:

x	0	4	8	12	16
$f(x)$	23	28	36	39	45

5.40 Find the value of $f(9)$ by applying Gauss's forward interpolation formula from the following data:

x	0	4	8	12	16
$f(x)$	15	25	34	37	42

5.41 Apply Gauss's forward interpolation formula to find the value of $f(12.2)$ from the following data:

x	10	11	12	13	14
$f(x)$	23967	28060	31788	35209	38368

5.42 Find the value of $f(9)$ by applying Gauss's forward interpolation formula from the following data:

x	0	4	8	12	16
$f(x)$	17	19	35	38	41

5.43 Use Gauss's forward interpolation formula to find y for $x = 10$ given that

x	0	4	8	12	16
y = f(x)	15	25	34	37	42

5.44 Use Gauss's backward interpolation formula to find the sales for the year 1966 given the following data:

Year	1931	1941	1951	1961	1971	1981
Sales (in millions)	5	7	12	17	23	31

5.45 Apply Gauss's backward interpolation formula and find the population of a city in 1946 based on the following data:

Year	1931	1941	1951	1961	1971
Population (in millions)	16	21	29	41	54

5.46 Use Gauss's backward interpolation formula to find the sales for the year 1966 based on the following data:

Year	1951	1961	1971	1981	1991
Sales (in millions)	23	32	43	52	61

5.47 Apply Gauss's backward interpolation formula to find the population of a city in 1986 based on the following data:

Year	1951	1961	1971	1981	1991	2001
Population (in millions)	15	21	25	29	47	61

5.48 Use Gauss's backward interpolation formula to find the sales for the year 1986 based on the following data:

Year	1951	1961	1971	1981	1991	2001
Sales (in millions)	1	3	6	11	17	23

5.49 Apply Bessel's interpolation formula to obtain y_{25}, given that $y_{20} = 515$, $y_{24} = 438$, $y_{28} = 348$ and $y_{32} = 249$.

5.50 Apply Bessel's interpolation formula to obtain y_{16}, given that $y_{15} = 0.345$, $y_{20} = 0.375$, $y_{25} = 0.478$ and $y_{30} = 0.653$.

5.51 Apply Bessel's interpolation formula to obtain $y_{1.6}$, given that $y_{1.5} = 0.345$, $y_{2.0} = 0.423$, $y_{2.5} = 0.512$ and $y_{3.0} = 0.756$.

5.52 Apply Bessel's interpolation formula to obtain y_{36}, given that $y_{21} = 19$, $y_{231} = 29$, $y_{41} = 43$ and $y_{51} = 54$.

5.53 Apply Bessel's interpolation formula to obtain $y_{1.4}$, given that $y_{1.25} = 1.0772$, $y_{1.5} = 1.1447$, $y_{1.75} = 1.2051$ and $y_{2.0} = 1.2599$.

5.54 Apply Bessel's interpolation formula to obtain $y_{0.644}$, given that $y_{0.64} = 1.89648$, $y_{0.65} = 1.91554$, $y_{0.66} = 1.93479$, and $y_{0.67} = 1.95424$.

5.55 Use Stirling's interpolation formula to find $y_{12.2}$ from the following table.

x	10	11	12	13	14
y = f(x)	24765	27876	30879	36543	39879

5.56 Use Stirling's interpolation formula to find $y_{1.22}$ from the following table.

x	0	0.5	1.0	1.5	2.0
y = f(x)	0	0.1910	0.3410	0.4330	0.4770

5.57 Use Stirling's interpolation formula to find $y_{22.6}$ from the following table.

x	20	21	22	23	24
y = f(x)	1.2123	1.3546	1.4879	1.5765	1.6987

5.58 Use Stirling's interpolation formula to find $y_{3.8}$ from the following table of data.

x	1	2	3	4	5
y = f(x)	0.12340	0.34560	0.87650	1.12346	1.34657

5.59 Use Stirling's interpolation formula to find $y_{3.25}$ from the following data.

x	2	2.5	3.0	3.5	4.0
y = f(x)	49225	48316	47236	45926	44306

5.60 Use Everett's interpolation formula to find the value of y when $x = 3.5$ from the following table.

x	1	2	3	4	5	6
y = f(x)	1.2567	1.4356	1.5678	1.6547	1.7658	1.8345

5.61 Use Everett's interpolation formula to find the value of y when $x = 6$ from the following table.

x	1	3	5	7	9	11
y = f(x)	−0.375	−2.947	−6.063	−2.331	24.857	105.165

5.62 Use Everett's interpolation formula to find the value of y when $x = 0.35$ from the following table.

x	0.1	0.2	0.3	0.4	0.5	0.6
y = f(x)	1.23900	1.12999	0.95294	0.70785	0.39469	0.01348

5.63 Use Everett's interpolation formula to find the value of y when $x = 0.35$ from the following table.

x	0.1	0.2	0.3	0.4	0.5	0.6
y = f(x)	2.4780	2.25997	1.90589	1.41569	0.78938	0.02696

5.64 Use Everett's interpolation formula to find the value of y when $x = 0.644$ from the following table.

x	0.61	0.62	0.63	0.64	0.65	0.66	0.67
y = f(x)	1.850431	1.858928	1.887610	1.906481	1.925541	1.944792	1.964237

5.65 Use Everett's interpolation formula to find the value of y when $x = 1.71$ from the following table.

x	1.4	1.5	1.6	1.7	1.8	1.9	2.0
y = f(x)	4.055200	4.481689	4.953032	5.473947	6.049647	6.685894	7.389056

5.66 Fit a cubic spline curve that passes through the points as shown below:

x	0	1	2	3
y	0	0.5	2	1.5

The natural end boundary conditions are: $y''(0) = y''(3) = 0$.

5.67 Apply natural cubic spline interpolation method to find y at $x = 1.5$. The data points are given below:

x	1	2	3	4	5
y	0	1	0	1	0

5.68 Develop a natural cubic spline for the following data:

x	3	4	5	6	7
y	3.7	3.9	3.9	4.2	5.7

Find $f'(3.4), f'(5.2)$ and $f'(5.6)$.

5.69 Find the zero of the function $y(x)$ from the following data:

x	1.0	0.8	0.6	0.4	0.2
y	−1.049	−0.0266	0.377	0.855	1.15

Use inverse interpolation with the natural cubic spline.

5.70 Fit a cubic spline curve for the following data with end conditions $y'(0) = 0.2$ and $y'(3) = -1$.

x	0	1	2	3
y	0	0.5	3.5	5

5.71 Construct a clamped cubic spline for the following data given that the slope of 0.2 at x_0 and a slope of 0.6 at x_n.

n	0	1	2	3	4
x	3	4	5	6	7
y	3.7	3.9	3.9	4.2	5.7

Find $f''(3.4), f'(5.2)$ and $f(5.6)$.

5.72 Fit the data in Table 5.31 with cubic spline and find the value at $x = 5$.

TABLE 5.31

i	1	2	3	4
x	3	4.5	7	9
y	2.5	1.0	2.5	0.5

5.73 Determine the cubic spline interpolation at $x = 2.6$ based on the data points given below:

x	0	1	2	3
y	1	1	0.5	0

Given the end conditions as $f'_{1,2}(0) = 0$ (zero slope).

6

CURVE FITTING, REGRESSION, AND CORRELATION

6.1 INTRODUCTION

In real life engineering practice, often a relationship is found to exist between two (or more) variables. For example: the experimental data for force (N) and velocity (m/s) from a wind tunnel experiment. A mechanical element/component is suspended in a wind tunnel and the force measured for various levels of wind velocity. This relationship can be visualized by plotting force versus velocity. It is frequently desirable to express this relationship in mathematical/analytical form by establishing an equation connecting the variables.

In order to determine an equation connecting the variables, it is often necessary to collect the data depicting the values of the variables under consideration.

For example, if x and y denote respectively the velocity and force from the wind tunnel experiment, then a sample of n individual would give the velocities $x_1, x_2, ..., x_n$ and the corresponding forces $y_1, y_2, ..., y_n$. When these points $(x_1, y_1), (x_2, y_2), ..., (x_n, y_n)$ are plotted on a rectangular coordinate system, the resulting set of points on the plot is called the *scatter diagram*. From such a scatter diagram, one can visualize a smooth curve approximating the given data points. Such a curve is known as an *approximating curve*. Figure 6.1(a) shows that the data appears to be approximated by a straight line and it clearly exhibits a *linear relationship* between the two variables. On the other hand Figure 6.1(b) shows a relationship that is not linear and in fact it is a nonlinear relationship between the variables. *Curve fitting* is the general problem of finding equations of approximating curves which best fit the given set of data.

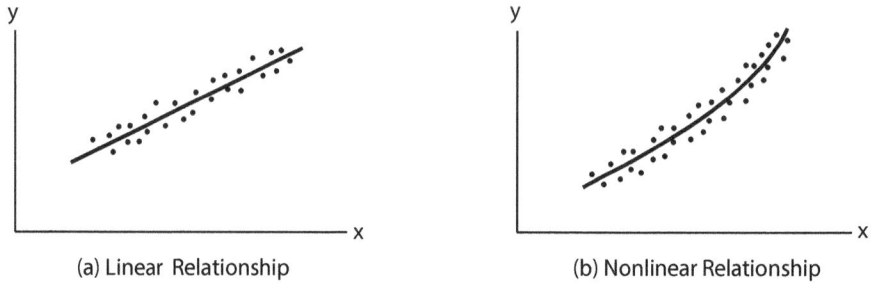

(a) Linear Relationship   (b) Nonlinear Relationship

FIGURE 6.1 Linear and nonlinear relationship.

6.1.1 Approximating Curves

Table 6.1 lists a few common approximating curves used in practice and their equations. The variables x and y are called the *independent* and *dependent variables* respectively. The scatter diagrams of the variables or transformed variables will help determine the curve to be used.

TABLE 6.1

No.	Equation	Description of the curve
1.	$y = a + bx$	Straight line
2.	$y = a + bx + cx^2$	Parabola or quadratic curve
3.	$y = a + bx + cx^2 + dx^3$	Cubic curve
4.	$y = a + bx + cx^2 + dx^3 + ex^4$	Quartic curve
5.	$y = a_0 + a_1 x + a_2 x^2 + \ldots + a_n x^n$	n^{th} degree curve
6.	$y = \dfrac{1}{c + mx}$ or $\dfrac{1}{y} = c + mx$	Hyperbola
7.	$y = ab^x$ or $\log y = \log a + x(\log b) = a_0 + b_0 x$	Exponential curve
8.	$y = ax^b$ or $\log y = \log a + b(\log x)$	Geometric curve (power function)
9.	$y = ab^x + c$	Modified exponential curve
10	$y = ax^b + c$	Modified geometric curve
11.	$y = pq^{b^x}$ or $\log y = \log p + b^x$ $\log q = ab^x + q$	Gompertz curve
12.	$y = pq^{b^x} + h$	Modified Gompertz curve
13.	$y = \dfrac{1}{ab^x + q}$ or $\dfrac{1}{y} = ab^x + q$	Logistic curve
14.	$y = be^{mx}$ or $y = b\,10^{mx}$	Exponential function

No.	Equation	Description of the curve
15.	$y = \dfrac{1}{mx+b}$	Reciprocal function
16.	$y = \alpha \dfrac{x}{\beta + x}$	Saturation-growth-rate equation

In Table 6.1, $a, b, c, d, e, a_0, a_1, a_2, \ldots, a_n, b_0, p, q, h, m, \alpha$ and β are all constant coefficients.

Linear Regression

Linear *regression* and *correlation* are two commonly used methods for examining the relationship between quantitative variables and for making predictions. In this chapter, we review linear equations with one independent variable, explain how to find the *regression equation*, the equation of the line that best fits a set of data points. We also examine the coefficient of determination that is a descriptive measure of the utility of the regression equation for making predictions. In addition, we discuss the linear correlation coefficient, which provides a descriptive measure of the strength of the linear relationship between the two quantitative variables.

6.2 LINEAR EQUATION

The general form of a *linear equation* with one independent variable can be written as

$$y = a + bx$$

where a and b are constants (fixed numbers), x is the independent variable, and y is the dependent variable. The graph of a linear equation with one independent variable is a *straight line*, or simply a *line*. Also, any nonvertical line can be represented by such an equation.

Linear equations with one independent variable occur frequently in applications of mathematics to many different fields, including the social sciences, engineering, and management as well as physical and mathematical sciences.

For a linear equation $y = a + bx$, the number a is the y-value of the point of intersection of the line and the y-axis. The number b measures the steepness of the line. The number b indicates how much the y-value changes when the x-value increases by 1 unit. Figure 6.2 illustrates these relationships.

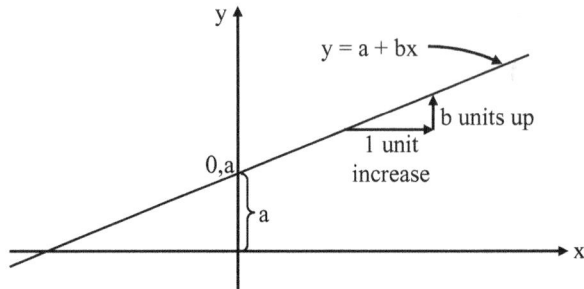

FIGURE 6.2 Graph of y = a + bx.

The numbers a and b have special names that reflect these geometric interpretations. For a linear equation $y = a + bx$, the number a is called the *y-intercept* and the number b is called the *slope*.

The graph of the linear equation $y = a + bx$ slopes upward if $b > 0$, slopes downward if $b < 0$, and is horizontal if $b = 0$, as shown in Figure 6.3.

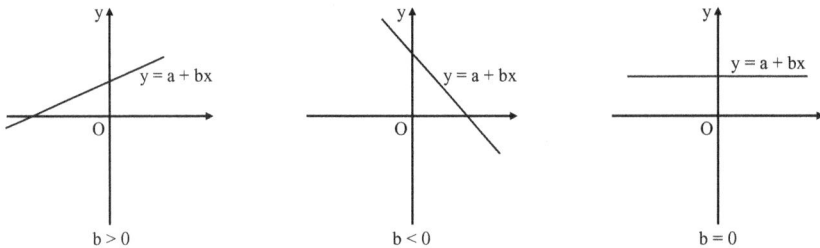

FIGURE 6.3 Graphical interpretation of slope.

6.3 CURVE FITTING WITH A LINEAR EQUATION

Curve fitting is a procedure in which a mathematical formula (equation) is used to best fit a given set of data points. The objective is to find a function that fits the data overall. Curve fitting is used when the values of the data points have some error, or scatter and require a curve fit to the points. Curve fitting can be accomplished with many types of functions and with polynomials of various orders.

Curve fitting using a linear equation (first degree polynomial) is the process by which an equation of the form

$$y = a + bx \tag{6.1}$$

is used to best fit the given data points. This can be accomplished by finding the constants a and b that give the smallest error when the data points are substituted in Eq. (6.1). If the data points consists of only two points, the constants can be obtained such that Equation (6.1) gives the exact values at the points. Figure 6.4 shows the straight line corresponding to the Eq. (6.1) and passing through the two points. When the data has more than two points, the constants a and b are determined such that the line has the best fit overall as shown in Figure 6.5.

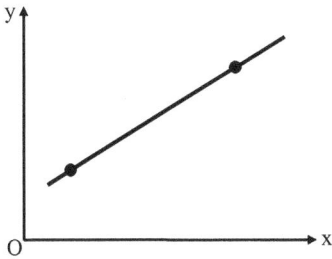

FIGURE 6.4 Straight line connecting two points.

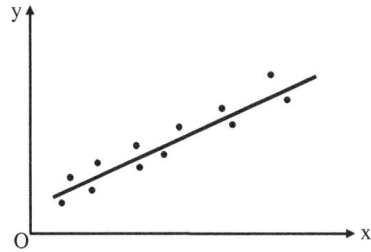

FIGURE 6.5 A straight line passing through many data points.

The procedure for obtaining the constants a and b that give the best fit requires a definition of best fit and an analytical procedure for deriving the constants a and b. The fitting between the given data points and an approximating linear function is obtained by first computing the error, also called the *residual*, which is the difference between a data point and the value of the approximating function, at each point. Figure 6.6 shows a linear function (straight line) that is used for curve fitting n points.

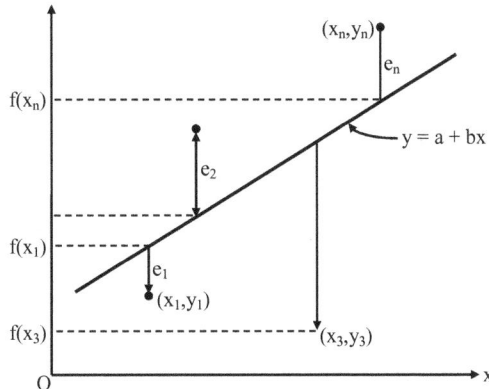

FIGURE 6.6 Curve-fitting points with a linear equation $y = a + bx$.

Thus, the *residual*, e is the discrepancy between the true value of y and the approximating value, $a + bx$, predicted by the linear equation.

6.4 CRITERIA FOR A *"BEST"* FIT

A criterion that measures how well the approximating function fits the given data can be determined by computing a total error E in terms of the residuals as

$$E = \sum_{i=1}^{n} e_i = \sum_{i=1}^{n} [y_i - (a + bx_i)] \qquad (6.2)$$

where n = total number of points.

However, this is an inadequate criterion, as illustrated in Figure 6.7, which shows that E is zero since $e_1 = -e_4$ and $e_2 = -e_3$.

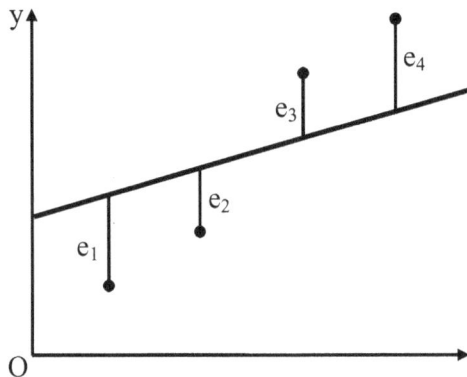

FIGURE 6.7 Straight line fit with $E = 0$.

One way to remove the effect of the signs might be to minimize the sum of the absolute values of the discrepancies:

$$E = \sum_{i=1}^{n} |e_i| = \sum_{i=1}^{n} |y_i - a - bx_i| \qquad (6.3)$$

Figure 6.8 shows why this criterion is also inadequate. For four points show, for the same set of points there can be several functions that give the same total error. E is the same for the two approximating lines in Figure 6.8.

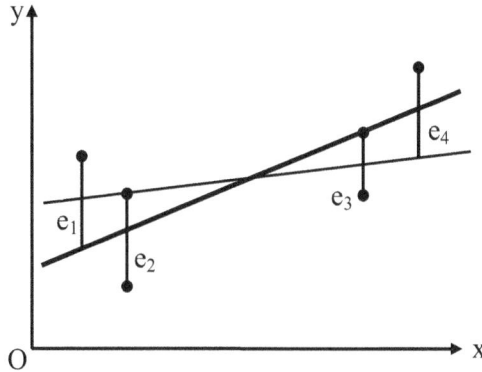

FIGURE 6.8 Two straight line fits with the same total error as per Eq. (6.3).

A third strategy for fitting a best line is the *minmax* criterion. In this technique, the straight line is chosen that minimizes the maximum distance that an individual point falls from the line. Again as shown in Figure 6.9, this technique gives undue influence to an outlier (a single point with a large error).

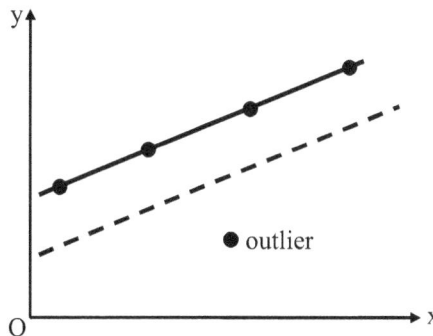

FIGURE 6.9 Minimization of the maximum error of any individual point.

A method that overcomes the shortcomings of the aforementioned approaches is to minimize the sum of the squares of the residuals:

$$S_r = \sum_{i=1}^{n} e_i^2 = \sum_{i=1}^{n} (y_i - a - bx_i)^2 \tag{6.4}$$

Equation (6.4) always gives a positive number of S_r (total error) and positive and negative residuals do not cancel each other. This criterion (Eq. 6.4) is called the *least squares* and has many advantages, including that it gives a unique line for a given set of data.

Equation (6.4) can be used to determine the coefficients a and b in the linear function $y = a + bx$ that yield the smallest total error. This is accomplished by using a procedure called linear least-squares regression, which is presented in the next section.

6.5 LINEAR LEAST-SQUARES REGRESSION

Linear least-squares regression is a method in which the coefficients a and b of a linear function $y = a + bx$ are determined such that the function has the best fit to a given set of data points. The best fit is defined as the smallest possible total error that is computed by adding the squares of the residuals according to Eq. (6.4).

For a given set of n data points (x_i, y_i), the overall error calculated by Eq. (6.4) is

$$S_r = \sum_{i=1}^{n} [y_i - (a + bx_i)]^2 \tag{6.5}$$

Since the values of x_i and y_i are known, S_r in Eq. (6.5) is a nonlinear function of two variables a and b. This function S_r has a minimum at the values of a and b, where the partial derivatives of S_r with respect to each variable is equal to zero.

Taking the partial derivatives and setting then equal to zero gives

$$\frac{\partial S_r}{\partial a} = -2 \sum_{i=1}^{n} (y_i - a - bx_i) = 0 \tag{6.6}$$

$$\frac{\partial S_r}{\partial b} = -2 \sum_{i=1}^{n} [(y_i - a - bx_i)x_i] = 0 \tag{6.7}$$

Equations (6.6) and (6.7) are a system to two linear equations for the two unknowns a and b, and can be rewritten in the form

$$na + \left[\sum_{i=1}^{n} x_i \right] b = \sum_{i=1}^{n} y_i \tag{6.8}$$

$$\left[\sum_{i=1}^{n} x_i \right] a + \left[\sum_{i=1}^{n} x_i^2 \right] b = \sum_{i=1}^{n} x_i y_i \tag{6.9}$$

Equations (6.8) and (6.9) are called the *normal equations* and can be solved simultaneously for

$$b = \frac{n\sum\limits_{i=1}^{n} x_i y_i - \left[\sum\limits_{i=1}^{n} x_i\right]\left[\sum\limits_{i=1}^{n} y_i\right]}{n\sum\limits_{i=1}^{n} x_i^2 - \left[\sum\limits_{i=1}^{n} x_i\right]^2} \tag{6.10}$$

or

$$b = \frac{\sum\limits_{i=1}^{n} x_i y_i - \dfrac{\left[\sum\limits_{i=1}^{n} x_i\right]\left[\sum\limits_{i=1}^{n} y_i\right]}{n}}{\sum\limits_{i=1}^{n} x_i^2 - \dfrac{\left[\sum\limits_{i=1}^{n} x_i\right]^2}{n}} \tag{6.11}$$

and

$$a = \frac{\left[\sum\limits_{i=1}^{n} x_i^2\right]\left[\sum\limits_{i=1}^{n} y_i\right] - \left[\sum\limits_{i=1}^{n} x_i y_i\right]\left[\sum\limits_{i=1}^{n} x_i\right]}{n\sum\limits_{i=1}^{n} x_i^2 - \left[\sum\limits_{i=1}^{n} x_i\right]^2} \tag{6.12}$$

or

$$a = \overline{y} - b\overline{x}$$

Since Eqs. (6.10) to (6.12) contain summations that are the same for a set of n points, they can also be written as

$$SS_{xx} = \Sigma(x_i - \overline{x})^2 = \Sigma x_i^2 - (\Sigma x_i)^2 / n$$

$$SS_{xy} = \Sigma(x_i - \overline{x})(y_i - \overline{y}) = \Sigma x_i y_i - (\Sigma x_i)(\Sigma y_i) / n$$

$$SS_{yy} = \Sigma(y_i - \overline{y})^2 = \Sigma y_i^2 - (\Sigma y_i)^2 / n$$

The regression equation for a set of n data points is

$$\hat{y} = a + bx$$

where

$$b = \frac{SS_{xy}}{SS_{xx}}$$

and

$$a = \frac{1}{n}(\Sigma y_i - b\Sigma x_i) = \overline{y} - b\overline{x}$$

6.6   LINEAR REGRESSION ANALYSIS

A regression model is a mathematical equation that describes the relationship between two or more variables. A *single regression* model includes only two variables: one independent and one dependent. The relationship between two variables in a regression analysis is expressed by a mathematical equation called a *regression equation* or *model*. A regression equation that gives a straight-line relationship between two variables is called a *linear regression model*; otherwise, it is called a *nonlinear regression model*. Figures 6.10(a) and (b) show a linear and a nonlinear relationship between independent variable and the dependent variable.

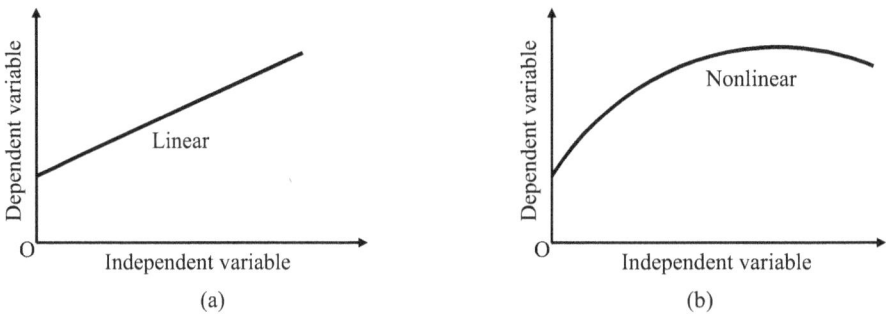

FIGURE 6.10   (a) Linear relationship. (b) Nonlinear relationship.

The equation of a linear relationship between two variables x and y is written as

$$y = a + bx \qquad (6.13)$$

where a gives the y-intercept and b represents the slope of the line.

In regression model, x is the independent variable and y is the dependent variable. The simple linear regression model for population is written as

$$y = A + Bx \qquad (6.14)$$

Equation (6.14) is called a *deterministic model*. It gives an *exact relationship* between x and y. However, in many instances the relationship between the variables is not exact. Therefore, the complete regression model is then written as

$$yA + Bx + \in \qquad (6.15)$$

where $\in$ is called *the random error term*. This regression model (Eq. 6.5) is called a *probabilistic model* (or a *statistical relationship*). The random error term $\in$ is included in the model to take into consideration of the following two phenomena:

a) Missing or omitted variables: The random error term $\in$ is included to capture the effect of all the missing or omitted variables that were not included in the model.

b) Random variation: The random error term $\in$ is included to capture the random variation.

In Eq. (6.15), A and B are the *population parameters*. The regression line obtained from Eq. (6.15) by using the population data is called the *population regression line*. The values of A and B combined is the population regression line is called the true values of the *y-intercept and slope*.

However, most often the population data is difficult to obtain. As a consequence, we almost always use the sample data and use the model given by Eq. (6.15). The values of the *y*-intercept and slope calculated from sample data on x and y are called the *estimated values* of A and B and are denoted by a and b.

The estimated regression model is then written as

$$\hat{y} = a + bx \tag{6.16}$$

where $\hat{y}$ is the *estimated* or *predicted value* of y for a given value of x. Equation (6.16) is called the *estimated regression model*. It gives the regression of y on x. A plot of paired observation is called a *scatter diagram* as shown in Figure 6.11.

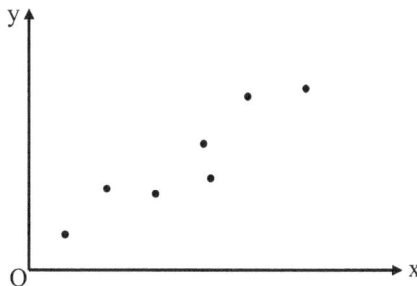

FIGURE 6.11 Scatter diagram.

To find the line that best fits the scatter of points, we minimize the *error sum of squares*, denoted by SSE, which is given by

$$\text{SSE} = \Sigma e^2 = \Sigma(y - \hat{y})^2 \tag{6.17}$$

where $e = y - \hat{y}$.

The least-squares method gives the values of a and b such that the sum of squared errors (SSE) is minimum.

For the least-squares regression line $\hat{y} = a + bx$ from Eqs. (6.10) to (6.17), we have

$$b = \frac{SS_{xy}}{SS_{xx}} \text{ and } a = \overline{y} - b\overline{x} \tag{6.18}$$

where

$$SS_{xy} = \Sigma xy - \frac{(\Sigma x)(\Sigma y)}{n}$$

$$SS_{xx} = \Sigma x^2 - \frac{(\Sigma x)^2}{n}$$

$$SS_{yy} = \Sigma(y - \overline{y}) = \Sigma y^2 - (\Sigma y)^2 / n \tag{6.19}$$

The least-squares regression line $\hat{y} = a + bx$ is also called the *regression of y on x*.

6.6.1 MATLAB Functions: *polyfit* and *polyval*

MATLAB has a built-in function *polyfit* that fits a least-square *n*th-order polynomial to data. It can be applied as in

$$>> p = \text{polyfit}(x, y, n)$$

where x and y are the vectors of the independent and the dependent variables, respectively, and n = the order of the polynomial. The function returns a vector p containing the polynomial's coefficients.

It should be noted here that it represents the polynomial using decreasing powers of x as in the following representation:

$$f(x) = p_1 x^n + p_2 x^{n-1} + p_3 x^{n-2} + \dots + p_n x + p_{n+1}$$

Since a straight line is a first-order polynomial, *polyfit* $(x, y, 1)$ will return the slope and the intercept of the best-fit straight line.

Another function, *polyval* can be used to compute a value using the coefficients. It has the general format:

$$>> y = \text{polyval}\,(p, x)$$

where p = the polynomial coefficients and y = the best-fit value at x.

EXAMPLE 6.1

Table 6.2 gives experimental data for force (N) and velocity (m/s) for an object suspended in a wind tunnel.

TABLE 6.2

Velocity, v(m/s)	10	20	30	40	50	60	70	80
Force F(N)	24	68	378	552	608	1218	831	1452

a) use the linear least-squares regression to determine the coefficients a and b in the function $\hat{y} = a + bx$ that best fits the data.

b) estimate the force when the velocity is 55 m/s.

Solution:

Here $n = 8$.

n	x	y	x^2	xy
1	10	24	100	240
2	20	68	400	1360
3	30	378	900	11340
4	40	552	1600	22080
5	50	608	2500	30400
6	60	1218	3600	73080
7	70	831	4900	58170
8	80	1452	6400	116160
Σ	360	5131	20400	312830

$$\bar{x} = \frac{\Sigma x}{n} = \frac{360}{8} = 45$$

$$\bar{y} = \frac{\Sigma y}{n} = \frac{5131}{8} = 641.375$$

From Eq. (6.16), we have

$$b = \frac{SS_{xy}}{SS_{xx}}$$

where

$$SS_{xy} = \Sigma xy - \frac{(\Sigma x)(\Sigma y)}{n} = 312830 - \frac{(360)(5131)}{8} = 81935$$

and

$$SS_{xx} = \Sigma x^2 - \frac{(\Sigma x)^2}{n} = 20400 - \frac{(360)^2}{8} = 4200$$

$$b = \frac{SS_{xy}}{SS_{xx}} = \frac{81935}{4200} = 19.5083$$

From Eq. (6.17), we have

$$a = \bar{y} - b\bar{x} = 641.375 - (19.5083)(45) = -236.50$$

Hence

$$\hat{y} = -236.50 + 19.5083x$$

b) The estimated value of the force when the velocity is 55 m/s, is given by

$$\hat{y} = a + bx = -236.50 + 19.5083(55) = 836.4583(N).$$

MATLAB Solution:

a)
```
>> x = [10 20 30 40 50 60 70 80];
>> y = [24 68 378 552 608 1218 831 1452];
>> a = polyfit(x, y, 1)
a =
19.5083  -236.5000
```

Hence, the slope is 19.5083 and the intercept is –236.50.

b) The MATLAB function, polyval can be used to compute a value using the coefficients. Therefore,

```
>>y = polyval (a, 55)
y =
    836.4583
```

Hence, the estimated value of the force when the velocity is 55 m/s is 836.4583(N).

6.7 INTERPRETATION OF *a* AND *b*

When b is positive, an increment in x will lead to an increase in y and a decrease in x will lead to a decrease in y. That is, when b is positive, the movements in x and y are in the same direction. Such a relationship between x and y is called a *positive linear relationship*. The regression line slopes upward from left to right.

Similarly, if the value of b is negative, an increase in x will cause a decrease in y and a decrease in x will cause an increase in y. The changes in x and y are in opposite directions. Such a relationship between x and y is called a *negative linear relationship*. The regression line slopes downward from left to right. Figure 6.12 shows these two relationships.

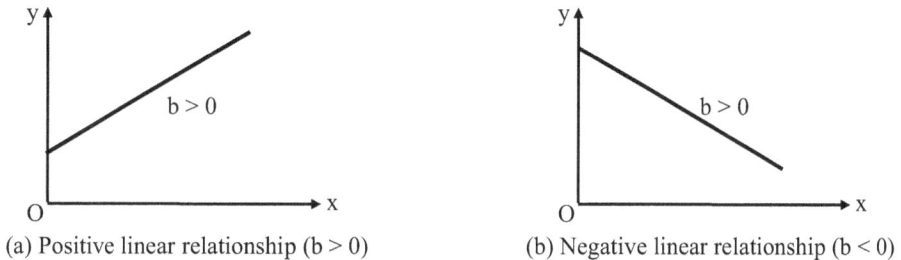

(a) Positive linear relationship (b > 0)   (b) Negative linear relationship (b < 0)

FIGURE 6.12 Positive and negative relationship between *x* and *y*.

Assumptions in the Regression Model

The linear regression analysis is based on the following assumptions:

1. The random error term $\in$ has a mean equal to zero for each x.
2. The errors associated with different observations are independent.
3. For any given x, the distribution of errors is normal.
4. The distribution of population errors for each x has the same (constant) standard deviation, which is denoted by σ_e as shown in Figures 6.13(a) and (b).

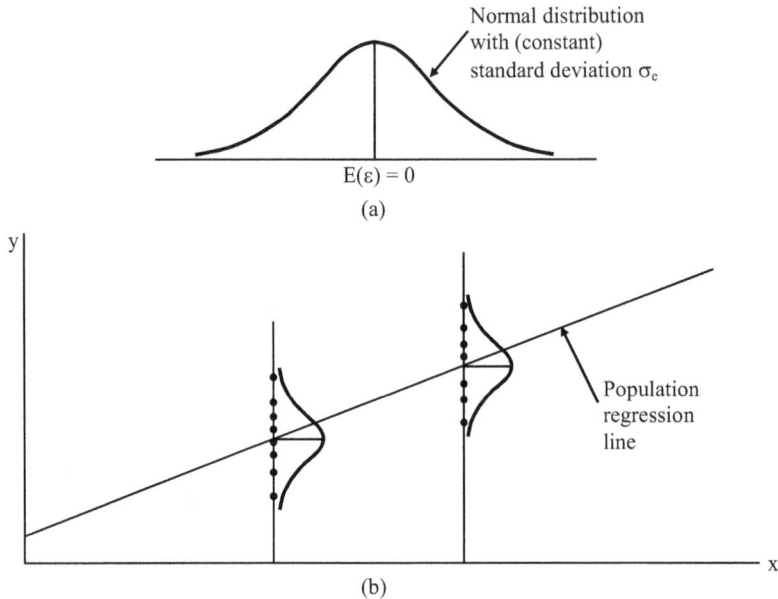

FIGURE 6.13 (a) Assumptions in the regression model. (b) Distribution on the regression line.

6.8   STANDARD DEVIATION OF RANDOM ERRORS

The standard deviation s_e measures the spread of the errors around the regression line as shown in Figure 6.13(b). The *standard deviation of errors* is calculated using

$$s_e = \sqrt{\frac{\text{SSE}}{n-2}}$$

where

$$\text{SSE} = \Sigma(y - \hat{y})^2$$

or

$$s_e = \sqrt{\frac{\text{SS}_{yy} - b\,\text{SS}_{xy}}{n-2}} \qquad (6.20)$$

where

$$\text{SS}_{yy} = \Sigma(y - \bar{y})^2 = \Sigma y^2 - \frac{(\Sigma y)^2}{n}$$

$$\text{SS}_{xy} = \Sigma xy - \frac{(\Sigma x)(\Sigma y)}{n} \qquad (6.21)$$

In Eq. (6.20), $(n - 2)$ represents the degrees of freedom for the regression model. The reason for $df = n - 2$ is that we lose one degree of freedom to calculate $\bar{x}$ and one for $\bar{y}$.

6.9 COEFFICIENT OF DETERMINATION

The *coefficient of determination*, denoted by r^2, represents the proportion of the *total sum of squares* that is explained by the use of the regression model. The computational formula for r^2 is given by

$$r^2 = b\frac{SS_{xy}}{SS_{yy}} \qquad 0 \leq r^2 \leq 1 \tag{6.22}$$

The *total sum of squares* denoted by SST is the total variation in the observed values of the response variable

$$SST = SS_{yy} = \Sigma(y - \bar{y})^2 = \Sigma y^2 - \frac{(\Sigma y)^2}{n} \tag{6.23}$$

which is the same formula we use for SS_{yy}.

The regression sum of squares, denoted by SSR, is the variation in the observed values of the response variable explained by the regression:

$$SSE = \Sigma(y - \hat{y})^2$$

$$SSR = SST - SSE$$

or $SST = SSR + SSE$ \hfill (6.24)

The ratio of SSR to SST gives the coefficient of determination. That is,

$$r^2 = \frac{SSR}{SST} = \frac{SST - SSE}{SST} = 1 - \frac{SSE}{SST} \tag{6.25}$$

This formula shows that we can also interpret the coefficient of determination as the percentage reduction obtained in the total squared error by using the regression equation instead of the mean, $\bar{y}$, to predict the observed values of the response variable.

The *coefficient of determination*, denoted by r^2, represents the portion of SST that is *explained* by the use of the regression model. Hence,

$$r^2 = \frac{b\,SS_{xy}}{SS_{yy}} \qquad (6.26)$$

and $\qquad 0 \le r^2 \le 1$

The value of r^2 is the proportion of the variation in y that is explained by the linear relationship between x and y. The coefficient of determination is the amount of the variation in y that is explained by the regression line. It is computed as

$$r^2 = \frac{\text{Explained variation}}{\text{Total variation}} = \frac{\Sigma(\hat{y} - \overline{y})^2}{\Sigma(y - \overline{y})^2}$$

The coefficient of determination, r^2, always lies between 0 and 1. A value of r^2 near 0 suggests that the regression equation is not very useful for making predictions, whereas a value of r^2 near 1 suggests that the regression equation is quite useful for making predictions.

EXAMPLE 6.2

For the data of Example 6.2, calculate the

    a)   standard deviation of errors, s_e

    b)   error sum of squares, SSE

    c)   total sum of squares, SST

    d)   regression sum of squares, SSR

    e)   the coefficient of determination, r^2

Solution:

Referring to Table 6.2, we have

$$n = 8,\ \Sigma x = 360,\ \Sigma y = 5131,\ \Sigma x^2 = 20400,\ \Sigma xy = 312830,\ \Sigma y^2 = 5104841$$

$$\overline{x} = \frac{\Sigma x}{n} = \frac{360}{8} = 45$$

$$\overline{y} = \frac{\Sigma y}{n} = \frac{5131}{8} = 641.375$$

$$SS_{xy} = \Sigma xy - \frac{(\Sigma x)(\Sigma y)}{n} = 312830 - \frac{(360)(5131)}{8} = 81935$$

$$SS_{xx} = \Sigma x^2 - \frac{(\Sigma x)^2}{n} = 20400 - \frac{(360)^2}{8} = 4200$$

$$SS_{yy} = \Sigma y^2 - \frac{(\Sigma y)^2}{n} = 5104841 - \frac{(5131)^2}{8} = 1813945.875$$

$$b = \frac{SS_{xy}}{SS_{xx}} = \frac{81935}{4200} = 19.5083$$

$$a = \overline{y} - bx = 641.375 - 19.5083x$$

a) The standard deviation of errors, s_e

$$s_e = \sqrt{\frac{SS_{yy} - bSS_{xy}}{n-2}} = \sqrt{\frac{1813945.875 - 19.5083(81935)}{8-2}} = 189.5304$$

b) The error sum of squares, SSE

$$SSE = \Sigma(y - \hat{y})^2 = 215530.5833$$

c) Total sum of squares, SST

$$SST = SS_{yy} = 1813945.875$$

d) The regression sum of squares, SSR

$$SSR = SST - SSE = 1813946.875 - 215530.5833 = 1598415.2917$$

e) The coefficient of determination, r^2

$$r^2 = \frac{b\,SS_{xy}}{SS_{yy}} = \frac{(19.5083)(81935)}{1813945.875} = 0.8812$$

6.10 LINEAR CORRELATION

Linear correlation coefficient is a measure of the relationship between two variables. Linear correlation coefficient measures how closely the points in a scatter diagram are spread around the regression line. The correlation coefficient calculated for the population is denoted by ρ and the one calculated for sample data is denoted by r. The linear correlation coefficient r measures the strength of the linear relationship between the paired x- and y-quantitative values in a sample. The linear correlation coefficient is sometimes referred to as the Pearson product moment correlation coefficient in honor of Karl Pearson (1857–1936), who originally developed it. Square of the correlation coefficient is equal to the coefficient of determination. The value of the *correlation coefficient* always lies in the range –1 to 1. Hence, $-1 \leq \rho \leq 1$ and $-1 \leq r \leq 1$.

If $r = 1$, it refers to a case of *perfect positive linear correlation* and all points in the scatter diagram lie on a straight line that slopes upward from left to right, as shown in Figure 6.14. If $r = -1$, the correlation is said to be *perfect negative linear correlation* and all points in the scatter diagram fall on a straight line that slopes downward from left to right, as shown in Figure 6.14(*b*).

When there is *no linear correlation* between the two variables and r is close to 0. Also, in this case, all the points are scattered all over the diagram as shown in Figure 6.14(*c*).

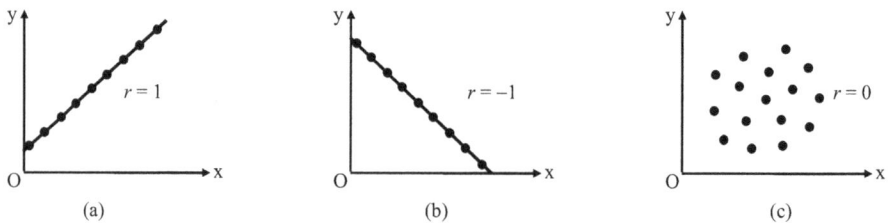

FIGURE 6.14 Linear correlation between two variables.

(a) Perfect positive linear correlation, $r = 1$

(b) Perfect negative linear correlation, $r = -1$

(c) No linear correlation, $r = 0$

Two variables are said to have a *strong positive linear correlation* when the correlation is positive and close to 1. If the correlation between the two variables is positive but close to zero, then the variables have a *weak positive linear correlation*. Similarly, when the correlation between two variables is negative and close to –1, then the variables are said to have a *strong negative linear correlation*. A *weak negative linear correlation* exists when the correlation between the variables is negative but close to zero. The four cases are shown in Figures 6.15(a) through (d). Figure 6.16 shows the various degrees of linear correlation.

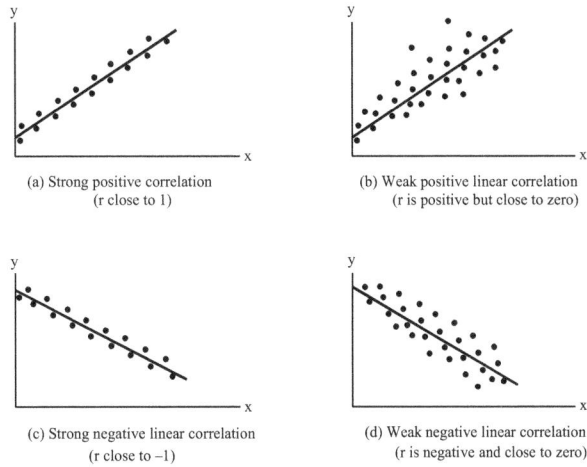

(a) Strong positive correlation
(r close to 1)

(b) Weak positive linear correlation
(r is positive but close to zero)

(c) Strong negative linear correlation
(r close to −1)

(d) Weak negative linear correlation
(r is negative and close to zero)

FIGURE 6.15 Linear correlation between variables.

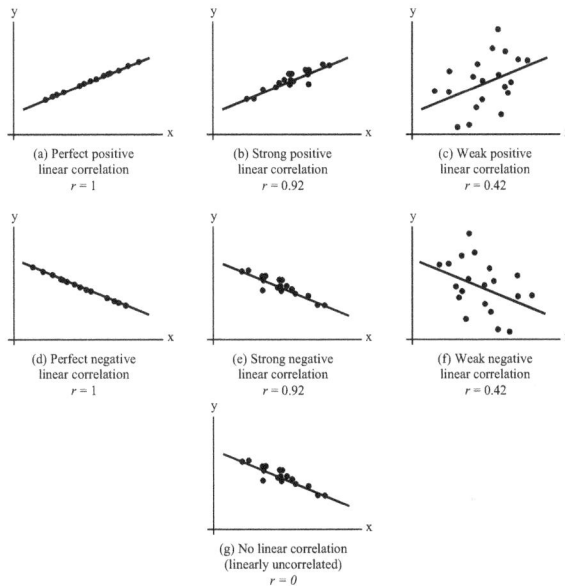

(a) Perfect positive
linear correlation
$r = 1$

(b) Strong positive
linear correlation
$r = 0.92$

(c) Weak positive
linear correlation
$r = 0.42$

(d) Perfect negative
linear correlation
$r = 1$

(e) Strong negative
linear correlation
$r = 0.92$

(f) Weak negative
linear correlation
$r = 0.42$

(g) No linear correlation
(linearly uncorrelated)
$r = 0$

FIGURE 6.16 Various degrees of linear correlation.

The *simple linear correlation,* denoted by r, measures the strength of the linear relationship between two variables for a sample and is calculated as

$$r = \frac{SS_{xy}}{\sqrt{SS_{xx}SS_{yy}}} \qquad (6.27)$$

It should be noted here that r and b calculated for the same sample will always have the same sign.

Properties of the Linear Correlation Coefficient r

1. The value of r is always between -1 and $+1$ inclusive. That is, $-1 \leq r \leq 1$.
2. The values of r do not change if all values of either variable are converted to a different scale.
3. The value of r is not affected by the choice of x or y. Interchange all x- and y- values, and the value of r will not change.
4. "r" measures the strength of a linear relationship. The magnitude of r indicates the strength of the linear relationship. A value of r close to -1 or to 1 indicates a strong linear relationship between the variables and that the variable x is a good linear predictor of the variable y. That is, the regression equation is extremely useful for making predictions. A value of r near 0 indicates at most a weak linear relationship between the variables and that the variable x is a poor linear predictor of the variable y. That is, the regression equation is either useless or not very useful for making predictions. It is not designed to measure the strength of a relationship that is not linear.
5. The sign of r suggests the type of linear relationship. A positive value of r suggests that the variables are positively correlated, meaning that y tends to increase linearly as x increases, with the tendency being greater the closer that r is to 1. A negative value of r suggests that the variables are negatively linearly correlated, meaning that y tends to decrease linearly as x increases, with the tendency being greater the closer that r is to -1.
6. r reflects the slope of the scatter plot. The linear correlation coefficient is positive when the scatter plot shows a positive slope and is negative when the scatter plot shows a negative slope.
7. The sign of r and the sign of the slope of the regression line are identical. If r is positive, so is the slope of the regression line. That is, the regression

line slopes upward. If r is negative, so is the slope of the regression line and the regression line slopes downward.

Explained and Unexplained Variation

The total variation is defined as $\Sigma(y - \overline{y})^2$, that is, the sum of the squares of the deviations of the values of y from the mean $\overline{y}$. This can be written as

$$\Sigma(y - \overline{y})^2 = \Sigma(y - \hat{y})^2 + \Sigma(\hat{y} - \overline{y})^2 \qquad (6.28)$$

where $\hat{y}$ is the value of y for given values of x as estimated from $\hat{y} = a + bx$, a measure of the scatter about the regression line of y on x.

The first term on the right side of Eq. (6.28) is called the *unexplained variation* while the second term is called the *explained variation*. The deviations $\hat{y} - \overline{y}$ have a definite pattern while the deviations $y - \hat{y}$ behave in a random or unpredictable manner. Similar results hold true for the variable x.

The ratio of the explained variation to the total variation is called the *coefficient of determination*. If there is zero explained variation, that is, the total variation is all unexplained, then this ratio is zero. If there is zero unexplained variation, i.e., the total variation is all explained, the ratio is one. In all other cases, the ratio lies between zero and one. The ratio is always nonnegative.

The quantity, r, is called the coefficient of correlation, and it is given by

$$r = \pm\sqrt{\frac{\text{Explained variation}}{\text{Total variation}}} = \pm\sqrt{\frac{\Sigma(\hat{y} - \overline{y})^2}{\Sigma(y - \overline{y})^2}} \qquad (6.29)$$

r varies between -1 and $+1$. The signs $\pm$ are used for positive linear correlation and negative error correlation respectively. 'r' is a dimensionless quantity. The coefficient of determination equals the square of the linear correlation coefficient.

EXAMPLE 6.3

Determine the correlation coefficient for the data given in Example 6.1.

Solution:

Refer to the solutions obtained earlier for Examples 6.1 and 6.2. We have

$$SS_{xy} = 81935, \quad SS_{xx} = 4200 \quad \text{and} \quad SS_{yy} = 1813945.875$$

Hence, $\quad r = \dfrac{SS_{xy}}{\sqrt{SS_{xx}SS_{yy}}} = \dfrac{81935}{\sqrt{(4200)(1813945.875)}} = 0.9387$

6.11 LINEARIZATION OF NONLINEAR RELATIONSHIPS

Linear regression provides a powerful technique for fitting a best line to data. There exists many situations in science and engineering that show the relationship between the quantities that are being considered is not linear. There are several examples of nonlinear functions used for curve fitting. A few of them were described in Table 6.1.

Nonlinear regression techniques are available to fit these equations in Table 6.1 to data directly. A simpler alternative is to use analytical manipulations to transform the equations into a linear form. Then linear regression can be used to fit the equations to data.

For instance, $y = bx^m$ can be linearized by taking its natural logarithm to give

$$\ell n\, y = \ell n\, b + m\ell nx \tag{6.30}$$

A plot of ℓny versus ℓnx will give a straight line with a slope of m and an intercept of ℓnb as shown in Figure 6.17.

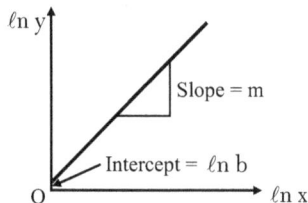

FIGURE 6.17 Linearized version of the exponential equation $y = bx^m$.

Many other nonlinear equations can be transformed into linear form in a similar way. Table 6.3 lists several such equations.

TABLE 6.3

No.	Nonlinear equation	Linear form	Relationship to $\hat{y} = a + b\hat{x}$	Values for least-squares regression
1.	$y = cx^m$	$ln(y) = m\,ln(x) + ln(c)$	$\hat{y} = ln(y)$, $\hat{x} = ln(x)$ $b = m$, $a = ln(c)$	$ln(x_i)$ and $ln(y_i)$
2.	$y = c\,e^{mx}$	$ln(y) = mx + ln(c)$	$\hat{y} = ln(y)$, $\hat{x} = x$ $b = m$, $a = ln(c)$	x_i and $ln(y_i)$
3.	$y = c\,10^{mx}$	$\log(y) = mx + \log c$	$\hat{y} = \log(y)$, $\hat{x} = x$ $b = m$, $a = \log(c)$	x_i and $ln(y_i)$
4.	$y = \dfrac{1}{mx + c}$	$\dfrac{1}{y} = mx + c$	$\hat{y} = \dfrac{1}{y}$, $\hat{x} = x$ $b = m$, $a = c$	x_i and $\dfrac{1}{y_i}$
5.	$y = \dfrac{mx}{c + x}$	$\dfrac{1}{y} = \dfrac{c}{mx} + \dfrac{1}{m}$	$\hat{y} = \dfrac{1}{y}$, $\hat{x} = \dfrac{1}{x}$ $b = \dfrac{c}{m}$, $a = \dfrac{1}{m}$	$\dfrac{1}{x_i}$ and $\dfrac{1}{y_i}$
6.	$xy^c = d$ Gas equation	$\log y = \dfrac{1}{c}\log d - \dfrac{1}{c}\log x$	$\hat{y} = \log y$, $\hat{x} = \log x$ $a = \dfrac{1}{c}\log d$, $b = -\dfrac{1}{c}$	$\log x_i$ and $\log y_i$
7.	$y = cd^x$	$\log y = \log c + x \log d$	$\hat{y} - \log y$, $\hat{x} = x$ $a = \log c$, $b = \log d$	x_i and $\log y_i$
8.	$y = c + d\sqrt{x}$	$y = c + d\hat{x}$ where $\hat{x} = \sqrt{x}$	$\hat{y} = y$ and $\hat{x} = \sqrt{x}$ $a = c$ and $b = d$	$\sqrt{x_i}$ and y_i

The curves in Figure 6.18 may be used as guides to some of the simpler variable transformations.

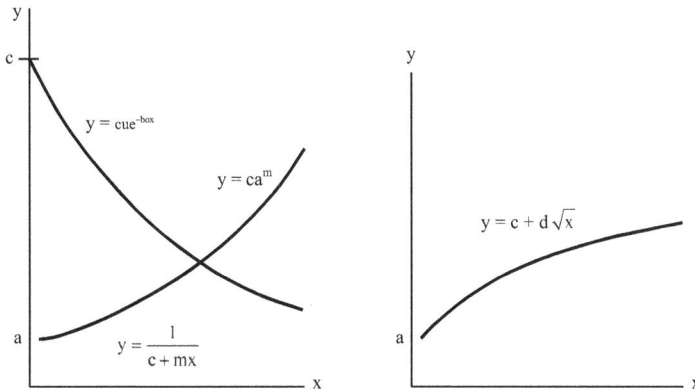

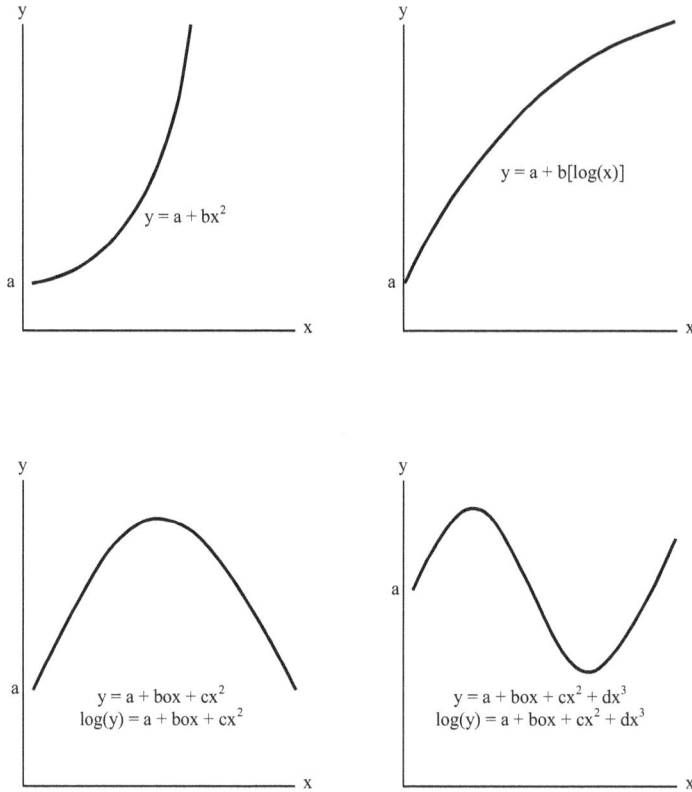

FIGURE 6.18 Nonlinear data curves.

EXAMPLE 6.4

Fit $y = cx^m$ (power function) to the data in Example 6.1 using a logarithmic transformation.

Solution:

The data can be set up in tabular form as shown in Table 6.4.

TABLE 6.4

i	x_i	y_i	log x_i	log y_i	$(\log x_i)^2$	$(\log x_i)(\log y_i)$
1	10	24	1.0000	1.3802	1.0000	1.3802
2	20	68	1.3010	1.8325	1.6927	2.3841
3	30	378	1.4771	2.5775	2.1819	3.8073
4	40	552	1.6021	2.7419	2.5666	4.3928
5	50	608	1.6990	2.7839	2.8865	4.7298
6	60	1218	1.7782	3.0856	3.1618	5.4867
7	70	831	1.8451	2.9196	3.4044	5.3870
8	80	1452	1.9031	3.1620	3.6218	6.0175
Σ	360	5131	12.606	20.483	20.516	33.585

The means are computed as

$$\overline{x} = \frac{\Sigma \log x}{n} = \frac{12.606}{8} = 1.5757$$

$$\overline{y} = \frac{\Sigma \log y}{n} = \frac{20.483}{8} = 2.5604$$

The slope and the intercept are then calculated using Eqs. (6.13), (6.14), (6.15), and (6.16).

$$b = \frac{SS_{xy}}{SS_{xx}} = \frac{n\Sigma(\log x_i)(\log y_i) - (\Sigma \log x_i)(\Sigma \log y_i)}{n\Sigma \log x_i^2 - (\Sigma \log x_i)^2}$$

$$= \frac{8(33.585) - (12.606)(20.483)}{8(20.516) - (12.606)^2} = 2.0055$$

$$a = \overline{y} - b\overline{x} = 2.5604 - 2.0055(1.5757) = -0.5997$$

The least-square fit is

$$\log y = -0.5997 + 2.0055 \log x$$

The fit along with the data is shown in Figure 6.18.

Transforming to the original coordinates, we have

$$c = 10^{-(0.5997)} = 0.2514 \quad \text{and} \quad m = 2.0055$$

Hence, the least-squares fit is

$$y = 0.2514\, x^{2.0055}$$

6.12 POLYNOMIAL REGRESSION

The least-squares procedure described in earlier sections can be readily extended to fit the data to a higher-order polynomial. Consider a second-order polynomial or quadratic:

$$y = a + bx + cx^2 + e \qquad (6.31)$$

The sum of the squares of the residuals is

$$S_r = \sum_{i=1}^{n} (y_i - a - bx_i - cx_i^2)^2 \qquad (6.32)$$

To generate the lease-squares fit, we take the first derivative of Eq. (6.32) with respect to each of the unknown coefficients of the polynomial.

$$\frac{\partial S_r}{\partial a} = -2\sum_{i=1}^{n} (y_i - a - bx_i - cx_i^2)$$

$$\frac{\partial S_r}{\partial b} = -2\sum_{i=1}^{n} x_i(y_i - a - bx_i - cx_i^2) \qquad (6.33)$$

$$\frac{\partial S_r}{\partial c} = -2\sum_{i=1}^{n} x_i^2(y_i - a - bx_i - cx_i^2)$$

Equation (6.33) is set equal to zero and rearranged to obtain the following set of normal equations:

$$na + \left(\sum_{i=1}^{n} x_i\right)b + \left(\sum_{i=1}^{n} x_i^2\right)c = \sum_{i=1}^{n} y_i$$

$$\left(\sum_{i=1}^{n} x_i\right)a + \left(\sum_{i=1}^{n} x_i^2\right)b + \left(\sum_{i=1}^{n} x_i^3\right)c = \sum_{i=1}^{n} x_iy_i \qquad (6.34)$$

$$\left(\sum_{i=1}^{n} x_i^2\right)a + \left(\sum_{i=1}^{n} x_i^3\right)b + \left(\sum_{i=1}^{n} x_i^4\right)c = \sum_{i=1}^{n} x_i^2y_i$$

Equation (6.34) is all linear equations in three unknowns: a, b and c. These coefficients a, b and c can be determined directly from the observed data.

The above procedure can be easily extended to an mth order polynomial as in

$$y = a + bx + cx^2 + dx^3 + \ldots + zx^m + e \qquad (6.35)$$

Hence, the determination of the coefficients of an mth order polynomial is equivalent to solving a system of $(m + 1)$ simultaneous linear equations.

The standard error in this case is given by

$$S_{y/x} = \sqrt{\frac{S_r}{n-(m+1)}} \tag{6.36}$$

The coefficient of determination, r^2, can be computed for a polynomial regression with Eq. (6.25).

EXAMPLE 6.5

Fit a second-order polynomial to the data in Table 6.5 and determine the total standard deviation, the standard error of the estimate and the correlation coefficient.

TABLE 6.5

x_i	0	1	2	3	4	5
y_i	2	8	14	27	41	61

Solution:

Tables 6.6 (a) and (b) shows the computations for an error analysis of the quadratic least-squares fit.

TABLE 6.6(a)

x_i	y_i	$x_i y_i$	x_i^2	$x_i^2 y_i$	x_i^3	x_i^4
0	2	0	0	0	0	0
1	8	8	1	8	1	1
2	14	28	4	56	8	16
3	27	81	9	243	27	81
4	41	164	16	656	64	256
5	61	305	25	1525	125	625
Σ 15	153	586	55	2488	225	979

TABLE 6.6(b)

x_i	y_i	$(y_i - \bar{y})^2$	$(y_i - a - bx_i - cx_i^2)^2$
0	2	552.3	0.2500
1	8	306.3	1.3391
2	14	132.3	0.6862
3	27	2.3	0.2951
4	41	240.3	0.5300
5	61	1260.3	0.1282
Σ 15	153	2493.50	3.2286

Hence, the simultaneous linear equations are

$$\begin{bmatrix} 6 & 15 & 55 \\ 15 & 55 & 225 \\ 55 & 225 & 979 \end{bmatrix} \begin{Bmatrix} a \\ b \\ c \end{Bmatrix} = \begin{Bmatrix} 153 \\ 586 \\ 2488 \end{Bmatrix}$$

Refer to Appendix C (Cramer's rule for solving a system of linear algebraic equations).

Here
$$D = \begin{vmatrix} 6 & 15 & 55 \\ 15 & 55 & 225 \\ 55 & 225 & 979 \end{vmatrix} = 3920$$

$$D_1 = \begin{vmatrix} 153 & 15 & 55 \\ 586 & 55 & 225 \\ 2488 & 225 & 979 \end{vmatrix} = 9800$$

$$D_2 = \begin{vmatrix} 6 & 153 & 55 \\ 15 & 586 & 225 \\ 55 & 2488 & 979 \end{vmatrix} = 9884$$

$$D_3 = \begin{vmatrix} 6 & 15 & 153 \\ 15 & 55 & 586 \\ 55 & 225 & 2488 \end{vmatrix} = 7140$$

Therefore,
$$a = \frac{D_1}{D} = \frac{9800}{3920} = 2.5$$

$$b = \frac{D_2}{D} = \frac{9884}{3920} = 2.5214$$

$$c = \frac{D_3}{D} = \frac{7140}{3920} = 1.8214$$

These equations can also be solved to determine the coefficients using MATLAB:

These equations can be solved to determine the coefficients. Here, we use MATLAB.

```
>>A = [6  15  55;  15  55  225;  55  225  979];
>>b = [153;  586;  2488];
    x = A/b
```

$$x = 2.5 \quad 2.5214 \quad 1.8214$$

or $\quad a = 2.5, b = 2.5214,$ and $c = 1.8214.$

Hence, the least-squares quadratic equation for this problem is

$$y = 2.5 + 2.5214x + 1.8214x^2$$

The standard error of the estimate is based on the regression polynomial given by Eq. (6.36), where $S_r = \Sigma(y - \hat{y})^2$. Here, we have

$$S_{y/x} = \sqrt{\frac{S_r}{n-(m+1)}} = \sqrt{\frac{3.229}{6-(2+1)}} = 1.0374$$

The coefficient of determination is given by Eq. (6.25)

$$r^2 = \frac{S_t - S_r}{S_t} = \frac{2493.5 - 3.229}{2493.5} = 0.9987$$

where $\qquad S_t = \Sigma(y_i - \overline{y})^2$

and $\qquad S_r = \Sigma(y_i - \hat{y})^2$

Therefore, the correlation coefficient is $r = \sqrt{0.9987} = 0.99935$. These results show that 99.935% of the original uncertainty has been explained by the model.

6.13 QUANTIFICATION OF ERROR OF LINEAR REGRESSION

Noting that the sum of the squares is defined as

$$S_r = \sum_{i=1}^{n}(y_i - a - bx_{1,i} - cx_{2,i})^2 \tag{6.37}$$

Equation (6.46) is similar to the equation

$$S_t = \sum_{i=1}^{n}(y_i - \overline{y})^2 \tag{6.38}$$

In Eq. (6.38), the squares of the residual represented the squares of the discrepancy between the data and a single estimate of the measure of central tendency (the mean). The squares of the residual represent the squares of the vertical distance between the data and another measure of central tendency (the straight line). If the spread of the points around the line is of similar magnitude along the entire range of data and the distribution of these points about the line is normal, then the least-squares regression will provide the best estimates of a and b. This is known as the *maximum likelihood principle*. Also, if these criteria are met, a standard deviation for the regression line can be determined as

$$S_{y/x} = \sqrt{\frac{S_r}{n-2}} \tag{6.39}$$

where $S_{y/x}$ is called the *standard error of the estimate*.

The difference between the S_t and S_r quantifies the improvement or error reduction due to describing the data in terms of a straight line rather than as an average value. The difference is therefore normalized to S_t to give

$$r^2 = \frac{S_t - S_r}{S_t} \tag{6.40}$$

where r^2 is called *the coefficient of determination* and r is the *correlation coefficient*. For a perfect fit, $S_r = 0$, and $r^2 = 1$, indicating that the line explains 100% of the variability of the data. For $r^2 = 0$, $S_r = S_t$ and the fit represents no improvement. An alternative formulation for r is given by

$$
r = \frac{n\sum\limits_{i=1}^{n}(x_i y_i) - \left(\sum\limits_{i=1}^{n} x_i\right)\left(\sum\limits_{i=1}^{n} y_i\right)}{\sqrt{\left[n\sum\limits_{i=1}^{n} x_i^2 - \left(\sum\limits_{i=1}^{n} x_i\right)^2\right]\left[n\sum\limits_{i=1}^{n} y_i^2 - \left(\sum\limits_{i=1}^{n} y_i\right)^2\right]}}
$$

$$
= \frac{\sum\limits_{i=1}^{n}(x_i y_i) - \left(\sum\limits_{i=1}^{n} x_i\right)\left(\sum\limits_{i=1}^{n} y_i\right)}{\sqrt{\left[\sum\limits_{i=1}^{n} x_i^2 - \left(\sum\limits_{i=1}^{n} x_i\right)^2\right]\left[\sum\limits_{i=1}^{n} y_i^2 - \left(\sum\limits_{i=1}^{n} y_i\right)^2\right]}} \tag{6.41}
$$

EXAMPLE 6.6

Determine (*a*) the total standard deviation, (*b*) the standard error of the estimate, and (*c*) the correlation coefficient for the data in Example 6.1.

Solution:

Table 6.7 shows the data and summation to compute the goodness-of-fit statistics. From Example 6.1, we have $a = -236.50$, $b = 19.5083$, $\bar{x} = 45$ and $\bar{y} = 641.375$.

TABLE 6.7

i	x_i	y_i	$a + bx_i$	$(y_i - \bar{y})^2$	$(y_i - a - bx_i)^2$
1	10	24	−41.4167	381151.8906	4279.3403
2	20	68	153.6667	328758.8906	7338.7778
3	30	378	348.7500	69366.3906	855.5625
4	40	552	543.8333	7987.8906	66.6944
5	50	608	738.9167	1113.8906	17139.1736
6	60	1218	934.0000	332496.3906	80656.0000
7	70	831	1129.0833	35957.6406	88853.6736
8	80	1452	1324.6667	657112.8906	16341.3611
Σ	360	5131	5131	1813945.875	215530.583

The standard deviation is given by

$$S_y = \sqrt{\frac{S_t}{n-1}}$$

where S_t is the total sum of the squares of the residuals between the data points and the mean.

Hence $\quad S_y = \sqrt{\dfrac{1813945.875}{8-1}} = 476.1746$

The standard error of the estimate is

$$S_{y/x} = \sqrt{\frac{S_r}{n-2}} = \sqrt{\frac{215530.583}{8-2}} = 164.1320$$

Since $S_{y/x} < S_y$, the linear regression model has merit.

The coefficient of determination r^2 is given by Eq. (6.41)

$$r^2 = \frac{S_t - S_r}{S_t} = \frac{1813945.875 - 215530.583}{1813945.875} = 0.8812$$

or $\quad r = \sqrt{0.8812} = 0.9387$

These results indicate that 93.87% of the original uncertainty has been explained by the linear model.

6.14 MULTIPLE LINEAR REGRESSION

Consider a function y which is a linear function of x_1 and x_2 as in

$$y = a + bx_1 + cx_2 + e \tag{6.42}$$

Equation (6.42) is quite useful in fitting experimental data where variable being studied is often a function of two other variables. For this two-dimensional case, the regression *line* becomes a *plane*. The best values of the coefficients are obtained by formulating the sum of the squares of the residuals:

$$S_r = \sum_{i=1}^{n} (y_i - a - bx_{1,i} - cx_{2,i})^2 \tag{6.43}$$

Differentiating Eq. (6.43) with respect to each of the unknown coefficients, we get

$$\frac{\partial S_r}{\partial a} = -2\sum_{i=1}^{n} (y_i - a - bx_{1,i} - cx_{2,i})$$

$$\frac{\partial S_r}{\partial b} = -2\sum_{i=1}^{n} x_{1,i}(y_i - a - bx_{1,i} - cx_{2,i})$$

$$\frac{\partial S_r}{\partial c} = -2\sum_{i=1}^{n} x_{2,i}(y_i - a - bx_{1,i} - cx_{2,i})$$

The coefficient giving the minimum sum of the squares of the residuals are obtained by setting the partial derivatives equal to zero and expressing the result in matrix form as

$$\begin{bmatrix} n & \sum_{i=1}^{n} x_{1,i} & \sum_{i=1}^{n} x_{2,i} \\ \sum_{i=1}^{n} x_{1,i} & \sum_{i=1}^{n} x_{1,i}^2 & \sum_{i=1}^{n} x_{1,i}x_{2,i} \\ \sum_{i=1}^{n} x_{2,i} & \sum_{i=1}^{n} x_{1,i}x_{2,i} & \sum_{i=1}^{n} x_{2,i}^2 \end{bmatrix} \begin{Bmatrix} a \\ b \\ c \end{Bmatrix} = \begin{Bmatrix} \sum_{i=1}^{n} y_i \\ \sum_{i=1}^{n} x_{1,i}y_i \\ \sum_{i=1}^{n} x_{2,i}y_i \end{Bmatrix} \tag{6.44}$$

EXAMPLE 6.7

The following data was generated from the equation $y = 7 + 3x_1 + 4x_2$. Use multiple linear regressions to fit this data.

x_1	0	1	2	1	4	7	2	0
x_2	0	2	1	3	5	1	3	4
y	7	18	17	22	39	32	25	23

Solution:

TABLE 6.8

y_i	$x_{1,i}$	$x_{2,i}$	$x_{1,i}^2$	$x_{1,i}x_{2,i}$	$x_{2,i}^2$	$x_{2,i}y_i$	$x_{1,i}y_i$
7	0	0	0	0	0	0	0
18	1	2	1	2	4	36	18
17	2	1	4	2	1	17	34
22	1	3	1	3	9	66	22
39	4	5	16	20	25	195	156
32	7	1	49	7	1	32	224
25	2	3	4	6	9	75	50
23	0	4	0	0	16	92	0
Σ 183	17	19	75	40	65	513	504

The summations required for Eq. (6.44) are computed in Table 6.8 as shown above. Substituting these values in Eq. (6.44), we get

$$\begin{bmatrix} 8 & 17 & 19 \\ 17 & 75 & 40 \\ 19 & 40 & 65 \end{bmatrix} \begin{Bmatrix} a \\ b \\ c \end{Bmatrix} = \begin{Bmatrix} 183 \\ 504 \\ 513 \end{Bmatrix}$$

which can be solved using MATLAB.

Refer to Appendix C (Cramer's rule for solving a system of linear algebraic equations).

$$D = \begin{vmatrix} 8 & 17 & 19 \\ 17 & 75 & 40 \\ 19 & 40 & 65 \end{vmatrix} = 6180$$

$$D_1 = \begin{vmatrix} 183 & 17 & 19 \\ 504 & 75 & 40 \\ 513 & 40 & 65 \end{vmatrix} = 43260$$

$$D_2 = \begin{vmatrix} 8 & 183 & 19 \\ 17 & 504 & 40 \\ 19 & 513 & 65 \end{vmatrix} = 18540$$

$$D_3 = \begin{vmatrix} 8 & 17 & 183 \\ 17 & 75 & 504 \\ 19 & 40 & 513 \end{vmatrix} = 24720$$

$$a = \frac{D_1}{D} = \frac{43260}{6180} = 7$$

$$b = \frac{D_2}{D} = \frac{18540}{6180} = 3$$

$$c = \frac{D_3}{D} = \frac{24720}{6180} = 4$$

MATLAB Program:

```
>>A = [8  17  19;  17  75  40;  19  40  65];
>>B = [183;  504;  513];
  x = A/B
  x = 7;  b = 3  and  c = 4
or          a =7,  b = 3,  and c = 4.
```

which is consistent with the original equation from which the data was derived.

6.15 WEIGHTED LEAST-SQUARES METHOD

Referring to the Sections 6.3 and 6.4 and assigning weights w_i to each error, e_i $(i = 1, 2, \dots, n)$ in Eq. (6.4) such that $\Sigma w_i = 1$.

Equation (6.5) can be written as

$$S_r = \sum_{i=1}^{n} w_i [y_i - (a + bx_i)]^2 \tag{6.45}$$

For S_r to be a minimum, we have

$$\frac{\partial S_r}{\partial a} = 0 \text{ and } \frac{\partial S_r}{\partial b} = 0 \tag{6.46}$$

We obtain the normal equations as

$$a + b\Sigma w_i x_i = \Sigma w_i y_i \tag{6.47}$$

$$a\Sigma w_i x_i + b\Sigma w_i x_i^2 = \Sigma w_i x_i y_i \tag{6.48}$$

The solution of Eqs. (6.47) and (6.48) gives the values for a and b. These values give the minimum of S_r with respect to the weight w_i.

Similarly, for a parabolic equation, we need to minimize

$$S_r = \Sigma w_i [y_i - (a + bx_i + cx_i^2)]^2 \qquad (6.49)$$

where $\qquad\qquad\qquad \Sigma w_i = 1.$

For S_r to be minimum, we have

$$\frac{\partial S_r}{\partial a} = 0, \ \frac{\partial S_r}{\partial b} = 0 \text{ and } \frac{\partial S_r}{\partial c} = 0 \qquad (6.50)$$

On simplification, we obtain the following normal equations to determine a, b, and c.

$$a + b\Sigma w_i x_i + c\Sigma w_i x_i^2 = \Sigma w_i y_i \qquad (6.51)$$

$$a\Sigma w_i x_i + b\Sigma w_i x_i^2 + c\Sigma w_i x_i^3 = \Sigma w_i x_i y_i \qquad (6.52)$$

$$a\Sigma w_i x_i^2 + b\Sigma w_i x_i^3 + c\Sigma w_i x_i^4 = \Sigma w_i x_i y_i \qquad (6.53)$$

6.16 ORTHOGONAL POLYNOMIALS AND LEAST-SQUARES APPROXIMATION

The previous sections considered the problem of least-squares approximation to fit a collections of data. This method is also applicable for continuous data.

6.17 LEAST-SQUARES METHOD FOR CONTINUOUS DATA

Let $y = f(x)$ be a continuous function on $[a, b]$ and it is to be approximated by the nth degree polynomial.

$$y = a_0 + a_1 x + a_2 x^2 + \ldots + a_n x^n \qquad (6.54)$$

Here the sum of the squares of residuals S is given by

$$S = \int_a^b w(x)[y - (a_0 x + a_2 x^2 + \ldots + a_n x^n)]^2 \, dx \qquad (6.55)$$

where $w(x)$ is a suitable weight function.

The necessary conditions for minimum S are given by

$$\frac{\partial S}{\partial a_0} = \frac{\partial S}{\partial a_1} = ... = \frac{\partial S}{\partial a_n} = 0 \tag{6.56}$$

Equation (6.56) give the normal equations as

$$-2\int_a^b w(x)[y - (a_0 + a_1 x + a_2 x^2 + ... + a_n x^n)]dx = 0$$

$$-2\int_a^b w(x)[y - (a_0 + a_1 x + a_2 x^2 + ... + a_n x^n)]x\, dx = 0$$

$$-2\int_a^b w(x)[y - (a_0 + a_1 x + a_2 x^2 + ... + a_n x^n)]x^2\, dx = 0$$

$$\vdots \qquad\qquad \vdots$$

$$-2\int_a^b w(x)[y - (a_0 + a_1 x + a_2 x^2 + ... + a_n x^n)]x^n\, dx = 0 \tag{6.57}$$

After simplification these equations reduce to

$$a_0\int_a^b w(x)\,dx + a_1\int_a^b xw(x)\,dx + ... + a_n\int_a^b x^n w(x)\,dx = \int_a^b w(x)y\,dx$$

$$a_0\int_a^b xw(x)\,dx + a_1\int_a^b x^2 w(x)\,dx + ... + a_n\int_a^b x^{n+1} w(x)\,dx = \int_a^b w(x)xy\,dx$$

$$a_0\int_a^b x^2 w(x)\,dx + a_1\int_a^b x^3 w(x)\,dx + ... + a_n\int_a^b x^{n+2} w(x)\,dx = \int_a^b w(x)x^2 y\,dx$$

$$\vdots \qquad\qquad \vdots$$

$$a_0\int_a^b x^n w(x)\,dx + a_1\int_a^b x^{n+1} w(x)\,dx + ... + a_n\int_a^b x^{2n} w(x)\,dx = \int_a^b w(x)x^n y\,dx \tag{6.58}$$

Since $w(x)$ and $y = f(x)$ are known, Eq. (6.58) form a system of linear equations with $(n + 1)$ unknowns $a_0, a_1, ..., a_n$. This system of equations possesses a unique solution. If

$$a_0 = a_0^*, a_1 = a_1^*, ..., a_n = a_n^*$$

is the solution for $a_0, a_1, ..., a_n$ then the approximate polynomial is given by

$$y = a_0^* + a_1^* x + a_2^* x^2 + ... + a_n^* x^n$$

EXAMPLE 6.8

Construct a least-squares quadrate approximation to the function $f(x) = \sin \pi x$ on $[0, 1]$.

Solution:

The normal equations for $P_2(x) = a_2x^2 + a_1x + a_0$ are

$$a_0 \int_0^1 1\,dx + a_1 \int_0^1 x\,dx + a_2 \int_0^1 x^2 dx = \int_0^1 \sin \pi x\,dx \tag{E.1}$$

$$a_0 \int_0^1 x\,dx + a_1 \int_0^1 x^2 dx + a_2 \int_0^1 x^3 dx = \int_0^1 x \sin \pi x\,dx \tag{E.2}$$

$$a_0 \int_0^1 x^2\,dx + a_1 \int_0^1 x^3 dx + a_2 \int_0^1 x^{43} dx = \int_0^1 x^2 \sin \pi x\,dx \tag{E.3}$$

Performing the integration gives

$$a_0 + \frac{1}{2}a_1 + \frac{1}{3}a_2 = \frac{2}{\pi} \tag{E.4}$$

$$\frac{1}{2}a_0 + \frac{1}{3}a_1 + \frac{1}{4}a_2 = \frac{1}{\pi} \tag{E.5}$$

$$\frac{1}{3}a_0 + \frac{1}{4}a_1 + \frac{1}{5}a_2 = \frac{\pi^2 - 4}{\pi^3} \tag{E.6}$$

Equations (E.4), (E.5), and (E.6) in three unknowns can be solved to obtain

$$a_0 = \frac{12\pi^2 - 120}{\pi^3} \approx -0.050465$$

and $\quad a_1 = -a_2 = \dfrac{720 - 60\pi^2}{\pi^3} \approx 4.12251$

Consequently, the least-squares polynomial approximation of degree 2 for $f(x) = \sin \pi x$ on $[0, 1]$ is $P_2(x) = -4.12251x^2 + 4.12251x - 0.050465$.

6.18 APPROXIMATION USING ORTHOGONAL POLYNOMIALS

In Section 6.19, a function is approximated as a polynomial containing the terms $1, x, x^2, ..., x^n$. These terms are called *base functions*, since, any function or even discrete data are approximated based on these functions.

Here, we assume that the base functions are some orthogonal polynomials $f_0(x), f_1(x), \ldots, f_n(x)$. Let the given function be approximated as

$$y = a_0 f_0(x) + a_1 f_1(x) + \ldots + a_n f_n(x) \tag{6.59}$$

where $f_i(x)$ is a polynomial in x of degree i. Then the residue is given by

$$S = \int_a^b w(x)[y - \{a_0 f_0(x) + a_1 f_1(x) + \ldots + a_n f_n(x)\}]^2 \, dx \tag{6.60}$$

For minimum S, the conditions are given by

$$\frac{\partial S}{\partial a_0} = 0, \frac{\partial S}{\partial a_1} = 0, \ldots, \frac{\partial S}{\partial a_n} = 0 \tag{6.61}$$

Equation (6.61) yield the following normal equations:

$$-2\int_a^b w(x)[y - \{a_0 f_0(x) + a_1 f_1(x) + \ldots + a_n f_n(x)\}] f_0(x) \, dx = 0$$

$$-2\int_a^b w(x)[y - \{a_0 f_0(x) + a_1 f_1(x) + \ldots + a_n f_n(x)\}] f_1(x) \, dx = 0$$

$$\vdots \qquad\qquad\qquad \vdots$$

$$-2\int_a^b w(x)[y - \{a_0 f_0(x) + a_1 f_1(x) + \ldots + a_n f_n(x)\}] f_n(x) \, dx = 0 \tag{6.62}$$

After simplification, the ith equation can be written as

$$a_0 \int_a^b w(x) f_0(x) f_i(x) \, dx + a_1 \int_a^b w(x) f_1(x) f_i(x) \, dx + \ldots$$

$$+ a_i \int_a^b w(x) f_i^2(x) \, dx + \ldots + a_n \int_a^b w(x) f_n(x) f_i(x) \, dx = \int_a^b w(x) y \, f_i(x) \, dx \tag{6.63}$$

$i = 0, 1, 2, \ldots, n$.

A set of polynomial $\{f_0(x), f_1(x), \ldots, f_n(x)\}$ is said to be *orthogonal* with respect to the weight function $w(x)$ if

$$\int_a^b f_i(x) f_j(x) w(x) \, dx = \begin{cases} 0, & \text{if } i \neq j \\ \int_b^a f_i^2(x) w(x) dx, & \text{if } i = j \end{cases} \tag{6.64}$$

Using Eq. (6.64), Eq. (6.63) can be written as

$$a_i \int_a^b w(x) f_i^2(x) \, dx = \int_a^b w(x) f_i(x) \, dx, \qquad i = 0, 1, 2, \ldots, n$$

Hence, $\qquad a_i = \dfrac{\int_a^b w(x)y \, f_i(x)\,dx}{\int_a^b w(x)f_i^2(x)\,dx}, \qquad i = 0, 1, 2, \dots, n$ $\qquad$ (6.65)

From Eq. (6.65), we can find the values of $a_0, a_1, \dots$, and the least-squares approximation is obtained by substituting these values in Eq. (6.59). However, the functions $f_0(x), f_1(x), \dots, f_n(x)$ are unknown. Several orthogonal functions are available in literature. A few of them are given in Table 6.9.

Any one of the orthogonal functions can be selected to fit a function dependent on the given problem.

TABLE 6.9 Some standard orthogonal polynomials

Name	$f_i(x)$	Interval	$w(x)$
Legendre	$P_n(x)$	$[-1, 1]$	1
Leguerre	$L_n(x)$	$[0, \infty]$	e^{-x}
Hermite	$H_n(x)$	$(-\infty, \infty)$	e^{-x^2}
Chebyshev	$T_n(x)$	$[-1, 1]$	$(1-x^2)^{-1/2}$

6.19 GRAM-SCHMIDT ORTHOGONALIZATION PROCESS

Let $f_i(x)$ be a polynomial in x of degree i and $\{f_i(x)\}$ be a given sequence of polynomials. Then the sequence of orthogonal polynomials $\{f_i^*(x)\}$ over the interval $[a, b]$ with respect to the weight function $w(x)$ can be generated by the following equation

$$f_i^*(x) = x^i - \sum_{r=0}^{i-1} a_{ir} f_r^*(x), \qquad i = 1, 2, \dots, n \qquad (6.66)$$

where the constants are a_{ir}, and $f_0^*(x) = 1$.

To obtain a_{ir}, we multiply Equation (6.66) with $w(x)f_n^*(x)$, $0 \le k \le i-1$ and integrating over $[a, b]$, we obtain

$$\int_a^b f_i^*(x)f_n^*(x)w(x)\,dx = \int_a^b x^i f_n^*(x)w(x)\,dx - \int_a^b \sum_{r=0}^{i-1} a_{ir} f_r^*(x)f_n^*(x)w(x)\,dx \quad (6.67)$$

Using the property of orthogonal polynomial, Equation (6.67) becomes

$$\int_a^b x^i f_n^*(x)w(x)\,dx - \int_a^b a_{in} f_n^{*2}(x)w(x)\,dx = 0$$

or
$$a_{in} = \frac{\int_a^b x^i f_n^*(x)\, w(x)\, dx}{\int_a^b f_n^{*2}(x)\, w(x)\, dx}, \qquad 0 \le n \le i-1 \qquad (6.68)$$

Hence, the set of orthogonal polynomials $\{f_i^*(x)\}$ are given by

$$f_0^*(x) = 1$$

$$f_i^*(x) = x^i - \sum_{r=0}^{i-1} a_{ir} f_r^*(x), \qquad i = 1, 2, \ldots., n$$

where
$$a_{ir} = \frac{\int_a^b x^i f_n^*(x)\, w(x)\, dx}{\int_a^b f_n^{*2}(x)\, w(x)\, dx} \qquad (6.69)$$

For the discrete data, the integral is replaced by summation.

EXAMPLE 6.9

Use the Gram-Schmidt orthogonalization process to find the first two orthogonal polynomials on $[-1, 1]$ with respect to the weight function $w(x) = 1$.

Solution:

Let $f_0^*(x) = 1$.

Hence $f_1^*(x) = x - a_{10} f_0^*(x)$

where
$$a_{10} = \frac{\int_{-1}^1 x\, dx}{\int_{-1}^1 dx} = 0.$$

or $f_1^*(x) = x.$

The second orthogonal polynomial is given by
$$f_2^*(x) = x^2 - a_{20} f_0^*(x) - a_{21} f_1^*(x)$$

where
$$a_{20} = \frac{\int_{-1}^1 x^2\, dx}{\int_{-1}^1 dx} = \frac{1}{3},$$

$$a_{21} = \frac{\int_{-1}^1 x^2 . x\, dx}{\int_{-1}^1 x^2\, dx} = 0$$

Hence, $f_2^*(x) = x^2 - \dfrac{1}{3} = \dfrac{1}{3}(3x^2 - 1)$

Thus, the first two orthogonal polynomials are

$$f_0^*(x) = 1, \ f_1^*(x) = x \text{ and } f_2^*(x) = \dfrac{1}{3}(3x^2 - 1)$$

6.20 FITTING A FUNCTION HAVING A SPECIFIED POWER

EXAMPLE 6.10

Determine the equation to the best fitting exponential curve of the form $y = ae^{bx}$ for the data given in Table 6.10.

TABLE 6.10

x	1	3	5	7	9
y	115	105	95	85	80

Solution:

Refer to Table 6.11.

TABLE 6.11

i	x_i	y_i	$\log y_i$	x_i^2	$x_i \log y_i$
1	1	115	2.0607	1	2.0607
2	3	105	2.0212	9	6.0636
3	5	95	1.9777	25	9.8886
4	7	85	1.9294	49	13.5059
5	9	80	1.9031	81	17.1278
Σ	25	480	9.8921	165	48.6466

Given $\quad y = ae^{bx}$

The normal equations are

$$\Sigma \log y_i = 5A + B \Sigma x_i \qquad\qquad (E.1)$$

$$\Sigma x_i \log y_i = A \Sigma x_i + B \Sigma x_i^2 \qquad\qquad (E.2)$$

where $A = \log a$ and $B = b \log e$.

Solving the two normal Eqs. (E.1) and (E.2), we get

$$A = 2.0802 \text{ and } B = -0.0203$$

Hence, $\qquad a = \text{antilog of } A = 10^{2.0802} = 120.2818$

and $\qquad b = B / \log e = \dfrac{-0.0203}{\log e} = -0.0075.$

Hence, the fitted equation is $y = 120.2818\, e^{-0.0075x}$.

EXAMPLE 6.11

For the data given in Table 6.12, find the equation to best fitting curve of the form $y = ab^x$.

TABLE 6.12

x	1	2	3	4	5
y	130	150	175	190	240

Solution:

The calculations are shown in Table 6.13.

TABLE 6.13

x	y	log y	x^2	x log y	$\hat{y}$ (estimated)
1	130	2.1139	1	2.1139	129.2062
2	150	2.1761	4	4.3522	149.5433
3	175	2.2430	9	6.7291	173.0814
4	190	2.2788	16	9.1150	200.3144
5	240	2.3802	25	11.9011	231.8555
Σ 15	885	11.192	55	34.2113	884.0108

The normal equations are obtained as follows:

$$y = ab^x \tag{E.1}$$

Taking logarithms (base 10) on both sides of the above Eq. (E.1), we get

$$\log y = \log a + x \log b \tag{E.2}$$

or $\qquad Y = A + Bx \tag{E.3}$

where $\qquad Y = \log y$, $A = \log a$ and $B = \log b$.

Hence, the normal equations are

$$\Sigma Y = nA + B\Sigma x \tag{E.4}$$

$$\Sigma xY = A\Sigma x + B\Sigma x^2 \tag{E.5}$$

Substituting the values from Table 6.9 (a) into Eqs. (E.4) and (E.5), we have

$$11.1920 = 5A + 15B \tag{E.6}$$

$$34.2113 = 15A + 55B \tag{E.7}$$

Solving Eqs. (E.6) and (E.7), we obtain

$$A = 2.0478 \quad \text{and} \quad B = 0.0635$$

Hence, $\qquad a = \text{antilog of } A = 10^{2.0478} = 111.6349$

$$b = \text{antilog of } B = 10^{0.0635} = 1.1574$$

Hence, the fitted equation is $y = 111.6349(1.1574)^x$. The estimated values of y (denoted by $\hat{y}$) are shown in the last column of Table 6.9 (a).

EXAMPLE 6.12

For the data given in Table 6.14, find the equation to best fitting curve of the form $xy^a = b$.

TABLE 6.14

x	200	150	100	60	40	10
y	1	1.5	1.8	2.4	4.1	6.5

Solution:

See Table 6.15.

TABLE 6.15

x	y	log x	log y	(log x)^2	(log x)(log y)	ŷ (estimated)
200	1	2.3010	0	5.2947	0	1.1762
150	1.5	2.1761	0.1761	4.7354	0.3832	1.4040
100	1.8	2.0000	0.2553	4.0000	0.5105	1.8019
60	2.4	1.7782	0.3802	3.1618	0.6761	2.4675
40	4.1	1.6021	0.6128	2.5666	0.9817	3.1668
10	6.5	1.0000	0.8129	1.0000	0.8129	7.4322
Σ 560	17.3	10.8573	2.2373	20.7585	3.3644	17.4485

Given

$$xy^a = b \qquad\qquad (E.1)$$

Taking logarithms (to the base 10) on both sides of the above Eq. (E.1), we get

$$\log x + a \log y = \log b \qquad\qquad (E.2)$$

$$\frac{1}{a}\log x + \log y - \frac{\log b}{a} \qquad\qquad (E.3)$$

The normal equations are given by

$$\Sigma Y = 6A + B\Sigma X \qquad\qquad (E.4)$$

$$\Sigma XY = A\Sigma X + B\Sigma X^2 \qquad\qquad (E.5)$$

where $Y = \log y$, $X = \log x$, $A = \dfrac{1}{a}\log b$ and $B = -1/a$.

Solving Eqs. (E.4) and (E.5), we obtain

$$A = 1.4865 \text{ and } B = -0.6154$$

Therefore, $a = -1/B = -1/-0.6154 = 1.6250$

and $b = $ antilog of $(aA) = 10^{(1.6250)(1.4865)} = 260.3529$

Hence, the fitted equation is $xy^{1.6250} = 260.3529$.

EXAMPLE 6.13

Fit the following data:

x	0	2	4	6
y	11	16	19	26

to a straight line by considering that the data (2, 16) and (4, 19) are more significant or reliable with weights 6 and 11 respectively.

Solution:

Weighted least-squares method.

Let the straight line be $y = a + bx$. The normal equations are

$$a\Sigma w_i + b\Sigma w_i x_i = \Sigma w_i y_i \tag{E.1}$$

and $$a\Sigma w_i x_i + b\Sigma w_i\, x_i^2 = \Sigma w_i x_i y_i \tag{E.2}$$

The values in Eqs. (E.1) and (E.2) are calculated as shown in Table 6.16.

TABLE 6.16

x	y	w	wx	wx^2	wy	wxy
0	11	1	0	0	11	0
2	16	6	12	24	96	192
4	19	11	44	176	209	836
6	26	1	6	36	26	156
Total		19	62	236	342	1184

The normal equations are

$$19a + 62b = 342 \tag{E.3}$$

and $$62a + 236b = 1184 \tag{E.4}$$

The solution of Eqs. (E.3) and (E.4) gives
$$a = 11.4125 \text{ and } b = 2.0188$$

Hence, $$y = 11.4125 + 2.0188x$$

Estimation of Error

x	y	w	Predicted y	Absolute error	(Absolute error)^2
0	11	1	11.4125	0.4125	0.1702
2	16	6	15.4500	0.5500	0.3025
4	19	11	19.4875	0.4875	0.2377
6	26	1	23.5250	2.4750	6.1256
			Sum of squares of errors		6.8329

EXAMPLE 6.14

Consider the Example 6.14 with the modified weights 300 and 50 instead of 6 and 11.

Solution:

The modified calculations are shown in Table 6.17.

TABLE 6.17

x	y	w	wx	wx^2	wy	wxy
0	11	1	0	0	11	0
2	16	30	60	120	480	960
4	19	50	200	800	950	3800
6	26	1	6	36	26	156
Total		82	266	956	1467	4916

The normal equations are

$$82a + 266b = 1467 \qquad \text{(E.1)}$$

and $\qquad 266a + 956b = 4916 \qquad$ (E.2)

The solution of Eqs. (E.1) and (E.2) gives

$$a = 12.4144 \text{ and } b = 1.6881$$

Hence, $\qquad y = 12.4144 + 1.6881x$

Estimation of Error

x	y	w	Predicted y	Absolute error	(Absolute error)^2
0	11	1	12.4144	1.4144	2.0004
2	16	30	15.7905	0.2096	0.0439
4	19	50	19.1666	0.1666	0.0277
6	26	1	22.5427	3.4573	11.9530
Sum of squares of errors					14.0250

It is noted that when the weights on $x = 2$ and $x = 4$ are increased then the absolute error in y are reduced at these points, but, the sum of squares of errors is increased due to the less importance of the data $(0, 11)$ and $(6, 26)$.

6.21 SUMMARY

In this chapter, we have reviewed the relationship between two variables in two ways: (1) by using the regression analysis, and (2) by computing the correlation coefficient. It was shown that the regression model can be used to evaluate the magnitude of change in one variable due to a certain change in another variable. The regression model also helps to predict the value of one variable for a given value of another variable. The correlation coefficient shows how strongly two variables are related. It does not, however, provide any information about the size of change in one variable as a result of a certain change in the other variable.

EXERCISES

6.1. Table 6.18 gives information on the monthly incomes (in hundreds of dollars) and monthly telephone bills (in dollars) for a random sample of 10 households.

TABLE 6.18

Income	16	45	35	31	30	14	40	15	36	40
Telephone bill	36	140	171	70	94	25	159	41	78	98

Use least-squares regression to determine the coefficients a and b in the function $y = a + bx$ that best fits the data.

6.2. The following Table 6.19 lists the annual incomes (in thousands of dollars) and amounts of life insurance (in thousands of dollars) of life insurance policies for six persons:

TABLE 6.19

Annual income	47	54	26	38	62	20
Life insurance	250	300	100	150	500	75

(a) Find the regression line $y = a + bx$ with annual income as an independent variable and amount of life insurance policy as a dependent variable.

(b) Determine the estimated value of life insurance of a person with an annual income of $50,000.

6.3. Find the least-squares regression line for the data on annual incomes and food expenditures of the seven households given in Table 6.20. Use income as an independent variable and food expenditure as a dependent variable. All data is given in thousands of dollars.

TABLE 6.20

Income: x	35	50	22	40	16	30	25
Expenditure: y	9	15	6	11	5	8	9

6.4. Table 6.21 gives data on age and crown-rump length for the foetuses. Use least-squares regression to determine the coefficients a and b in the function $y = a + bx$ that best fits the data:

TABLE 6.21

x	10	10	13	13	18	19	19	23	25	28
y	66	66	108	106	160	165	176	227	234	279

6.5. The following data in Table 6.22 refers to the number of hours that 10 students studied for a math test and their scores on the test:

TABLE 6.22

Hours studied	1	17	22	12	7	4	14	10	9	4
Test score	21	83	90	60	45	38	74	66	59	32

(a) Find the equation of the least-squares line that approximates the regression of the test scores on the number of hours studied.

(b) Determine the average test score of a person who studied 15 hours for the test.

6.6. The following Table 6.23 shows the first two grades, denoted by x and y respectively, of 10 students on two midterm examinations in applied statistics. Find the least-squares regression line of y on x.

TABLE 6.23

Grade on first midterm examination (x)	60	50	80	80	70	60	100	40	90	70
Grade on second midterm examination (y)	80	70	70	90	50	80	95	60	80	60

6.7. The following Table 6.24 shows ages x and systolic blood pressure y of 12 men.

(a) Determine the least-squares regression equation of y on x

(b) Estimate the blood pressure of a man whose age is 45 years.

TABLE 6.24

Age (x)	56	42	72	36	63	47	55	49	38	42	68	60
Blood pressure (y)	147	125	160	118	149	128	150	145	115	140	152	155

6.8. Table 6.25 shows the respective weight x and y of a sample of 12 fathers and their oldest sons. Find the least-squares regression line of y on x.

TABLE 6.25

Weight of father, x (kg)	65	63	67	64	68	62	70	66	68	67	69	71
Weight of son, y (kg)	68	66	68	65	69	66	67	65	70	67	68	70

6.9. Find the least-squares regression line for the data on annual incomes and food expenditures of seven households given in Table 6.26. Use income as independent variable and food expenditure as a dependent variable. The income and food-expenditures are in thousands of dollars.

TABLE 6.26

Income x	35	49	21	29	15	28	25
Food expenditure y	9	15	7	10	5	8	8.5

6.10. A car manufacturing company wanted to investigate how the price of one of its car models depreciates with age. The company took a sample of eight cars of this model and collected the following information on the ages (in years) and prices (in hundreds of dollars) of these cars as shown in Table 6.27.

TABLE 6.27

Age	8	3	6	9	2	5	6	3
Price	16	74	40	19	120	36	33	86

(a) Find the regression line $\hat{y} = a + bx$ with price as a dependent variable and age as independent variable.

(b) Give a brief interpretation of the values of a and b calculated in part (a).

(c) Predict the price of a 7-year old car of this model.

(d) Estimate the price of an 4-year old car of this model.

For Exercises 6.28 through 6.37 do the following:

Fit a least-squares regression line of the form $\hat{y} = a + bx$ for the data given in Tables 6.18 through 6.27 respectively. Assume x as the independent variable and y as the dependent variable.

(a) Give a brief interpretation of the values of a and b calculated in $\hat{y} = a + bx$.

(b) Compute the standard deviation of the sample errors, s_e.

(c) Compute the error sum of squares, SSE.

(d) Compute the total sum of squares, SST.

(e) Compute the regression sum of squares, SSR.

(f) Compute the coefficient of determination, r^2.

(g) Compute the correlation coefficient, r.

6.11. For the data given in Table 6.18.

6.12. For the data given in Table 6.19.

6.13. For the data given in Table 6.20.

6.14. For the data given in Table 6.21.

6.15. For the data given in Table 6.22.

6.16. For the data given in Table 6.23.

6.17. For the data given in Table 6.24.

6.18. For the data given in Table 6.25.

6.19. For the data given in Table 6.26.

6.20. For the data given in Table 6.27.

6.21. Fit $y = bx^m$ (power function) in Exercise 6.18 using a logarithmic transformation.

6.22. Fit $y = bx^m$ (power function) to the data in Exercise 6.19 using a logarithmic transformation.

6.23. Fit $y = bx^m$ (power function) to the data in Exercise 6.20 using a logarithmic transformation.

6.24. Fit $y = bx^m$ (power function) to the data in Exercise 6.21 using a logarithmic transformation.

6.25. Fit $y = bx^m$ (power function) to the data in Exercise 6.22 using a logarithmic transformation.

6.26. Determine the coefficient of the polynomial $y = a + bx + cx^2$ that best fit the data given in the following table.

x	1	3	5	7	10
y	2.1	5.1	5.45	6.12	6.62

Determine the standard error of the estimate and correlation coefficient.

6.27. The following data were collected in an experiment to study the relationship between shear strength in kPa (y) and curing temperature in °C (x).

x	1.38	1.40	1.46	1.48	1.52	1.53
y	5.392	5.612	5.671	5.142	4.481	4.129

(a) Fit a least-squares quadratic model of the form $y = a + bx + cx^2$ to the above data.

(b) Using the equation, compute the residuals.

(c) Compute the error sum of squares and total sum of squares.

(d) Compute the error variance estimate.

(e) Compute the coefficient of determination.

6.28. The following data were collected in an experiment to study the relationship between the number of kilograms of fertilizer (x) and the yield of tomatoes in bushels (y).

x	5	10	30	40	50
y	32	42	54	50	42

(a) Fit a least-squares quadratic model of the form $y = a + bx + cx^2$ to the above data.

(b) Using this equation, compute the regression sum of squares
$$\sum_{i=1}^{n} (\hat{y}_i - \bar{y})^2 .$$

(c) Compute the error sum of squares $\sum_{i=1}^{n} (y_i - \hat{y}_i)^2$ and total sum of squares $\sum_{i=1}^{n} (y_i - \bar{y})^2 .$

(d) Compute the error variance estimate $(b) + (c)$.

(e) Compute the coefficient of determination, r^2.

6.29. Fit a least-square parabola $y = a + bx + c^2$ to the following data:

x	0	1	2	3	4	5	6
y	2.4	2.1	3.2	5.6	9.3	14.6	21.9

Determine the coefficient of determination.

6.30. The following table gives the data collected in an experiment to study the relationship between the stopping distance $d(m)\_$ of an automobile travelling at speeds v(km/hr) at the instant the danger is sighted.

(a) Fit a least-squares parabola of the form $d = a + bv + cv^2$ to the data.

(b) Determine the coefficient of determination.

Speed v(km/hr)	32	48	64	80	96	112
Stopping distance d(m)	16.5	27.5	19.5	24.5	29.3	34.2

6.31. Use multiple linear regression fit of the form $y = a + bx_1 + cx_2$ for the following data:

x_1	0	1	1	2	2	3	3	4	4
x_2	0	1	2	1	2	1	2	1	2
y	15	18	12.8	25.7	20.4	35	30	45.3	40.1

Compute the coefficients, the standard error of the estimate, and the correlation coefficient.

6.32. Use multiple linear regression fit of the form $y = a + bx_1 + cx_2$ for the following data:

x_1	0	0	1	2	1	1.5	3	3	−1
x_2	0	1	0	1	2	1	2	3	−1
y	1	6	4	−4	−2	−1.5	−12	−15	17

Compute the coefficients, the standard error of estimate, and the correlation coefficient.

6.33. Use multiple linear regression fit of the form $y = a + bx_1 + cx_2$ for the following data:

x_1	0	0	1	1	2	3	0	2	1	4
x_2	0	1	0	1	2	0.5	2	3	4	1
y	3	8	7	12	21	15	13	26	27	24

Compute the coefficients, the standard error of estimate, and the correlation coefficient.

6.34. Use multiple linear regression fit of the form $y = a + bx_1 + cx_2$ for the following data:

x_1	0	0	1	1	2	0	1	2	1	1
x_2	0	1	0	1	0	2	2	1	3	1
y	23	15	19	10	15	5	0	5	−10	0

Compute the coefficients, the standard error of estimate, and the correlation coefficient.

6.35. Use multiple linear regression fit of the form $y = a + bx_1 + cx_2$ for the following data:

x_1	0	0	1	1	2	0	1	2	1	3
x_2	0	1	0	1	0	2	2	1	3	1
y	29	10	23	4	19	−10	−16	−2	−36	−8

Compute the coefficients, the standard error of estimate, and the correlation coefficient.

6.36. For the data given in Table 6.28, find the equation to the best fitting exponential curve of the form $y = ae^{bx}$.

TABLE 6.28

x	1	2	3	4	5
y	100	90	80	75	70

6.37. For the data given in Table 6.29, find the equation to the best fitting exponential curve of the form $y = ae^{bx}$.

TABLE 6.29

x	2	3	4	5	6
y	3.8	5.8	7.8	8.8	9.8

6.38. For the data given in Table 6.30, find the equation to the best fitting exponential curve of the form $y = ae^{bx}$.

TABLE 6.30

x	2.2	3	4	6	7
y	31	38	45	68	84

6.39. For the data given in Table 6.31, find the equation to the best fitting exponential curve of the form $y = ab^x$.

TABLE 6.31

x	1	2	3	4	5
y	22	8	3	1	0.35

6.40. For the data given in Table 6.32, find the equation to the best fitting exponential curve of the form $y = ab^x$.

TABLE 6.32

x	2	4	6	8	10
y	3	13	32	57	91

6.41. For the data given in Table 6.33, find the equation to the best fitting exponential curve of the form $y = ab^x$.

TABLE 6.33

x	1	3	5	7	9
y	3	2	1.3	0.72	0.43

6.42. For the data given in Table 6.34, find the equation to the best fitting exponential curve of the form $y = xy^a = b$.

TABLE 6.34

x	190	134	89	55	37	8.9
y	0.97	1.14	1.32	1.63	1.92	3.5

6.43. For the data given in Table 6.35, find the equation to the best fitting exponential curve of the form $y = xy^a = b$.

TABLE 6.35

x	2	3	5	7	9	11
y	1.25	1.21	1.16	1.14	1.11	1.10

6.44. For the data given in Table 6.36, find the equation to the best fitting exponential curve of the form $y = xy^a = b$.

TABLE 6.36

x	232	178	99	66	51

y	1.1	1.3	1.8	2.2	2.5

6.45. Find a nonlinear relationship of the form $y = a + b \log x$ for the data given in Table 6.37. Determine the linear correlation coefficient.

TABLE 6.37

x	1.2	4.7	8.3	20.9
y	0.6	5.1	6.9	10

6.46. Fit the following data to a straight line $y = a + bx$ by considering the weights as given in the table. Compute the sum of squares of errors.

x	1	17	22	12	7	4	14	10	9	4
y	21	83	90	60	45	38	74	66	59	32
w	5	1	7	1	1	8	1	11	1	4

6.47. Fit the following data to a straight line $y = a + bx$ by considering the weights as given in the table. Compute the sum of squares of errors.

x	16	45	35	31	30	14	40	15	36	40
y	50	134	107	95	90	44	120	47	110	120
w	1	6	1	3	1	7	1	11	1	16

6.48. Fit the following data to a straight line $y = a + bx$ by considering the weights as given in the table. Compute the sum of squares of errors.

x	47	54	26	38	62	20
y	250	300	100	150	500	75
w	1	5	1	3	1	7

6.49. Fit the following data to a straight line $y = a + bx$ by considering the weights given in the table. Compute the sum of squares of errors.

x	35	50	22	40	16	30	25
y	9	15	6	11	5	8	9
w	1	2	1	2	1	3	4

6.50. Fit the following data to a straight line $y = a + bx$ by considering the weights given in the table. Compute the sum of squares of errors.

x	10	10	13	13	18	19	19	23	25	28
y	66	66	108	106	160	165	176	227	234	279
w	2	1	3	1	4	1	5	1	6	1

6.51. Fit the following data to a straight line $y = a + bx$ by considering the weights given in the table. Compute the sum of squares of errors.

x	1	17	22	12	7	4	14	10	9	4
y	21	83	90	60	45	38	74	66	59	32
w	1	2	1	3	1	2	1	4	1	5

6.52. Fit the following data to a straight line $y = a + bx$ by considering the weights given in the table. Compute the sum of squares of errors.

x	1	17	22	12	7	4
y	21	83	90	60	45	38
w	5	1	7	1	1	8

6.53. Fit the following data to a straight line $y = a + bx$ by considering the weights given in the table. Compute the sum of squares of errors.

x	16	45	35	31	30
y	50	134	107	95	90
w	1	6	1	3	1

6.54. Fit the following data to a straight line $y = a + bx$ by considering the weights given in the table. Compute the sum of squares of errors.

x	47	54	26	38	62	20
y	250	300	100	150	500	75
w	1	4	1	2	1	7

6.55. Fit the following data to a straight line $y = a + bx$ by considering the weights given in the table. Compute the sum of squares of errors.

x	35	50	22	40
y	9	15	6	11
w	2	3	4	2

6.56. Fit the following data to a straight line $y = a + bx$ by considering the weights given in the table. Compute the sum of squares of errors.

x	10	10	13	13	18
y	66	66	108	106	160
w	5	2	3	7	4

6.57. Fit the following data to a straight line $y = a + bx$ by considering the weights given in the table. Compute the sum of squares of errors.

x	1	17	22	12	7
y	21	75	96	60	45
w	1	2	1	3	1

6.58. Construct a least-squares quadratic approximation to the function $y = e^x$ on $[0, 1]$.

6.59. Construct a least-squares quadratic approximation to the function $y = x \ln x$ on $[1, 3]$.

6.60. Construct a least-squares quadratic approximation to the function $y = x^3$ on $[0, 2]$.

6.61. Construct a least-squares quadratic approximation to the function $y = \dfrac{1}{x}$ on $[1, 3]$.

6.62. Construct a least-squares quadratic approximation to the function $y = x^2 + 3x + 2$ on $[0, 1]$.

6.63. Use the Gram-Schmidt orthogonalization process to construct $\phi_0(x)$, $\phi_1(x)$, $\phi_2(x)$, and $\phi_3(x)$ for the interval $[0, 1]$.

6.64. Use the Gram-Schmidt orthogonalization process to construct $\phi_0(x)$, $\phi_1(x)$, $\phi_2(x)$, and $\phi_3(x)$ for the interval $[0, 2]$.

6.65. Use the Gram-Schmidt orthogonalization process to construct $\phi_0(x)$, $\phi_1(x)$, $\phi_2(x)$, and $\phi_3(x)$ for the interval $[1, 3]$.

*N*UMERICAL *I*NTEGRATION

7.1 INTRODUCTION

If $F(x)$ is a differentiable function whose derivative is $f(x)$, then we can evaluate the definite integral I as

$$I = \int_a^b f(x) = F(b) - F(a), \; F'(x) = f(x) \tag{7.1}$$

Equation (7.1) is known as the *fundamental theorem of calculus*. Most integrals can be evaluated by the formula given by Eq. (7.1) and there exists many techniques for making such evaluations. However, in many applications in science and engineering, most integrals cannot be evaluated because most integrals do not have antiderivatives $F(x)$ expressible in terms of elementary functions.

In other circumferences, the integrands could be empirical functions given by certain measured values. In all these instances, we need to resort to numerical methods of integration. It should be noted here that, sometimes, it is difficult to evaluate the integral by analytical methods. *Numerical integration* (or *numerical quadrature,* as it is sometimes called) is an alternative approach to solve such problems. As in other numerical techniques, it often results in approximate solution. The integration can be performed on a continuous function or a set of data.

The integration given by Eq. (7.1) is shown in Figure 7.1. The integration shown in Figure 7.1 is called *closed* since the function values at the two points (a, b) where the limits of integration are located are used to find the integral. In *open* integration, information on the function at one or both limits of integration is not required.

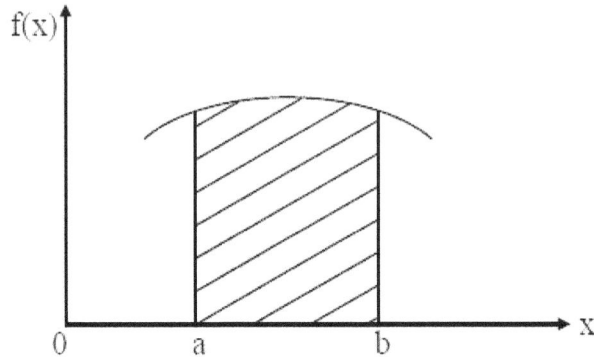

FIGURE 7.1

The range of integration $(b - a)$ is divided into a *finite* number of intervals in numerical integration. The integration techniques consisting of equal intervals are based on formulas known as *Newton-Cotes closed quadrature formulas*.

In this chapter, we present the following methods of integration with illustrative examples:

1. Trapezoidal rule
2. Simpson's 1/3 rule
3. Simpson's 3/8 rule
4. Boole's and Weddle's rules

7.1.1 Relative Error

Suppose we are required to evaluate the definite integral

$$I = \int_a^b f(x)\, dx$$

In numerical integration, we approximate $f(x)$ by a polynomial $f(x)$ of suitable degree. Then, we integrate $f(x)$ within the limits (a, b). That is,

$$\int_a^b f(x)\, dx \cong \int_a^b \phi(x)\, dx$$

Here the exact value if

$$I = \int_a^b f(x)\, dx$$

Approximate value $= \int_a^b \phi(x)\,dx$.

The difference $\left[\int_a^b f(x)\,dx - \int_a^b \phi(x)\,dx \right]$

is called the *error of approximation* and

$$\frac{\left[\int_a^b f(x)\,dx - \int_a^b \phi(x)\,dx \right]}{\int_a^b f(x)\,dx}$$

is called the *relative error of approximation*.

Hence, relative error of approximation $= \dfrac{\text{exact values } - \text{ approximate value}}{\text{exact value}}$

7.2 NEWTON-COTES CLOSED QUADRATURE FORMULA

The general form of the problem of numerical integration may be stated as follows:

Given a set of data points (x_i, y_i), $i = 0, 1, 2, \ldots, n$ of a function $y = f(x)$, where $f(x)$ is not explicitly known. Here, we are required to evaluate the definite integral

$$I = \int_a^b y\,dx \tag{7.2}$$

Here, we replace $y = f(x)$ by an interpolating polynomial $\phi(x)$ in order to obtain an approximate value of the definite integral of Eq. (7.2).

In what follows, we derive a general formula for numerical integration by using Newton's method forward difference formula. Here, we assume the interval (a, b) is divided into n-equal subintervals such that

$$h = \frac{b-a}{n}$$

$$a = x_0 < x_2 < x_3 \ldots < x_n = b \tag{7.3}$$

with $\quad x_n = x_0 + n\,h$

where $\quad h =$ the internal size

$n =$ the number of subintervals

a and $b =$ the limits of integration with $b > a$.

Hence, the integral in Eq. (7.2) can be written as

$$I = \int_{x_0}^{x_n} y\, dx \tag{7.4}$$

Using Newton's method forward interpolation formula, we have

$$I = \int_{x_0}^{x_n} \left[y_0 + p\,\Delta y_0 + \frac{p(p-1)}{2!}\Delta^2 y_0 + \frac{p(p-1)(p-2)}{3!}\Delta^3 y_0 + ... \right] dx$$

where $x = x_0 + ph$ \tag{7.5}

$$= h \int_0^n \left[y_0 + p\Delta y_0 + \frac{p^2 - p}{2}\Delta^2 y_0 + \frac{p^3 - 3p^2 + 2p}{6}\Delta^3 y_0 + ... \right] dp \tag{7.6}$$

Hence, after simplification, we get

$$I = \int_{x_0}^{x_n} y\, dn = nh \left[y_0 + \frac{n}{2}\Delta y_0 + \frac{n(2n-3)}{12}\Delta^2 y_0 + \frac{n(n-2)^2}{24}\Delta^3 y_0 + ... \right] \tag{7.7}$$

The formula given by Eq. (7.7) is known as the *Newton-Cotes closed quadrature formula*. From the general formula (Eq. [7.7]), we can derive or deduce different integration formulae by substituting $n = 1, 2, 3, ...,$ etc.

7.3   TRAPEZOIDAL RULE

In this method, the known function values are joined by straight lines. The area enclosed by these lines between the given end points is computed to approximate the integral as shown in Figure 7.2.

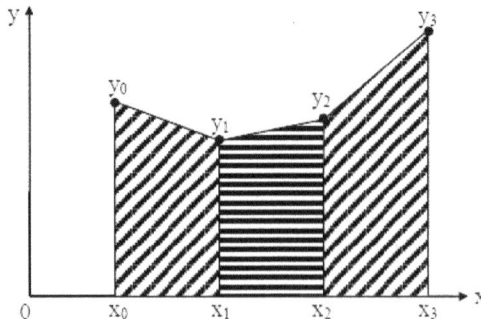

FIGURE 7.2

Each subinterval with the line approximation for the function forms a trapezoid as shown in Figure7.2. The area of each trapezoid is computed by multiplying the interval size h by the average value of the function value in that subinterval. After the individual trapezoidal areas are obtained, they are all added to obtain the overall approximation to the integral.

Substituting $n = 1$ in Eq. (7.7) and considering the curve $y = f(x)$ through the points (x_0, y_0) and (x_1, y_1) as a straight line (a polynomial of first degree so that the differences of order higher than first become zero), we get

$$I_1 = \int_{x_0}^{x_1} y\,dx = h\left[y_0 + \frac{1}{2}\Delta y_0\right] = \frac{h}{2}\left[y_0 + \frac{1}{2}(y_1 - y_0)\right] = \frac{h}{2}(y_0 + y_1) \qquad (7.8)$$

Similarly, we have

$$I_2 = \int_{x_1}^{x_2} y\,dx = \frac{h}{2}(y_1 + y_2)$$

$$I_3 = \int_{x_2}^{x_3} y\,dx = \frac{h}{2}(y_2 + y_3)$$

and so on. See Figure 7.3.

In general, we have

$$I_n = \int_{x_{n-1}}^{x_n} y\,dx = \frac{h}{2}(y_{n-1} + y_n) \qquad (7.9)$$

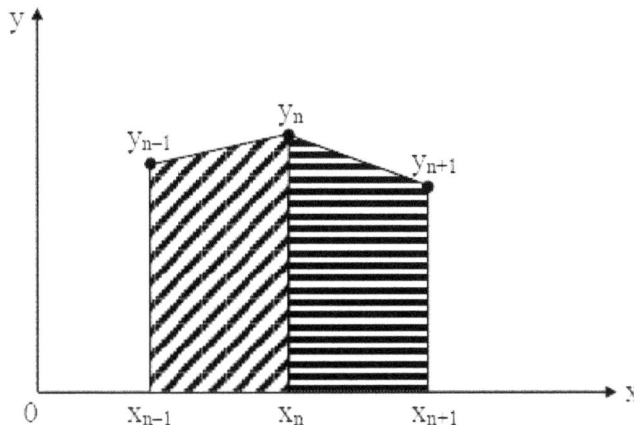

FIGURE 7.3

Adding all the integrals (Eq. [7.8], Eq.[7.9]) and using the interval additive property of the definite integrals, we obtain

$$I = \sum_{i=1}^{n} I_i = \int_{x_0}^{x_n} y \, dx = \frac{h}{2}[y_0 + 2(y_1 + y_2 + y_3 + \ldots + y_{n-1}) + y_n] = \frac{h}{2}[X + 2I] \quad (7.10)$$

where X = sum of the end points

         I = sum of the intermediate ordinates

Equation (7.10) is known as the *trapezoidal rule*.

    Summarizing, the trapezoidal rule signifies that the curve $y = f(x)$ is replaced by n-straight lines joining the points (x_n, y_n), i = 0, 1, 2, 3, …, n. The area bounded by the curve $y = f(x)$, the ordinates $x = x_0$, $x = x_n$ and the x-axis is then approximately equivalent to the sum of the areas of the n-trapezoids so obtained.

7.3.1   Error Estimate in Trapezoidal Rule

Let $y = f(x)$ be a continuous function with continuous derivatives in the interval $[x_0, x_n]$. Expanding y in a Taylor's series around $x = x_0$, we get

$$\int_{x_0}^{x_1} y \, dx = \int_{x_0}^{x_1} \left[y_0 + (x - x_0) y_0' + \left(\frac{x - x_0}{2!} \right)^2 y_0'' + \ldots \right] dx$$

$$hy_0 + \frac{h^2}{2} y_0' + \frac{h^3}{6} y_0'' + \frac{h^4}{24} y_0''' + \ldots \quad (7.11)$$

Likewise, $\frac{h}{2}(y_0 + y_1) = \frac{h}{2}(y_0 + y(x_0 + h)) = \frac{h}{2}\left[y_0 + y + hy_0' + \frac{h^2}{2} y_0'' + \ldots \right]$

$$= hy_0 + \frac{h^2}{2} y_0' + \frac{h^4}{4} y_0'' + \frac{h^4}{12} y_0''' + \ldots \quad (7.12)$$

Hence, the error e_1 in (x_0, x_1) is obtained from Eqs. (7.11) and (7.12) as

$$e_1 = \int_{x_0}^{x_1} y \, dx - \frac{h}{2}(y_0 + y_1) = \frac{-1}{12} h^3 y_0'' + \ldots$$

In a similar way, we can write

$$e_2 = \int_{x_1}^{x_2} y\,dx - \frac{h}{2}(y_1 + y_2) = \frac{-1}{12}h^3 y_1'' + ...$$

$$e_3 = \frac{-1}{12}h^3 y_2'' + ...$$

$$e_4 = \frac{-1}{12}h^3 y_3'' + ... \qquad (7.13)$$

and so on.

In general, we can write

$$e_n = \frac{-1}{12}h^3 y_{n+1}'' + ...$$

Hence, the total error E in the interval (x_0, x_n) can be written as

$$E = \sum_{n=1}^{n} e_n = \frac{-h^3}{12}\left[y_0'' + y_1'' + y_2'' + ... + y_{n-1}''\right] \qquad (7.14)$$

If $y''(\overline{x})$ is the largest value of the n quantities in the right-hand side of Eq. (7.14), then we have

$$E = \frac{-1}{12}h^3 n y''(\overline{x}) = -\frac{(b-a)}{12}h^2 y''(\overline{x}) \qquad (7.15)$$

Now, since $h = \dfrac{b-a}{n}$, the total error in the evaluation of the integral of Eq. (7.2) by the trapezoidal rule is of the order of h^2.

EXAMPLE 7.1

Evaluate the integral $\int_0^{1.2} e^x dx$, taking six intervals by using trapezoidal rule up to three significant figures.

Solution:

$$a = 0, b = 1.2, n = 6$$

$$h = \frac{b-a}{n} = \frac{1.2-0}{6} = 0.2$$

x	0	0.2	0.4	0.6	0.8	1.0	1.2
$y = f(x)$	0	1.221	1.492	1.822	2.226	2.718	3.320
	y_0	y_1	y_2	y_3	y_4	y_5	y_6

The trapezoidal rule can be written as

$$I = \frac{h}{2}[(y_0 + y_6) + 2(y_1 + y_2 + y_3 + y_4 + y_5)]$$

$$I = \frac{0.2}{2}[(1 + 3.320) + 2(1.221 + 1.492 + 1.822 + 2.226 + 2.718)]$$

$$I = 2.3278 \approx 2.328.$$

The exact value is $= \int_0^{1.2} e^x dx = 2.3201.$

EXAMPLE 7.2

Evaluate $\int_0^{12} \frac{dx}{1+x^2}$ by using trapezoidal rule, taking $n = 6$, correct to give significant figures.

Solution:

$$f(x) = \frac{1}{1+x^2}$$

$$a = 0, b = 12$$

$$h = \frac{b-a}{n} = \frac{12-0}{6} = 2$$

x	0	2	4	6	8	10	12
$y = f(x)$	1	$\frac{1}{5}$	$\frac{1}{17}$	$\frac{1}{37}$	$\frac{1}{65}$	$\frac{1}{101}$	$\frac{1}{145}$
	1.00000	0.20000	0.05882	0.02703	0.01538	0.00990	0.00690
y	y_0	y_1	y_2	y_3	y_4	y_5	y_6

The trapezoidal rule can be written as

$$I = \frac{h}{2}[(y_0 + y_6) + 2(y_1 + y_2 + y_3 + y_4 + y_5)]$$

$$I = \frac{2}{2}[(1 + 0.00690) + 2(0.2 + 0.05882 + 0.02703 + 0.01538 + 0.00990)]$$

$$I = 1.62916.$$

The exact value is

$$\int_0^{12} \frac{1}{1+x^2}\, dx = \tan^{-1} x \Big|_0^{12} = 1.48766.$$

EXAMPLE 7.3

Evaluate $\int_2^6 \log_{10} x\, dx$ by using trapezoidal rule, taking $n = 8$, correct to five decimal places.

Solution:

$$f(x) = \log_{10} x$$

$$a = 2,\ b = 6,\ n = 8$$

$$h = \frac{b-a}{n} = \frac{6-2}{8} = \frac{1}{2} = 0.5$$

x	2	2.5	3.0	3.5	4.0	4.5	5.0	5.5	6.0
f(x)	0.30103	0.39794	0.47712	0.54407	0.60206	0.65321	0.69897	0.74036	0.77815
	y_0	y_1	y_2	y_3	y_4	y_5	y_6	y_7	y_8

The trapezoidal rule is

$$I = \frac{h}{2}[(y_0 + y_8) + 2(y_1 + y_2 + y_3 + y_4 + y_5 + y_6 + y_7)]$$

$$I = \frac{0.5}{2}[(0.30103 + 0.77815) + 2(0.39794 + 0.47712 + 0.54407$$
$$+ 0.60206 + 0.65321 + 0.69897 + 0.74036 + 0.77815)]$$
$$I = 2.32666.$$

The exact value is given by

$$\int_2^6 \log x\, dx = [x\log x - x]_2^6 = 6.06685$$

7.4   SIMPSON'S 1/3 RULE

In Simpson's 1/3 rule, the function is approximated by a second degree polynomial between successive points. Since a second degree polynomial contains three constants, it is necessary to know three consecutive function values forming two intervals as shown in Figure 7.4.

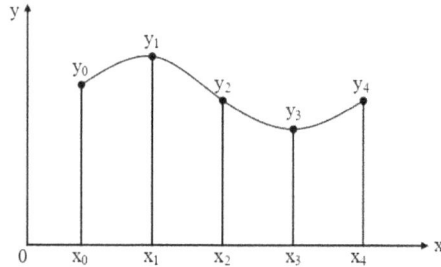

FIGURE. 7.4

Consider three equally spaced points x_0, x_1, and x_2. Since the data are equally spaced, let $h = x_{n+1} - x_n$ (see Figure 7.5).

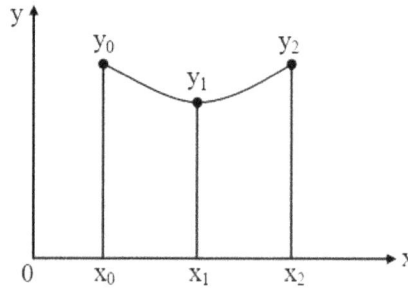

FIGURE. 7.5

Substituting $n = 2$ in Eq. (7.7) and taking the curve through the points (x_0, y_0), (x_1, y_1) and (x_2, y_2) as a polynomial of second degree (parabola) so that the differences of order higher than two vanish, we obtain

$$I_1 = \int_{x_0}^{x_1} y\,dx = 2h\left[y_0 + 4y_0 + \frac{1}{6}\Delta^2 y_0\right] = \frac{h}{3}[y_0 + 4y_1 + y_2] \qquad (7.16)$$

Similarly, $I_2 = \int_{x_2}^{x_4} y\,dx = \frac{h}{3}[y_2 + 4y_3 + y_4]$

$$I_3 = \int_{x_4}^{x_6} y\,dx = \frac{h}{3}[y_4 + 4y_5 + y_6] \qquad (7.17)$$

and so on.

In general, we can write

$$I_n = \int_{x_{2n-2}}^{x_{2n}} y\,dx = \frac{h}{3}[y_{2n-2} + 4y_{2n-1} + y_{2n}] \qquad (7.18)$$

Summing up all the above integrals, we obtain

$$I = \int_{x_0}^{x_n} y \, dx = \frac{h}{3}[y_0 + 4(y_1 + y_3 + y_5 + \dots + y_{2n-1}) + 2(y_2 + y_4 + y_6 + \dots$$

$$+ y_{2n-2}) + y_{2n}]$$

$$= \frac{h}{3}[X + 40 + 2E] \qquad (7.19)$$

where $X =$ sum of end ordinates

$O =$ sum of odd ordinates

$E =$ sum of even ordinates

Equation (7.19) is known as *Simpson's 1/3 rule*. Simpson's 1/3 rule requires the whole range (the given interval) must be divided into even number of equal subintervals.

7.4.1 Error Estimate in Simpson's 1/3 Rule

Expanding $y = f(x)$ around $x = x_0$ by Taylor's series, we obtain

$$\int_{x_0}^{x_2} y \, dx = \int_{x_0}^{x_0+2h}\left[y_0 + (x - x_0)y_0^i + \frac{(x-x_0)^2}{2!} y_0^{ii} + \dots \right] dx$$

$$= 2h y_0 + \frac{4h^2}{2!} y_0^i + \frac{8h^3}{3!} y_0^{ii} + \frac{16h^4}{4!} y_0^{iii} + \frac{32h^5}{5!} y_0^{iv} + \dots$$

$$= 2h y_0 + 2h^2 y_0^i + \frac{4}{3} h^3 y_0^{ii} + \frac{2h^4}{3} y_0^{iii} + \frac{4}{15} y_0^{iv} + \dots \qquad (7.20)$$

In addition, we have

$$\frac{h}{3}[y_0 + 4y_1 + y_2] = \frac{h}{3}\left[\begin{array}{l} y_0 + 4\left(y_0 + hy_0^i + \frac{h^2}{2!} y_0^{ii} + \dots \right) \\ + (y_0 + hy_0^i + \frac{4h^2}{2!} y_0^{ii} + \frac{8h^3}{3!} y_0^{iii} + \dots \end{array}\right]$$

$$= 2h y_0 + 2h^2 y_0^i + 4\frac{h^3}{3} y_0^{ii} + \frac{2h^4}{3} y_0^{iii} + \frac{5h^5}{18} y_0^{iv} + \dots \qquad (7.21)$$

Hence, from Eqs. (7.20) and (7.21), the error in the subinterval (x_0, x_2) is given by

$$e_1 = \int_{x_0}^{x_2} y\, dx - \frac{h}{3}(y_0 + 4y_1 + y_2) = \left(\frac{4}{15} - \frac{5}{18}\right)h^5 y_0^{iv} + ...$$

$$= \frac{-h^5}{90} y_0^{iv} + ... \cong \frac{-h^5}{90} y_0^{iv} \qquad (7.22)$$

Likewise, the errors in the subsequent intervals are given by

$$e_2 = \frac{-h^5}{90} y_2^{iv}$$

$$e_3 = \frac{-h^5}{90} y_4^{iv} \qquad (7.23)$$

and so on.

Hence, the total error E is given by

$$E = \sum e_n = \frac{-h^5}{90}\left[y_0^{iv} + y_2^{iv} + y_4^{iv} + ... + y_{2n-2}^{iv}\right] \cong \frac{-nh^5}{90} h_0^{iv}(\overline{x})$$

or $$E = \frac{-(b-a)}{180} h^2 y^{iv}(\overline{x}) \qquad (7.24)$$

where $y^{iv}(\overline{x})$ = largest value of the fourth-order derivatives

$h = (b-a)/n$

The error in Simpson's 1/3 rule can be written as

$$e = \frac{-nh^5}{180} f^{iv}(\xi) = \frac{-(b-a)^5}{2880 n^4} f^{iv}(\xi)$$

where $a = x_0 < \xi < x_n = $ b (for n subintervals of length h).

EXAMPLE 7.4

Evaluate the integral $\int_0^{1.2} e^x dx$, taking $n = 6$ using Simpson's 1/3 rule.

Solution:

$f(x) = e^X$

$a = 0, b = 1.2, n = 6$

$$h = \frac{b-a}{n} = \frac{1.2-0}{6} = 0.2$$

x	0	0.2	0.4	0.6	0.8	1.0	1.2
y = f(x)	1.0	1.22140	1.49182	1.82212	2.22554	2.71828	3.32012
	y_0	y_1	y_2	y_3	y_4	y_5	y_6

The Simpson's rule is

$$I = \frac{h}{2}[(y_0 + y_6) + 4(y_1 + y_3 + y_5) + 2(y_2 + y_4)]$$

$$I = \frac{0.2}{3}[(1 + 3.32012) + 4(1.22140 + 1.82212 + 2.71828)$$
$$+ 2(1.49182 + 2.22554)]$$

$$I = \frac{0.2}{3}[(4.32012) + 4(5.7618) + 2(3.71736)]$$

$$I = 2.320136 \approx 2.32014$$

The exact value is = 2.3201.

EXAMPLE 7.5

Evaluate $\int_0^{12} \frac{dx}{1+x^2}$ by using Simpson's 1/3 rule, taking $n = 6$.

Solution:

$$f(x) = \frac{1}{1+x^2}$$
$$a = 0, b = 12, n = 6$$
$$h = \frac{b-a}{n} = \frac{12-0}{6} = 2$$

x	0	2	4	6	8	10	12
y = f(x)	1	0.2	0.05882	0.02703	0.01538	0.0099	0.0069
	y_0	y_1	y_2	y_3	y_4	y_5	y_6

The Simpson's 1/3 rule is

$$I = \frac{h}{3}[(y_0 + y_6) + 4(y_1 + y_3 + y_5) + 2(y_2 + y_4)]$$

$$I = \frac{2}{3}[(1+0.0069) + 4(0.2 + 0.02703 + 0.0099) + 2(0.05882 + 0.01538)]$$

$$I = 1.40201.$$

EXAMPLE 7.6

Evaluate $\int_{2}^{6} \log_{10} x \, dx$ by using Simpson's 1/3 rule, taking $n = 6$.

Solution:

$$f(x) = \log_{10}x$$

$$a = 2, \text{b} = 6, n = 6$$

$$h = \frac{b-a}{n} = \frac{6-2}{6} = \frac{2}{3}$$

x	2 = 6/3	8/3	10/3	12/3 = 4	14/3	16/3	18/3 = 6
y = f(x)	0.30103	0.42597	0.52288	0.60206	0.66901	0.72700	0.77815
	y_0	y_1	y_2	y_3	y_4	y_5	y_6

The Simpson's 1/3 rule is

$$I = \frac{h}{3}[(y_0 + y_6) + 4(y_1 + y_3 + y_5) + 2(y_2 + y_4)]$$

$$I = \frac{2/3}{3}[(0.30103 + 0.77815) + 4(0.42597 + 0.60206 + 0.72700)$$
$$+ 2(0.52288 + 0.66901)]$$

$$I = 2.32957.$$

7.5 SIMPSON'S 3/8 RULE

Putting $n = 3$ in Eq. (7.7) and taking the curve through (x_n, y_n), $n = 0, 1, 2, 3$ as a polynomial of degree three such that the differences higher than the third order vanish, we obtain

$$I_1 = \int_{x_0}^{x_3} y \, dx = 3h\left[y_0 + \frac{3}{2}\Delta y_0 + \frac{3}{2}\Delta^2 y_0 + \frac{1}{8}\Delta^3 y_0\right]$$

$$= \frac{3}{8}h[h_0 + 3y_1 + 3y_2 + y_3] \tag{7.25}$$

Similarly, we get

$$I_2 = \int_{x_3}^{x_6} y\,dx = \frac{3}{8}h[y_3 + 3y_4 + 3y_5 + y_6]$$

$$I_3 = \int_{x_6}^{x_9} y\,dx = \frac{3}{8}h[y_6 + 3y_7 + 3y_8 + y_9] \qquad (7.26)$$

and so on.

Finally, we have

$$I_n = \int_{x_{3n-3}}^{x_{3n}} y\,dx = \frac{3}{8}h[y_{3n-3} + 3y_{3n-2} + 3y_{3n-1} + y_{3n}] \qquad (7.27)$$

Summing up all the expressions above, we obtain

$$I = \int_{x_0}^{x_{3n}} y\,dx = \frac{h}{8}[y_0 + 3(y_1 + y_2 + y_4 + y_5 + y_7 + y_8 + \ldots + y_{3n-2} + y_{3n-1})$$

$$+ 2(y_3 + y_6 + y_9 + \ldots + y_{3n-3}) + y_{3n}] \qquad (7.28)$$

Equation (7.28) is called the *Simpson's 3/8 rule*. Here, the number of subintervals should be taken as multiples of 3. Simpson's 3/8 rule is not as accurate as Simpson's 1/3 rule. The dominant term in the error of this formula is $\frac{-3}{80}y^5 y^{iv}(\overline{x})$. Simpson's 3/8 rule can be applied when the range (a, b) is divided into a number of subintervals, which must be a multiple of 3. The error in Simpson's 3/8 rule is $e = \frac{-nh^5}{80}f^{iv}(\xi)$, where $x_0, \xi x_n$ (for n subintervals of length h).

EXAMPLE 7.7

Evaluate the integral $\int_0^{1.2} e^x dx$, by using Simpson's 3/8 rule and taking seven ordinates.

Solution:

$$n + 1 = 7 \Rightarrow n = 6$$

The points of division are

$$0, \frac{1}{6}, \frac{2}{6}, \frac{3}{6}, \frac{4}{6}, \frac{5}{6}, 1, \qquad h = \frac{1}{6}$$

x	0	1/6	2/6	3/6	4/6	5/6	1
y = f(x)	1	1.18136	1.39561	1.64872	1.94773	2.30098	2.71828
	y_0	y_1	y_2	y_3	y_4	y_5	y_6

The Simpson's 3/8's rule is

$$I = \frac{3h}{8}[(y_0 + y_6) + 3(y_1 + y_2 + y_4 + y_5) + 2(y_3)]$$

$$I = \frac{3(1/6)}{8}[(1 + 2.71828) + 3(1.18136 + 1.39561 + 1.94773 + 2.30098) + 2(1.64872)]$$

$$I = 1.71830$$

EXAMPLE 7.8

Evaluate $\int_0^{12} \frac{dx}{1+x^2}$ by using Simpson's 3/8 rule and taking seven ordinates.

Solution:

$$n + 1 = 7 \Rightarrow n = 6, h = 2.$$

The points of division are

$$0, 2, 4, 6, 8, 10, 12$$

x	0	2	4	6	8	10	12
y = f(x)	1	0.2	0.05882	0.02703	0.01538	0.00990	0.00690
	y_0	y_1	y_2	y_3	y_4	y_5	y_6

The Simpson's 3/8's rule is

$$I = \frac{3}{8}h[(y_0 + y_6) + 3(y_1 + y_2 + y_4 + y_5) + 2(y_3)]$$

$$I = \frac{3}{8}2[(1 + 0.00690) + 3(0.2 + 0.05882 + 0.01538 + 0.00990) + 2(0.02703)]$$

$$I = 1.43495.$$

EXAMPLE 7.9

Repeat Example 7.6 by using Simpson's 3/8 rule, taking $n = 6$, correct to five decimal places.

Solution:

The points of division are

$$2, \frac{8}{3}, \frac{10}{3}, \frac{12}{3}, \frac{14}{3}, \frac{16}{3}, \frac{18}{3}$$

x	6/3	8/3	10/3	12/3	14/3	16/3	18/3
y = f(x)	0.30103	0.42597	0.52288	0.60206	0.66901	0.727	0.77815
	y_0	y_1	y_2	y_3	y_4	y_5	y_6

Here $\quad h = \dfrac{2}{3}$

The Simpson's 3/8's rule is

$$I = \frac{3.h}{8}\left[(y_0 + y_6) + 3(y_1 + y_2 + y_4 + y_5) + 2(y_3)\right]$$

$$I = \frac{3(2/3)}{8}\left[(0.30103 + 0.77815) + 3(0.42597 + 0.52288 + 0.66901 \right.$$
$$\left. + 0.72700) + 2(0.60206)\right]$$

$$I = 2.32947.$$

7.6 BOOLE'S AND WEDDLE'S RULES

7.6.1 Boole's Rule

Substituting $n = 4$ in Eq. (7.7) and taking the curve through (x_n, y_n), $n = 0, 1, 2, 3, 4$ as a polynomial of degree 4, so that the difference of order higher than four vanish (or neglected), we obtain

$$\int_{x_0}^{x_4} y\, dx = 4h\left[y_0 + 2\Delta y_0 + \frac{5}{3}\Delta^2 y_0 + \frac{2}{3}\Delta^3 y_0 + \frac{7}{90}\Delta^4 y_0\right]$$

$$= \frac{2h}{45}\left[7y_0 + 32y_1 + 12y_2 + 32y_3 + 7y_4\right] \tag{7.29}$$

Likewise,
$$\int_{x_4}^{x_8} y\,dx = \frac{2h}{45}(7y_4 + 32y_5 + 12y_6 + 32y_7 + 7y_8)$$

and so on.

Adding all the above integrals from x_0 to x_n, where n is a multiple of 4, we obtain

$$I = \int_{x_0}^{x_n} y\,dx = \frac{2h}{45}[7y_0 + 32(y_1 + y_3 + y_5 + y_7 + ...) + 12(y_2 + y_6 + y_{10} + ...)$$

$$+14(y_4 + y_8 + y_{12} + ...) + 7y_n] \qquad (7.30)$$

Equation (7.30) is known as *Boole's rule*. It should be noted here that the number of subintervals should be taken as a multiple of 4.

The leading term in the error of formula can be shown as

$$\frac{-8}{945}h^7 y^{vi}(\bar{x}).$$

7.6.2 Weddle's Rule

Substituting $n = 6$ in Eq. (7.7) and taking the curve $y = f(x)$ through the point (x_n, y_n), $n = 0, 1, 2, 3, 4, 5, 6$ as a polynomial of degree 6 so that the differences of order higher than 6 are neglected, we obtain

$$\int_{x_0}^{x_6} y\,dx = 6h\left[y_0 + 3\Delta y_0 + \frac{9}{2}\Delta^2 y_0 + 4\Delta^3 y_0 + \frac{123}{60}\Delta^4 y_0\right.$$

$$\left. + \frac{11}{20}\Delta^5 y_0 + \frac{41}{140}\Delta^6 y_0 \right]$$

$$= \frac{3h}{10}[y_0 + 5y_1 + y_2 + 6y_3 + y_4 + 5y_5 + y_6] \qquad (7.31)$$

Approximating $\frac{41}{140}\Delta^6 y_0$ as $\frac{3}{10}\Delta^6 y_0$, we have, similarly, we can write

$$\int_{x_6}^{x_{12}} y\,dx = \frac{3h}{10}[y_6 + 5y_7 + y_8 + 6y_9 + y_{10} + 5y_{11} + y_{12}] \qquad (7.32)$$

and so on.

Adding all the above integrals from x_0 to x_n, where x is a multiple of 6, we obtain

$$\int_{x_0}^{x_n} y\,dx = \frac{3h}{10}[y_0 + 5(y_1 + y_5 + y_7 + y_{11} + ...) + (y_2 + y_4 + y_8 + y_{10} + ...)$$

$$+6(y_3 + y_9 + y_{15} + ...) + 2(y_6 + y_{12} + y_{18} + ...) + y_n] \quad (7.33)$$

Equation (7.33) is known as *Weddle's rule*. Weddle's rule was found to be more accurate than most of the other rules. The error estimate is given by $\frac{-h^7}{140} y^{vi}(\bar{x})$. In Weddle's rule, the number of subintervals should be taken as multiple of 6.

A summary of the Newton-Cotes formulas and their errors is presented in Table 7.1.

TABLE 7.1 Summary of Newton-Cotes formula.

No.	Integral	Name	Integration formula	Error
1.	$\int_{x_0}^{x_1} y\,dx$	Trapezoidal Rule	$\frac{h}{2}[y_0 + y_1]$	$\frac{-h^3}{12} y''(\bar{x})$
2.	$\int_{x_0}^{x_2} y\,dx$	Simpson's 1/3 Rule	$\frac{h}{3}[y_0 + 4y_1 + y_2]$	$\frac{-h^5}{90} y^{iv}(\bar{x})$
3.	$\int_{x_0}^{x_3} y\,dx$	Simpson's 3/8 Rule	$\frac{3h}{8}[y_0 + 3y_1 + 3y_2 + y_3]$	$\frac{-3h^5}{80} y^{iv}(\bar{x})$
4.	$\int_{x_0}^{x_4} y\,dx$	Boole's Rule	$\frac{2h}{45}[7y_0 + 32y_1 + 12y_2 + 32y_3 + 7y_4]$	$\frac{-8}{945} h^7 y^{vi}(\bar{x})$
5.	$\int_{x_0}^{x_6} y\,dx$	Weddle's Rule	$\frac{3h}{10}[y_0 + 5y_1 + y_2 + 6y_3 + y_4 + 5y_5 + y_6]$	$\frac{-h^7}{140} y^{vi}(\bar{x})$

EXAMPLE 7.10

Evaluate the integral $\int_0^{1.2} e^x dx$ by using Boole's rule using exactly five functional evaluations and correct to five significant figures.

Solution:

Taking $h = \frac{1.2}{4}$ and applying Boole's rule, we have

$$\int_0^{1.2} f(x)\,dx = \frac{2h}{45}[7y_0 + 32y_1 + 12y_2 + 32y_3 + 7y_4]$$

$$\int_0^{1.2} f(x)\,dx = \frac{2 \times 0.3}{45}[7f(0) + 32f(0.3) + 12f(0.6) + 32f(0.9) + 7f(1.2)]$$

x	0	0.3	0.6	0.9	1.2
$y = f(x)$	1	1.34986	1.82212	2.45960	3.32012
	y_0	y_1	y_2	y_3	y_4

$$\int_0^{1.2} f(x)\, dx = 0.01333[7x1 + 32 \times 1.34986 + 12 \times 1.82212 + 32$$
$$\times 2.45960 + 7 \times 3.32012]$$

$$\int_0^{1.2} f(x)\, dx = 2.31954.$$

EXAMPLE 7.11

Evaluate the integral $\int_0^{12} \dfrac{dx}{1+x^2}$ by using Boole's rule using exactly five functional evaluations and correct to five significant figures.

Solution:

x	0	3	6	9	12
$y = f(x)$	1	0.1	0.02703	0.01220	0.00690

The Boole's rule is

$$\int_0^{12} f(x)\, dx = \frac{2h}{45}[7f(0) + 32f(3) + 12f(6) + 32f(9) + 7f(12)]$$

$$I = \frac{2x3}{45}[7 \times (1) + 32 \times (0.1) + 12 \times (0.02703) + 32 \times (0.01220) + 7 \times (0.00690)]$$

$$I = 1.46174.$$

EXAMPLE 7.12

Evaluate the integral $\int_0^{1.2} e^x dx$ by using Weddle's rule and taking $n = 6$, correct to five significant figures.

Solution:

$$f(x) = e^x; a = x_0 = 0; b = x_n = 1.2; n = 6$$

$$h = \frac{1.2 - 0}{6} = 0.2$$

The Weddle's rule is

$$I = \frac{3h}{10} \left[y_0 + 5y_1 + y_2 + 6y_3 + y_4 + 5y_5 + y_6 \right]$$

x	0	0.2	0.4	0.6	0.8	1	1.2
y = f(x)	1	1.2214	1.4918	1.8221	2.2255	2.7183	3.3201
	y_0	y_1	y_2	y_3	y_4	y_5	y_6

$$I = \frac{3(0.2)}{10} [1 + 5(1.2214) + 1.4918 + 6(1.8221) + 2.2255 + 5(2.7183)$$
$$+ 3.3201]$$

$$I = 2.32011 \approx 2.3201.$$

EXAMPLE 7.13

Evaluate the integral $\int_0^{12} \dfrac{dx}{1+x^2}$ by using Weddle's rule and taking $n = 6$, correct up to five significant figures.

Solution:

$$a = 0; b = 12; n = 6$$

$$h = \frac{b-a}{n} = \frac{12-0}{6} = 2$$

x	0	2	4	6	8	10	12
y = f(x)	1	0.2	0.05882	0.02703	0.01538	0.00990	0.00690
	y_0	y_1	y_2	y_3	y_4	y_5	y_6

The Weddle's rule is

$$I = \frac{3h}{10} \left[y_0 + 5y_1 + y_2 + 6y_3 + y_4 + 5y_5 + y_6 \right]$$

$$I = \frac{3 \times 2}{10} [1 + 5 \times 0.2 + 0.05882 + 6 \times 0.02703 + 0.01538 + 5 \times 0.00990 + 0.00690]$$

$$I = 1.37567.$$

EXAMPLE 7.14

Repeat Example 7.6 by using Weddle's rule, taking $n = 6$, correct to five decimal places.

Solution:

$$a = 2; b = 6; n = 6$$

$$h = \frac{b - a}{n} = \frac{6 - 2}{6} = \frac{2}{3}$$

x	6/3	8/3	10/3	12/3	14/3	16/3	18/3
y = f(x)	0.30103	0.42597	0.52288	0.60206	0.66901	0.727	0.77815
	y_0	y_1	y_2	y_3	y_4	y_5	y_6

The Weddle's rule is

$$I = \frac{3h}{10} [y_0 + 5y_1 + y_2 + 6y_3 + y_4 + 5y_5 + y_6]$$

$$I = \frac{3(2/3)}{10} [0.30103 + 5 \times 0.42597 + 0.52288 + 6 \times 0.60206 + 0.66901 \\ + 5 \times 0.727 + 0.77815]$$

$$I = 2.32966.$$

EXAMPLE 7.15

Repeat Example 7.6 by Boole's rule, using exactly five functional evaluations and correct to five significant figures.

Solution:

We use five functional evaluations here.

Taking $h = 1$ and applying Boole's rule, we have

$$I = h \frac{2}{45} [7f(2) + 32f(3) + 12f(4) + 32f(5) + 7f(6)]$$

$$I = \frac{2}{45} [7 \times 0.30103 + 32 \times 0.47712 + 12 \times 0.60206 + 32 \times 0.69897 \\ + 7 \times 0.77815]$$

x	2	3	4	5	6
y = f(x)	0.30103	0.47712	0.60206	0.69897	0.77815

$$I = 2.32950.$$

7.7 ROMBERG'S INTEGRATION

Romberg's integration employs a successive error reduction technique. It applies the trapezoidal rule with different interval sizes in order to obtain some preliminary approximations to the integral to start with. The method starts with the preliminary approximations obtained by the trapezoidal rule and then applies the Richardson extrapolation procedure which refines these values successfully to a single more accurate approximation.

7.7.1 Richardson's Extrapolation

The Richardson extrapolation is a simple method for improving the accuracy of certain numerical procedures, including the finite difference approximations and in numerical integration.

Assume that we have an approximate means of computing some quantity G. In addition, assume that the result depends on a parameter h. Let us denote the approximation by $g(h)$, then we have $G = g(h) + E(h)$, where $E(h)$ denotes the error.

Richardson extrapolation can remove the error, provided that it has the form $E(h) = ch^p$, where c and p are constants.

We begin by computing $g(h)$ with some value of h, say $h = h_1$. In this case, we have

$$G = g(h_1) + ch_1^p \tag{7.34}$$

Repeating the calculations with $h = h_2$, such that

$$G = g(h_2) + ch_2^p \tag{7.35}$$

Now, eliminating c and solving for G from Eqs. (7.34) and (7.35), we get

$$G = \frac{\left(\dfrac{h_1}{h_2}\right)^p g(h_2) - g(h_1)}{\left(\dfrac{h_1}{h_2}\right)^p - 1} \tag{7.36}$$

Equation (7.36) is called the *Richardson extrapolation formula*.

It is general practice to use $h_2 = \dfrac{h_1}{2}$ and in this case Eq. (7.36) becomes

$$G = \frac{2^p g\left(\dfrac{h_1}{2}\right) - g(h_1)}{2^p - 1} \tag{7.37}$$

7.7.2 Romberg Integration Formula

As mentioned earlier, Romberg's integration provides a simple modification to the approximate quadrature formula obtained with the aid of finite difference method in order to obtain their better approximations.

Consider as an example to improve the value of the integral

$$I = \int_a^b y\, dx = \int_a^b f(x)\, dx \tag{7.38}$$

by the trapezoidal rule.

We can evaluate Eq. (7.38) by means of the *trapezoidal rule*, namely

$$I = \sum_{i=1}^{n} I_i = \int_{x_0}^{x_n} y\, dx = \frac{h}{2}[y_0 + 2(y_1 + y_2 + y_3 + \dots + y_{n-1}) + y_n]$$

$$= \frac{h}{2}[X + 2I] \tag{7.39}$$

where X = sum of end ordinates

and I = sum of intermediate ordinates.

Equation (7.39) signifies that the curve $y = f(x)$ is replaced by n straight lines joining the points (x_i, y_i), $i = 0, 1, 2, 3, \dots, n$. The area bounded by the curve $y = f(x)$ the ordinates $x = x_0$, $x = x_n$ and the x-axis is then approximated equivalent to the sum of the areas of the n-trapeziums so obtained.

Now, we evaluate Eq. (7.38) by means of two different widths, h_1 and h_2, in order to obtain the approximate values I_1 and I_2, respectively. The corresponding errors E_1 and E_2 are given by

$$E_1 = \frac{(b-a)h_1^2}{12}y''(\bar{x})$$

$$E_2 = \frac{-(b-a)}{12}y''(\bar{\bar{x}}) \tag{7.40}$$

Noting that $y''(\overline{\overline{x}})$ is also the largest value of $y''(x)$, we can assume that the quantities $y''(\overline{x})$ and $y''(\overline{\overline{x}})$ are nearly equal.

Hence, we can write

$$\frac{E_1}{E_2} = \frac{h_1^2}{h_2^2} = \frac{E^2}{E_2 - E_1} = \frac{h_2^2}{h_2^2 - h_1^2} \tag{7.41}$$

Noting, now that $I = I_1 - E_1 = I_2 - E_2$, we have

$$E_2 - E_1 = I_1 - I_2 \tag{7.42}$$

From Eqs. (7.41) and (7.42), we have

$$E_2 = \frac{h_2^2}{h_2^2 - h_1^2}(E_2 - E_1) = \frac{h_2^2}{h_2^2 - h_1^2}(I_1 - I_2)$$

$$I = I_2 - E_2 = \frac{I_1 h_2^2 - I_2 h_1^2}{h_2^2 - h_1^2} \tag{7.43}$$

Equation (7.43) gives a better approximation for I.

In order to compute I, we let $h_1 = h$ and $h_2 = h/2$ such that Eq. (7.43) gives

$$I = \frac{I_1 \left(\dfrac{h^2}{4}\right) - I_2 h^2}{\dfrac{h^2}{4} - h^2} = \frac{4I_2 - I_1}{3} = I_2 + \frac{I_2 - I_1}{3}$$

or $\qquad I\left(h, \dfrac{h}{2}\right) = \dfrac{4I\left(\dfrac{h}{2}\right) - I(h)}{3}$ $\qquad\qquad$ (7.44)

If we apply the trapezoidal rule several times successively halving h, every time the error is reduced by a factor 1/4. The above computation is continued with two successive values are very close to each other. This refinement of Richardson's method is known as the *Romberg integration*. The values of the integral in the Romberg integration can be tabulated in the following scheme.

Romberg Integration Scheme

$$
\begin{array}{l}
I(h) \\
\qquad I\left(h,\dfrac{h}{2}\right) \\
I\left(\dfrac{h}{2}\right) \qquad\qquad I\left(h,\dfrac{h}{2},\dfrac{h}{4}\right) \\
\qquad I\left(\dfrac{h}{2},\dfrac{h}{4}\right) \qquad\qquad I\left(h,\dfrac{h}{2},\dfrac{h}{4},\dfrac{h}{8}\right) \\
I\left(\dfrac{h}{4}\right) \qquad\qquad I\left(\dfrac{h}{2},\dfrac{h}{4},\dfrac{h}{8}\right) \\
\qquad I\left(\dfrac{h}{4},\dfrac{h}{8}\right) \\
I\left(\dfrac{h}{8}\right)
\end{array}
$$

where
$$
I\left(h,\frac{h}{2}\right)=\frac{1}{3}\left[4I\left(\frac{h}{2}\right)-I(h)\right]
$$

$$
I\left(\frac{h}{2},\frac{h}{4}\right)=\frac{1}{3}\left[4I\left(\frac{h}{4}\right)-I\left(\frac{h}{2}\right)\right]
$$

$$
\vdots \qquad\qquad\qquad \vdots
$$

$$
I\left(h,\frac{h}{2},\frac{h}{4}\right)=\frac{1}{3}\left[4I\left(\frac{h}{2},\frac{h}{4}\right)-I\left(h,\frac{h}{2}\right)\right]
$$

$$
I\left(\frac{h}{2},\frac{h}{4},\frac{h}{8}\right)=\frac{1}{3}\left[4I\left(\frac{h}{4},\frac{h}{8}\right)-I\left(\frac{h}{2},\frac{h}{4}\right)\right]
$$

$$
I\left(h,\frac{h}{2},\frac{h}{4},\frac{h}{8}\right)=\frac{1}{3}\left[4I\left(\frac{h}{2},\frac{h}{4},\frac{h}{8}\right)-I\left(h,\frac{h}{2},\frac{h}{4}\right)\right] \tag{7.45}
$$

The computations are continued until the successive values are close to each other. The general extrapolation formula used in this scheme is

$$
R_{i,j}=\frac{4^{j-1}R_{i,j-1}-R_{i-1,j-1}}{4^{j-1}-1}, \qquad i>1, j=2,3,\ldots,I \tag{7.46}
$$

A pictorial representation of Eq. (7.46) is shown below:

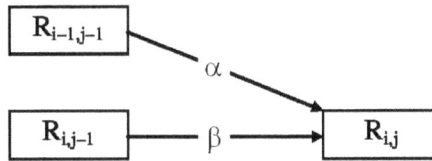

where the multipliers α and β depend on j in the following manner:

j	2	3	4	5	6
α	−1/3	−1/15	−1/63	−1/255	−1/1023
β	4/3	16/15	64/63	256/255	1024/1023

EXAMPLE 7.16

Apply Romberg's integration to find $\int_0^\pi f(x)\,dx$, where $f(x) = \sin x$.

Solution:

From the recursive trapezoidal rule in Eq. (7.9), we have

$$R_{1,1} = I(\pi) = \frac{\pi}{2}[f(0) + f(\pi)] = 0$$

$$R_{2,1} = I\left(\frac{\pi}{2}\right) = \frac{1}{2}I(\pi) + \frac{\pi}{2}f\left(\frac{\pi}{2}\right) = 1.5708$$

$$R_{3,1} = I\left(\frac{\pi}{4}\right) = \frac{1}{2}I\left(\frac{\pi}{2}\right) + \frac{\pi}{4}\left[f\left(\frac{\pi}{4}\right) + f\left(\frac{3\pi}{4}\right)\right] = 1.8961$$

$$R_{4,1} = I\left(\frac{\pi}{8}\right) = \frac{1}{2}I\left(\frac{\pi}{4}\right) + \frac{\pi}{8}\left[f\left(\frac{\pi}{8}\right) + f\left(\frac{3\pi}{8}\right) + f\left(\frac{5\pi}{8}\right) + f\left(\frac{7\pi}{8}\right)\right] = 1.9742$$

Using the extrapolation formula in Eq. (7.46), we obtain the following table:

$$\begin{bmatrix} R_{1,1} \\ R_{2,1} & R_{2,2} \\ R_{3,1} & R_{3,2} & R_{3,3} \\ R_{4,1} & R_{4,2} & R_{4,3} & R_{4,4} \end{bmatrix} = \begin{bmatrix} 0 \\ 1.5708 & 2.0944 \\ 1.8961 & 2.0046 & 1.9986 \\ 1.9742 & 2.0003 & 2.0000 & 2.0000 \end{bmatrix}$$

The above table shows that the procedure has converged. Hence, $\int_0^\pi \sin x\,dx = R_{4,4} = 2.0000$, which is of course, the exact result.

EXAMPLE 7.17

Apply Romberg's integration method to find $\int_0^{1.2}\left(\dfrac{1}{1+x}\right)dx$ correct to five decimal places.

Solution:

$$f(x) = \frac{1}{1+x}$$

Let $\quad h = 0.6, 0.3,$ and 0.15 or $h = 0.6,\ h/2 = 0.3$ and $h/4 = 0.15.$

x	0	0.15	0.30	0.40	0.60	0.75	0.90	1.05	1.20
$y = f(x)$	1	0.86957	0.76923	0.71429	0.62500	0.57143	0.52632	0.48780	0.45455

Applying trapezoidal rule for $h = 0.6$, we obtain

$$I(h) = I(0.6) = I_1 = \frac{0.6}{2}[(1 + 0.45455) + 2(0.6250)] = 0.81132$$

For $h = \dfrac{0.6}{2} = 0.3$, we obtain

$$I\left(\frac{h}{2}\right) = I(0.3) = I_2 = \frac{0.3}{2}[(1 + 0.45455) + 2(0.76923 + 0.6250 + 0.52632)]$$

$$= 0.79435$$

For h $= \dfrac{0.6}{4} = 0.15$, we have

$$I\left(\frac{h}{4}\right) = I(0.15) = I_3 = \frac{0.15}{2}[(1 + 0.45455) + 2(0.86957 + 0.76923 + 0.71429)$$

$$+ \frac{0.15}{2}[2(0.6250 + 0.57143 + 0.52632 + 0.48780)] = 0.78992$$

Now $I\left(h,\dfrac{h}{2}\right) = I(0.6, 0.3)$

Therefore, $I(0.6, 0.3) = \dfrac{1}{3}[4I(0.3) - I(0.6)] = \dfrac{1}{3}[4(0.79435) - 0.81132] = 0.78864$

In a similar manner, we obtain

$$I\left(\frac{h}{2}, \frac{h}{4}\right) = I(0.3, 0.15) = \frac{1}{3}[4I(0.15) - I(0.3)]$$

$$= \frac{1}{3}[4(0.78992 - 0.79435)] = 0.78846$$

Hence $\quad I\left(h, \dfrac{h}{2}, \dfrac{h}{4}\right) = I(0.6, 0.3, 0.15)$

or $\qquad I(0.6, 0.3, 0.15) = \dfrac{1}{3}[4I(0.15, 0.3) - I(0.3, 0.6)]$

$$= \frac{1}{3}[4(0.78846) - 0.78864] = 0.78832$$

The computations are summarized in the table below:

0.81132		
	0.7864	
0.79435		0.78832
	0.78846	
0.78992		

Hence $\displaystyle\int_0^{1.2} \frac{1}{1+x}\,dx = 0.78832$ correct to five decimal places.

EXAMPLE 7.18

Apply Romberg's integration method to find $\displaystyle\int_0^1 \frac{dx}{1+x^2}$, correct to four decimal places. Take $h = 0.5, 0.25.$ and $0.125.$

Solution:

Applying the trapezoidal rule, for $h = 0.25$, we obtain

x	0	0.5	1
$y = f(x) = \dfrac{1}{(1+x^2)}$	1	0.8	0.5

Hence, $\quad I = \displaystyle\int_0^1 \frac{1}{1+x^2} = \frac{0.5}{2}[1 + 2(0.8) + 0.5] = 0.775$

For $h = 0.25$, we have

x	0	0.25	0.5	0.75	1
$y = f(x) = \dfrac{1}{(1+x^2)}$	1	0.9412	0.8	0.64	0.5

Hence, $I = \int_0^1 \dfrac{dx}{1+x^2} = \dfrac{0.25}{2}[1 + 2(0.9412 + 0.8 + 0.64) + 0.5] = 0.7848$

Similarly, when $h = 0.125$, we find $I = 0.7848$.

Applying Eq. (7.46), we obtain the table as follows:

0.5	0.775		
0.25	0.7828	0.7854	
0.125	0.7848	0.7855	0.7855

Hence, $I = \int_0^1 \dfrac{dx}{1+x^2} = 0.7855$ correct to four decimal places.

7.8 SUMMARY

In this chapter, we have presented the various techniques on numerical integration. Integration methods such as the trapezoidal rule, Simpson's 1/3 rule, Simpson's 3/8 rule, and Boole's and Weddle's rules and their composite versions, and Romberg's integration were presented with illustrative examples. These methods use uniformly spaced-based points.

EXERCISES

7.1. Evaluate $\int_0^1 \cos x^2 dx$ by taking eight subintervals using the trapezoidal rule.

7.2. Use the trapezoidal rule to evaluate $\int_0^1 x^3 dx$, corresponding five subintervals.

7.3. Compute the following integral numerically using the trapezoidal rule:

$$I = \int_0^1 e^x dx$$

Use (a) $n = 1$, (b) $n = 2$, (c) $n = k$, and (d) $n = 8$. The exact value of $I = 1.7183$. Compare your computed results in each case with the exact result.

7.4. Evaluate $\int_0^1 \frac{dx}{1+x^2}$ using the trapezoidal rule. Take $h = 0.25$.

7.5. Determine the area bounded by the curve $f(x) = xe^{2x}$ and the x-axis between $x = 0$ and $x = 1$ using the trapezoidal rule with an interval size of $(a)\ h = 0.5$, $(b)\ h = 0.1$. Determine the relative error in each case given that the exact value of the integral $I = 2.09726$.

7.6. Evaluate $\int_1^5 \log_{10} x\ dx$, taking eight subintervals correct to four decimal places by the trapezoidal rule.

7.7. Evaluate $\int_1^7 \sin x^2\ dx$ by taking seven ordinates using the trapezoidal rule.

7.8. Evaluate $\int_0^\pi t \sin t\ dt$ using the trapezoidal rule.

7.9. Repeat Exercise 7.9 using Simpson's 1/3 rule.

7.10. Repeat Exercise 7.2 using Simpson's 1/3 rule taking $h = 0.25$.

7.11. Compute the integral $I = \int_0^1 e^x dx$ using Simpson's rule with $n = 8$ intervals rounding off the results to four digits.

7.12. Evaluate $\int_0^{0.6} e^x dx$, taking $n = 6$, correct to five significant figures by Simpson's 1/3 rule.

7.13. Evaluate $\int_0^{\pi/2} \sqrt{\cos x}\ dx$ by Simpson's 1/3 rule taking $n = 6$.

7.14. Evaluate $\int_4^{5.2} \log x\ dx$ by taking seven grid points and using Simpson's 1/3 rule.

7.15. Repeat Exercise 7.15 using Simpson's 1/3 rule.

7.16. Evaluate $\int_0^1 \frac{dx}{1+x^2}$ by taking six equal parts using Simpson's 1/3 rule.

7.17. Evaluate $\int_0^6 \frac{dx}{1+x^2}$ by using Simpson's 3/8 rule.

7.18. Repeat Exercise 7.24 using Simpson's 3/8 rule taking h = 1/6.

7.19. Evaluate $\int_0^1 \frac{1}{1+x^2}$, by taking seven ordinates, using Simpson's 3/8 rule.

7.20. Evaluate $\int_0^1 \sqrt{\sin x + \cos x}\, dx$ correct to two decimal places using Simpson's 3/8 rule.

7.21. Evaluate $\int_2^6 \dfrac{1}{\log e^x}\, dx$ by using Simpson's 3/8 rule.

7.22. Evaluate $\int_4^{5.2} \log x\, dx$ by taking seven grid points. Using Simpson's 3/8 rule.

7.23. Evaluate $\int_0^{\pi/2} e^{\sin x}\, dx$ correct to four decimal places using Simpson's 3/8 rule.

7.24. Repeat Exercise 7.24 using Simpson's 3/8 rule.

7.25. Evaluate the integral $\int_0^1 1 + e^{-x} \sin 4x$ using Boole's rule with h = 1/4.

7.26. Repeat Exercise 7.25 using Boole's rule.

7.27. Repeat Exercise 7.22 using Weddle's rule taking h = 1/6.

7.28. Repeat Exercise 7.25 using Weddle's rule.

7.29. Evaluate $\int_4^{5.2} \log_e x\, dx$ using Weddle's rule. Take $n = 6$.

7.30. Evaluate $\int_4^{5.2} \log x\, dx$ by taking seven grid points. Use Boole's and Weddle's rules.

7.31. Evaluate $\int_0^{1/2} \dfrac{dx}{\sqrt{1-x^2}}$ using Weddle's rule.

7.32. Evaluate $\int_0^2 \dfrac{1}{1+x^2}\, dx$ by using Weddle's rule taking twelve intervals.

7.33. Use Romberg's integration method to evaluate $\int_4^{5.2} \log x\, dx$, given that

x	4	4.2	4.4	4.6	4.8	5.0	5.2
$\log_e^2$	1.3863	1.4351	1.4816	1.5260	1.5686	1.6094	1.4684

7.34. Use Romberg's integration method to compute $\int_0^1 \dfrac{1}{1+x}$ with $h = 0.5$, 0.25, and 0.125. Hence, find $\log_e^2$ correct to four decimal places.

7.35. Approximate the integral $f(x) = \int_0^1 xe^{-x}dx$ using Romberg's integration with accuracy of $n = 8$ intervals. Round off results to six digits.

7.36. Use Romberg's integration to evaluate $\int_0^{\sqrt{\pi}} 2x^2 \cos x^2 dx$.

7.37. Evaluate $\int_0^2 (x^5 + 3x^3 - 2)dx$ by Romberg's integration.

7.38. Estimate $\int_0^\pi f(x)dx$ as accurately as possible, where $f(x)$ is defined by the data:

x	0	$\pi/4$	$\pi/2$	$3\pi/4$	π
$f(x)$	1	0.3431	0.25	0.3431	1

7.39. Use Romberg's integration method to compute $R_{3,3}$ for the following integrals:

(a) $\int_0^1 x^2 e^{-x}dx$ **(b)** $\int_1^{1.5} x^2 \ln x\, dx$

(c) $\int_0^{\pi/4} (\cos x)^2\, dx$ **(d)** $\int_0^{\pi/4} e^{3x} \sin 2x\, dx$

7.40. Use Romberg's integration method to find $R_{3,3}$ for the integral $\int_0^{\pi/4} x^2 \sin x\, dx$.

7.41. Apply Romberg's integration method to find $\int_1^5 f(x)\,dx$ for the following data:

x	1	2	3	4	5
$y = f(x)$	2.4142	2.6734	2.8974	3.0976	3.2804

7.42. Apply Romberg's integration method to find $\int_0^1 x^{1/3}dx$.

NUMERICAL SOLUTION OF ORDINARY DIFFERENTIAL EQUATIONS

8.1 INTRODUCTION

Numerical methods are becoming more and more important in engineering applications, simply because of the difficulties encountered in finding exact analytical solutions but also, because of the ease with which numerical techniques can be used in conjunction with modern high-speed digital computers. Several numerical procedures for solving initial value problems involving first-order ordinary differential equations are discussed in this chapter.

In spite of the fact that the error analysis is an important part of any numerical procedure, the discussion in this chapter is limited primarily to the use of the procedure itself. The theory of errors and error analysis is sometimes fairly complex and goes beyond the intended scope of this chapter.

An *ordinary differential equation* is one in which an ordinary derivative of a dependent variable y with respect to an independent variable x is related in a prescribed manner to x, y, and lower derivatives. The most general form of an ordinary differential equation of nth order is given by

$$\frac{d^n y}{dx^n} = f\left(x, y, \frac{dy}{dx}, \frac{d^2 y}{dx^2}, \ldots, \frac{d^{n-1} y}{dx^{n-1}}\right) \tag{8.1}$$

The Eq. (8.1) is termed *ordinary* because there is only one independent variable.

To solve an equation of the type (Eq. [8.1]), we also require a set of conditions. When all the conditions are given at one value x and the solution proceeds from that value of x, we have an *initial-value problem*. When the conditions are given at different values of x, we have a *boundary-value problem*.

A general solution of an ordinary differential equation (Eq. [8.1]) would be a relation between y, x, and n arbitrary constants which is of form

$$f(x, y, c_1, c_2, ..., c_n) = 0 \qquad (8.2)$$

If particular values are given to the constants c_n in Eq. (8.2), then the resulting solution is called a *particular solution*. There are many analytical methods available for finding the solution of the Eq. (8.1). However, there exist a large number of ordinary differential equations in science and engineering, whose solutions cannot easily be obtained by the well-known analytical methods. For such ordinary differential equations, we can obtain an approximate solution of a given ordinary differential equations using numerical methods under the given initial conditions.

Any ordinary differential equation can be replaced by a system of first-order differential equations (which involve only first derivatives). The single first-order ordinary differential equation with an initial value is a special case of Eq. (8.1). It is described by

$$\frac{dy}{dx} = f(x,y) \qquad y = y_0 \text{ at } x = x_0 \qquad (8.3)$$

The description in Eq. (8.3) consists of the differential equation itself and a given solution y_0 at initial location x_0. We then obtain the solution y as x ranges from its initial value to some other value.

The general solution of Eq. (8.3) can be obtained in two forms:

1. the values of y as a power series in independent variable x
2. as a set of tabulated values of x and y.

There are two categories of methods to solve ordinary differential equations:

1. one-step methods or single-step methods
2. step-by-step methods or marching methods

In one-step methods or single-step methods, the information about the curve represented by an ordinary differential equation at one point is utilized and the solution is not iterated. In step-by-step methods or the marching methods, the next point on the curve is evaluated in short steps ahead, for equal intervals of width h of the independent variable, by performing iterations till the desired level of accuracy is obtained.

In general, we divide the interval (a, b) on which the solution is derived into a finite number of subintervals by the points $a = x_0 < x_1 < x_2, \ldots < x_n = b$, called the *mesh points*. This is done by setting up $x_n = x_0 + nh$.

The existence of the uniqueness of the solution to an initial value problem in (x_0, b) is based on Lipschitz theorem. Lipschitz theorem states that:

a) If $f(x, y)$ is a real function defined and continuous in (x_0, b), $y \in (-\infty, +\infty)$, where x_0 and b are finite.

b) There exists a constant $k > 0$ called *Lipschitz constant*, such that for any two values $y = y_1$ and $y = y_2$

$$|f(x, y_1) - (f(x, y_2)| < k|k_1 - k_2|$$

where $x \in (x_0, b)$, then for any $y(x_0) = y_0$, the initial value problem (Eq. [8.3]), has unique solution for $x \in (x_0, b)$.

Also, there are two types of methods, *explicit* and *implicit*, can be used to compute the solution at each step. *Explicit* methods are those methods that use an *explicit* formula for calculating the value of the dependent variable at the next value of the independent variable. In an explicit method, the right-hand side of the equation only has all known quantities. Therefore, the next unknown value of the dependent variable, y_{n+1}, is calculated by evaluating an expression of the form:

$$y_{n+1} = F(x_n, x_{n+1}, y_n) \tag{8.4}$$

where x_n, y_n and x_{n+1} are all known quantities.

In *implicit* methods, the equation used for computing y_{n+1} from the known x_n, y_n and y_{n+1} has the form:

$$y_{n+1} = F(x_n, x_{n+1}, y_{n+1}) \tag{8.5}$$

Here, the unknown y_{n+1} appears on both sides of the equation. Generally speaking, the right-hand side of Eq. (8.3c) is nonlinear. Therefore, the

Eq. (8.5) must be solved for y_{n+1} using suitable numerical methods. In general, implicit methods give better accuracy over explicit methods at the expense of additional effort.

In this chapter, we present among the one-step or single-step methods, Picard's method of successive approximations, Taylor series methods were presented. Euler's method, modified Euler's method, and Runge-Kutta methods of order two and four, the Adam-Moulton predictor-corrector method and Milne's predictorcorrector methods were presented among the step-by-step methods or the marching methods. All these methods will be illustrated with worked examples.

8.2 ONE-STEP METHODS OR SINGLE-STEP METHODS

In the single-step explicit method, the approximate solution (x_{n+1}, y_{n+1}) is computed from the known solution at point (x_n, y_n) using

$$x_{n+1} = x_n + h \tag{8.6}$$

$$y_{n+1} = y_n + (\text{slope})\, h \tag{8.7}$$

This is illustrated in Figure 8.1. Here in Eq. (8.6), h is the step size and the *slope* is a constant that estimates the value of $\dfrac{dy}{dx}$ in the interval from x_n to x_{n+1}.

The numerical solution starts at the point where the initial value is known corresponding to $n = 1$ and point (x_1, y_1). Then, n is increased to $n = 2$, and the solution at the next point, (x_2, y_2) is computed using Eqs. (8.6) and (8.7). This procedure is repeated for $n = 3$ and so on until the points cover the whole domain of the solution.

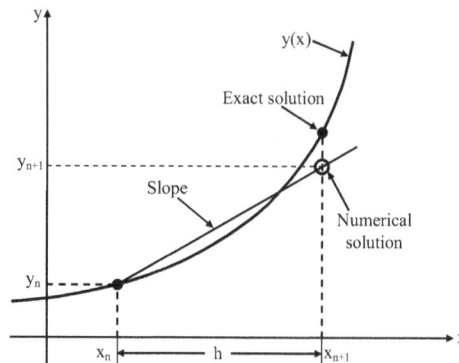

FIGURE 8.1 Single-step explicit methods.

8.2.1 Picard's Method of Successive Approximation

Consider the differential equation given by Eq. (8.3), namely, $\dfrac{dy}{dx} = f(x, y)$ with the initial condition $y(x_0) = y_0$.

Integrating this Eq. (8.3) between x_0 and x, we obtain

$$\int_{x_0}^{x} dy = \int_{x_0}^{x} f(x, y) \, dx$$

or $\quad\quad y - y_0 = \int_{x_0}^{x} f(x, y) \, dx$

or $\quad\quad y = y_0 + \int_{x_0}^{x} f(x, y) \, dx \quad\quad\quad\quad\quad\quad\quad\quad (8.8)$

Equation (8.8) is called the *integral equation* since the dependent variable y in the function $f(x, y)$ on the right-hand side occurs under the sign of integration.

Here, the problem of solving the differential equation (8.3) has been reduced to solving the integral Eq. (8.8). The first approximation y_1 of y can be obtained by replacing y by y_0 in $f(x_0, y_0)$ in Eq. (8.8).

Therefore $\quad\quad y_1 = y_0 + \int_{x_0}^{x} f(x, y_0) \, dx \quad\quad\quad\quad\quad\quad (8.9)$

Similarly, the second approximation is obtained as

$$y_2 = y_0 + \int_{x_0}^{x} f(x, y_1) \, dx \quad\quad\quad\quad\quad\quad (8.10)$$

Likewise, $\quad\quad y_3 = y_0 + \int_{x_0}^{x} f(x, y_2) \, dx$

$$y_4 = y_0 + \int_{x_0}^{x} f(x, y_3) \, dx \quad\quad\quad\quad\quad\quad (8.11)$$

and so on.

Continuing this process, we obtain $y_5, y_6, \ldots, y_{n+1}$, or

$$y_{n+1} = y_0 + \int_{x_0}^{x} f(x, y_{n-1}) \, dx \quad\quad\quad\quad\quad\quad (8.12)$$

The sequence of $\{y_n\}$, $n = 1, 2, 3, \ldots$ converges to the exact solution provided that the function $f(x, y)$ is bounded in some region in the neighborhood of (x_0, y_0) and satisfies the Lipschitz condition. That is, there exists a constant k such that $| f(x, y) - f(x, \bar{y}) | \le k | y - \bar{y} |$, for all x. The process of iteration is concluded when the values of y_{n-1} and y_n are approximately equal.

EXAMPLE 8.1

Use Picard's method of successive approximation to find the value of y when $x = 0.1$, given that $y = 1$ when $x = 0$ and $\dfrac{dy}{dx} = 3x + y^2$.

Solution:

Here $\dfrac{dy}{dx} = f(x, y) = 3x + y^2$, $x_0 = 0$, $y_0 = 1$.

From Eq. (8.9)

$$y_1 = y_0 + \int_{x_0}^{x} f(x, y_0) dx = y_0 + \int_0^x (3x + y_0^2) dx = 1 + \int_0^x (3x + 1) dx = \frac{3}{2}x^2 + x + 1 \tag{E.1}$$

From Eq. (8.10)

$$y_2 = y_0 + \int_{x_0}^{x} f(x, y_1) dx = y_0 + \int_0^x (3x + y_1^2) dx$$

$$= 1 + \int_0^x \left(\frac{9}{4}x^4 + 3x^3 + 4x^2 + 5x + 1 \right) dx$$

$$= \frac{9}{20}x^5 + \frac{3}{4}x^4 + \frac{4}{3}x^3 + \frac{5}{2}x^2 + x + 1 \tag{E.2}$$

From Eq. (8.11)

$$y_3 = y_0 + \int_{x_0}^{x} f(x, y_2) dx = 1 + \int_0^x \left[\frac{81}{400}x^{10} + \frac{27}{40}x^9 + \frac{141}{80}x^8 + \frac{17}{4}x^7 + \frac{1157}{180}x^6 \right.$$

$$\left. + \frac{136}{15}x^5 + \frac{125}{12}x^4 + \frac{23}{3}x^3 + 6x^2 + 5x + 1 \right] dx$$

$$= \frac{81}{4400}x^{11} + \frac{27}{400}x^{10} + \frac{47}{240}x^9 + \frac{17}{32}x^8 + \frac{1157}{1260}x^7$$

$$+ \frac{68}{45}x^6 + \frac{25}{12}x^5 + \frac{23}{12}x^4 + 2x^3 + \frac{5}{2}x^2 + x + 1 \tag{E.3}$$

When $x = 0.1$, Eqs. (E.1), (E.2), and (E.3) respectively give

$$y_0 = 1$$

$$y_1 = 1.1150$$

$$y_3 = 1.1272$$

EXAMPLE 8.2

Use Picard's method of successive approximation to find the value of y for the following:

a) $\dfrac{dy}{dt} = 2y, y(0) = 1$

b) $\dfrac{dy}{dx} = 2x - y, y(0) = 1.$

Solution:

a) The stated initial value problem is equivalent to the integral equation

$$y(x) = 1 + \int_0^x 2y(t)dt$$

Hence $\qquad y_{j+1}(x) = 1 + \int_0^x 2y_j(t)dt$

Using $y_0(x) = 1$, we find

$$y_1(x) = 1 + \int_0^x 2dt = 1 + 2x$$

$$y_2(x) = 1 + \int_0^x 2(1 + 2t)dt = 1 + 2x + 2x^2$$

$$y_3(x) = 1 + \int_0^x 2\left(1 + 2t + 2t^2\right)dt = 1 + 2x + 2x^2 + \frac{4x^3}{3}$$

In general, we have

$$y_j(x) = 1 + 2x + 2x^2 + \frac{4x^3}{3} + \dots + \frac{(2x)^j}{j!} = \sum_{\ell=0}^{j} \frac{2(x)^\ell}{\ell!}$$

These are the partial sums for the power series expansion of $y = e^2 x$. Hence, the solution of our initial value problem is $y = e^{2x}$.

b) The equivalent integral equation is

$$y(x) = 1 + \int_0^x [2t - y(t)]dt$$

Hence, $\qquad y_{j+1}(x) = 1 + \int_0^x \left[2t - y_j(t)\right]dt$

Taking $y_0(x) = 1$, we have

$$y_1(x) = 1 + \int_0^x (2t - 1)dt = 1 + x^2 - x$$

$$y_2(x) = 1 + \int_0^x \left(2t - \left[1 + t^2 - t \right] \right) dt = 1 + \frac{3x^2}{2} - x - \frac{x^3}{3}$$

$$y_3(x) = 1 + \int_0^x \left(2t - \left[1 + 3t^2/2 - t - t^3/3 \right] \right) dt = 1 + \frac{3x^2}{2} - x - \frac{x^3}{2} + \frac{x^4}{4.3}$$

$$y_4(x) = 1 + \int_0^x \left(2t - \left[1 + 3t^2/2 - t - t^3/2 + t^4/4.3 \right] \right) dt$$

$$= 1 + \frac{3x^2}{2} - x - \frac{x^3}{2} + \frac{x^4}{4.2} - \frac{x^5}{5.4.3}$$

Therefore $y_j(x) = 1 + x + \dfrac{3x^2}{2!} - \dfrac{3x^3}{3!} + \dfrac{4x^4}{4!} - \ldots + (-1)^j \dfrac{3x^j}{j!} + (-1)^{j+1} \dfrac{2x^{j+1}}{(j+1)!}$

$$= [2x - 2] + 3 \left[\sum_{\ell=0}^{j} (-1)^\ell \frac{x^\ell}{\ell!} \right] + (-1)^{j+1} \frac{2x^{j+1}}{(j+1)!}$$

$$= [2x - 2] + 3 \left[\sum_{\ell=0}^{j} \frac{(-x)^\ell}{\ell!} \right] + (-1)^{j+1} \frac{2x^{j+1}}{(j+1)!}$$

The iterates $y_j(x)$ converge to the solution $y(x) = [2x - 2] + 3e^{-x}$ for the initial value problem.

8.2.2 Taylor's Series Method

Consider the differential equation

$$\frac{dy}{dx} = f(x,y) \quad \text{with } y(x_0) = y_0 \tag{8.13}$$

Let $y = y(x)$ be a continuously differentiable function satisfying the Eq. (8.13). Expanding y in terms of the Taylor's series around the point $x = x_0$, we obtain

$$y = y_0 + \frac{(x - x_0)}{1!} y_0' + \frac{(x - x_0)^2}{2!} y_0'' + \frac{(x - x_0)^3}{3!} y_0''' + \ldots. \tag{8.14}$$

Now, substituting $x = x_1 = x_0 = h$, in Eq. (8.14), we get

$$f(x_1) = y_1 = y_0 + \frac{h}{1!} y_0' + \frac{h^2}{2!} y_0'' + \frac{h^3}{3!} y_0''' + \dots \tag{8.15}$$

Finally, we obtain

$$y_{n+1} = y_n + \frac{h}{1!} y_n' + \frac{h^2}{2!} y_n'' + \frac{h^3}{3!} y_n''' + \dots \tag{8.16}$$

Equation (8.16) can be written as

$$y_{n+1} = y_n + \frac{h}{1!} y_n' + \frac{h^2}{2!} y_n'' + O(h^3) \tag{8.17}$$

where $O(h^3)$ represents all the terms containing the third and higher power of h. The local truncation error in the solution is kh^3 where k is a constant when the terms containing the third and higher powers of h are ignored. It should be noted here that the Taylor's series method is applicable only when the derivatives of $f(x, y)$ exist and the value of $(x - x_0)$ in the expansion of $y = f(x)$ near x_0 must be very small so that the series converges. Taylor's series method is a single-step method and works well as long as the successive derivatives can be calculated easily.

The truncation error, due to the terms neglected in the series is given by

$$E = \frac{1}{(n+1)!} y^{(n+1)}(\xi) h^{n+1} \quad x < \xi < x + h \tag{8.17a}$$

Using the finite difference approximation

$$y_{n+1}(\xi) = \frac{y^n(x+h) - y^n(x)}{h} \tag{8.17b}$$

or $\quad E = \frac{h^n}{(n+1)!} [y^n(x+h) - y^n(x)] \tag{8.17c}$

Equation (8.17c) is in more usable form and could be incorporated in the algorithm to monitor the error in each integration step.

If the series in Eq. (8.17) is truncated after the term h^k, then the truncation error can be written as

$$T_e = \frac{h^{k+1}}{(k+1)!} f^{(k+1)}(p) \quad x_k < p < x_k + h \tag{8.17d}$$

EXAMPLE 8.3

Use the second-order Taylor series method on $(2, 3)$ for the initial value problem $\dfrac{dy}{dx} = -xy^2$, $y(2) = 1$. Take $h = 0.1$. Compare the results obtained with the exact solution of $y = \dfrac{2}{x^2 - 2}$.

Solution:

For $f(x, y) = -xy^2$, the first partial derivatives are $f_x = -y^2$ and $f_y = -2xy$.

Hence, the second-order Taylor's series method (Eq. [8.17]) becomes

$$y_{n+1} = y_n + h\left\{-x_n y_n^2 + \frac{h}{2}\left[-y_n^2 + (-2x_n y_n)(x_n y_n^2)\right]\right\}$$

$$= y_n + hy_n^2\left\{-x_n + \frac{h}{2}[-1 + 2x_n^2 y_n]\right\}$$

Taking $h = 0.1$ and starting with $x_0 = 2$, $y_0 = 1$, we get

$n = 0$:
$$y(x_1) = y(2.1) = y_1 = y_0 + hx_0^2\left\{-t_0 + \frac{h}{2}[-1 + 2x_0^2 y_0]\right\}$$

$$y(x_1) = 1 + 0.1(1)^2\{-2 + 0.05[-1 + 2(2)^2 1]\} = 0.8350$$

$n = 1$:
$$y(x_2) = y(2.2) = y_2 = y_1 + hx_1^2\left\{-x_1 + \frac{h}{2}[-1 + 2x_1^2 y_1]\right\}$$

$$y(x_2) = 0.8350 + 0.1(0.8350)^2\{-2.1 + 0.05[-1 + 2(2.1)^2(0.8350)]\} = 0.71077$$

The resulting approximations of $y(2.0)$, $y(2.1)$,...., $y(3.0)$ are shown in Table 8.1 along with the exact values and the relative error, E_n.

TABLE 8.1 Second-order Taylor's series method for $\dfrac{dy}{dx} = -xy^2$, $y(2) = 1$.

x_n	Exact $y(x_n)$	Using $h = 0.1$	
		$y_n[0.1]$	$E_n[0.1]$
$x_0 = 2.0$	1	1	0
2.1	0.8299	0.835	−0.0051
2.2	0.7042	0.7108	−0.0065
2.3	0.6079	0.6145	−0.0066
2.4	0.5319	0.5380	−0.0061

x_n	Exact $y(x_n)$	Using $h = 0.1$	
		$y_n[0.1]$	$E_n[0.1]$
2.5	0.4706	0.4761	−0.0055
2.6	0.4202	0.4250	−0.0049
2.7	0.3781	0.3823	−0.0043
2.8	0.3425	0.3462	−0.0037
2.9	0.3120	0.3153	−0.0033
$x_F = 3.0$	0.2857	0.2886	−0.0029

EXAMPLE 8.4

Use the Taylor series method to solve the equation $\dfrac{dy}{dx} = 3x + y^2$ to approximate y when $x = 0.1$, given that $y = 1$ when $x = 0$.

Solution:

Here $(x_0, y_0) = (0, 1)$ and $y^1 = \dfrac{dy}{dx} = 3x + h^2$

From Eq. (8.17)

$$y_{n+1} = y_n + \frac{h}{1!}y_n^i + \frac{h^2}{2!}y_n^{ii} + \frac{h^3}{3!}y_n^{iii} + \frac{h^4}{4!}y_n^{iv} + \dots$$

$y^i = 3x + y^2$ $\qquad$ y^i at $(x_0) = y^i$ at $(0) = 1$

$y^{ii} = 3 + 2yy^i$ $\qquad$ y^{ii} at $x_0 = 3 + 2(1)(1) = 5$

$y^{iii} = 2(y^i)^2 + 2yy^{ii}$ $\qquad$ y^{iii} at $x_0 = 2(1)^2 + 2(1)(5) = 12$

$y^{iv} = 6y^iy^{ii} + 2yy^{iii}$ $\qquad$ y^{iv} at $x_0 = 6(1)(5) + 2(1)(12) = 54$

Hence, the required Taylor series in Eq. (8.17) becomes

$$y = 1 + x + \frac{5}{2!}x^2 + \frac{12}{3!}x^3 + \frac{54}{4!}x^4 + \dots = 1 + x + \frac{5}{2}x^2 + 2x^3 + \frac{9}{4}x^4 + \dots$$

When $x = 0.1$, we have

$$y = 1 + 0.1 + \frac{5}{2}(0.1)^2 + 2(0.1)^3 + \frac{9}{4}(0.1)^4 + \dots$$

$$= 1 + 0.1 + 0.025 + 0.002 + 0.00022 + \dots = 1.12722$$

EXAMPLE 8.5

Use the fourth-order Taylor series method with a single integration step to determine $y(0.2)$. Given that

$$\frac{dy}{dx} + 4y = x^2, \quad y(0) = 1$$

The analytical solution of the differential equation is

$$y = \frac{31}{32}e^{-4x} + \frac{1}{4} + x^2 - \frac{1}{8}x + \frac{1}{32}$$

Compute also the estimated error and compare it with the actual error.

Solution:

The Taylor series solution up to and including the term with h^4 is given by

$$y_{n+1} = y_n + \frac{h}{1!}y_n^i + \frac{h^2}{2!}y_n^{ii} + \frac{h^3}{3!}y_n^{iii} + \frac{h^4}{4!}y_n^{iv} \qquad \text{(E.1)}$$

or $\quad y(h) = y(0) + hy^i(0) + \dfrac{h^2}{2!}y^{ii}(0) + \dfrac{h^3}{3!}y^{iii}(0) + \dfrac{h^4}{4!}y^{iv}(0)$

The given differential equation is

$$\frac{dy}{dx} + 4y = x^2$$

or $\quad y^i = -4y + x^2$

Differentiating the above equation gives

$$y^{ii} = -4y^i + 2x = 16y - 4x^2 + 2x$$
$$y^{iii} = 16y^i - 8x + 2 = -64y + 16x^2 - 8x + 2$$
$$y^{iv} = -64y^i + 32x - 8 = 256y - 64x^2 + 32x - 8$$

Hence, $\quad y^i(0) = -4(1) = -4$
$$y^{ii}(0) = 16(1) = 16$$
$$y^{iii}(0) = -64(1) + 2 = -62$$
$$y^{iv}(0) = 256(1) - 8 = 248$$

For $h = 0.2$, Eq. (E.1) becomes

$$y^i(0.2) = 1 + (-4)(0.2) + \frac{1}{2!}(16)(0.2)^2 + \frac{1}{3!}(-62)(0.2)^3 + \frac{1}{4!}(248)(0.2)^4 = 0.4539$$

According to Eq. (8.17c), the approximate truncation error is given by

$$E = \frac{h^n}{(n+1)!}[y_n(x+h) - y_n(x)]$$

or $\quad E = \dfrac{h^4}{(n+1)!}[y^n(x+h) - y^n(x)] \quad$ for $n = 4$

$$= \frac{h^4}{5!}[y^{(4)}(0.2) - y^{(4)}(0)]$$

where $\quad y^{(4)}(0) = 248$

$\qquad y^{(4)}(0.2) = 256(0.4539) - 64(0.2)^2 + 32(0.2) - 8 = 112.04$

Hence, $\quad E = \dfrac{(0.2)^4}{5!}[112.04 - 248] = -0.0018$

The analytical solution gives

$$y(0.2) = \frac{31}{32}e^{-4(0.2)} + \frac{1}{4}(0.2)^2 - \frac{1}{8}(0.2) + \frac{1}{32} = 0.4515$$

Hence, the actual error is $0.4515 - 0.4539 = -0.0024$.

8.3 STEP-BY-STEP METHODS OR MARCHING METHODS

In explicit multistep methods, the solution y_{n+1}, at the next point is calculated from an explicit formula. For instance, if three prior points are used, the next unknown value of the dependent variable, y_{n+1}, is computed by evaluating an expression of the form:

$$y_{n+1} = F(x_{n-2}, y_{n-2}, x_{n-1}, y_{n-1}, x_n, y_n, x_{n+1}) \qquad (8.18)$$

Equation (8.18) is of explicit form since the right-hand side of the equation has only all known quantities. In implicit multistep methods, the unknown y_{n+1} appears on both sides of the equation, which needs to be solved using numerical methods.

8.3.1 Euler's Method

Euler's method (also called the *forward Euler method*) is a single-step, explicit method for solving a first-order ordinary differential equation. The method uses Eqs. (8.6) and (8.7), where the value of the *slope* in Eq. (8.7) is the slope of $y(x)$ at point (x_n, y_n). This slope is computed from the differential equation:

$$slope = \left.\frac{dy}{dx}\right|_{x=x_n} = f(x_n, y_n) \qquad (8.19)$$

Euler's explicit method is illustrated schematically in Figure 8.2. Euler's method assumes that for a short distance h near (x_n, y_n), the function $y(x)$ has a constant slope equal to the slope at (x_n, y_n). Based on this assumption, the next point of the numerical solution (x_{n+1}, y_{n+1}) is obtained by:

$$x_{n+1} = x_n + h \tag{8.20}$$

$$y_{n+1} = y_n + f(x_n, y_n)h \tag{8.21}$$

The error in this method depends on the value of h and is smaller for smaller h.

Equation (8.21) can be derived in several ways.

Consider the differential equation

$$\frac{dy}{dx} = f(x,y) \tag{8.22}$$

with the initial condition $y(x_0) = y_0$.

Integrating Eq. (8.22), we obtain

$$y = y_0 + \int_{x_0}^{x} f(x,y)\, dx \tag{8.23}$$

Suppose we want to obtain an approximate value of y say y_n when $x = x_n$. We divide the interval $[x_0, x_n]$ into n subintervals of equal length, say, h, with the division point $x_0, x_1, x_2, \ldots, x_n$, where $x = x_r = x_0 = rh$, $r = 1, 2, 3, \ldots$.

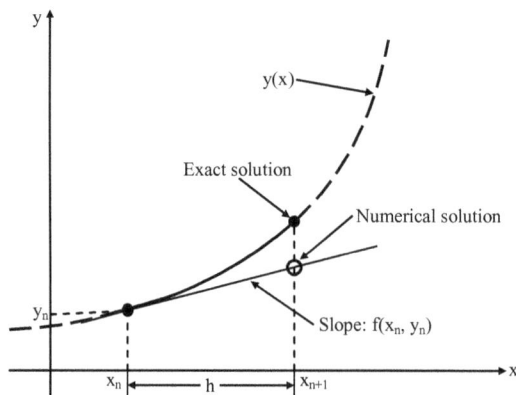

FIGURE 8.2 Euler's explicit method.

Then, from Eq. (8.23), we have

$$y_1 = y_0 + \int_{x_0}^{x_1} f(x,y)\,dx \qquad (8.24)$$

Assuming that $f(x, y) = f(x_0, y_0)$ in $x_0 \le x \le x_1$, the Eq. (8.24) leads to

$$y_1 = y_0 + h f(x_0, y_0) \qquad (8.25)$$

Equation (8.25) is called the *Euler's formula*.

Similarly, for the range $x_1 \le x \le x_2$, we have

$$y_2 = y_1 + \int_{x_1}^{x_2} f(x,y)\,dx = y_1 + h f(x_1, y_1)$$

and for the range $x_2 \le x \le x_3$, we get

$$y_3 = y_2 + h f(x_2, y_2)$$

and so on.

Finally, we obtain

$$y_{n+1} = y_n + h f(x_n, y_n), \qquad n = 0, 1, 2, 3, \dots. \qquad (8.26)$$

Euler's method accumulates large error as the process proceeds. The process is known to be very slow and in order to obtain reasonable accuracy, the value of h needs to be smaller. It can be shown that the error in Euler's method is $O(h)$, i.e., the error tends to zero as $h \to 0$, for $x = x_n$ fixed. The local truncation error of Euler's explicit method is $O(h^2)$. The global truncation error $O(h)$. The total numerical error is the sum of the global truncation error and the round-off error. The truncation error can be reduced by using smaller h (step size). However, if h becomes too small such that round-off errors become significant, the total error might increase.

EXAMPLE 8.6

Use Euler's method to solve the following differential equation

$$\frac{dy}{dx} = -t\, y^2,\ y(2) = 1 \text{ and } 2 < x < 3 \text{ with } h = 0.1.$$

Compare the results with exact solution from $y = \dfrac{2}{x^2 - 2}$.

Solution:

Euler's formula given by Eq. (8.26) is

$$y_{n+1} = y_n + hf(x_n, y_n) \qquad n = 0, 1, 2, 3, \ldots.$$

or $\qquad y_{n+1} = y_n + h\left[-t_n \, y_n^2\right] \approx y(x_{n+1}), \, x_{n+1} = 2 + (n+1)h$

Starting with $x_0 = 2$ and $y_0 = 1$ and taking $h = 0.1$, we get

$$n = 0: \quad y_1 = y_0 - h\left[x_0 \, y_0^2\right] = 1 - 0.1[2(1)^2] = 0.8 \approx y(2.1)$$

$$n = 1: \quad y_2 = y_1 - h\left[x_1 \, y_1^2\right] = 0.8 - 0.1[2.1(0.8)^2] = 0.6656 \approx y(2.2)$$

$$n = 2: \quad y_3 = y_2 - h\left[x_2 \, y_2^2\right] = 0.6656 - 0.1[2.2(0.6656)^2] = 0.5681 \approx y(2.3)$$

$$n = 3: \quad y_4 = y_3 - h\left[x_3 \, y_3^2\right] = 0.5681 - 0.1[2.3(0.5681)^2] = 0.4939 \approx y(2.4)$$

The results are tabulated for $x_n = 2, 2.1, 2.2, \ldots, 3$ in the $h = 0.1$ in Table 8.2. The exact values of $y(x_n)$ were obtained from the solution of $y(n)$ were also shown in the table. That is, $y(x_n) = \dfrac{2}{x_n^2 - 2}$.

TABLE 8.2 Euler's method values for $\dfrac{dy}{dx} = -xy^2, y(2) = 1$.

x_n	Exact $y(x_n)$	Using $h = 0.1$	
		$y_n[0.1]$	$E_n[0.1]$
$x_0 = 2.0$	1	1	0
2.1	0.8299	0.8000	0.0299
2.2	0.7042	0.6656	0.0386
2.3	0.6079	0.5681	0.0398
2.4	0.5319	0.4939	0.0380
2.5	0.4706	0.4354	0.0352
2.6	0.4202	0.3880	0.0322
2.7	0.3781	0.3488	0.0292
2.8	0.3425	0.3160	0.0265
2.9	0.3120	0.2880	0.0240
$x_F = 3.0$	0.2857	0.2640	0.0217

In the above table the error, $E_n = y(x_n) - y_n$.

EXAMPLE 8.7

Apply Euler's method to approximate the solution of the initial value problem $\frac{dy}{dt} = -2t\,y^2$ with $y(0) = 1$ in the interval $0 \le t \le 0.5$, using $h = 0.1$ and compute the error and the percentage error. The exact solution is $y = \frac{1}{(t^2 + 1)}$.

Solution:

Here, Eq. (8.26) becomes

$$y_{n+1} = y_n + h\,f(x_n, y_n)$$

Since $h = 0.1$ and $f(x_n, y_n) = -2t_n y_n^2$, we have

$$y_{n+1} = y_n - 2h\,t_n\,y_n^2 \quad n = 0, 1, 2, \ldots .$$

For $h = 0.1$, we set $n = 0$ and compute

$$n = 0: \quad y_1 = y_0 - 2(0.1)\,t_0 y_0^2 = 1 - 2(0.1)(0)(1)^2 = 1$$

which will be our estimate to the value $y(0.1)$. Continuing, we obtain

$$n = 1: \quad y_2 = y_1 - 2(0.1)\,t_1 y_1^2 = 1 - 2(0.1)(0.1)(1)^2 = 0.98$$

$$n = 2: \quad y_3 = y_2 - 2(0.1)\,t_2 y_2^2 = 0.98 - 2(0.1)(0.2)(0.98)^2 = 0.9416$$

and so on.

The exact value is given by $y = \frac{1}{t^2 + 1}$.

Table 8.3 gives a summary of the results computed for $0 \le t \le 0.5$.

$$\text{Error} = \text{Exact value} - y_n \text{ (from Euler's method)}$$

$$\text{Percentage error} = \frac{|\text{error}|}{\text{exact value}} \times 100$$

From Table 8.3, we note that Euler's method works quite well and the errors are acceptable in many practical applications.

TABLE 8.3

n	t	y_n	Exact value	Error	Percentage error
0	0	1	1	0	0
1	0.1	1	0.9901	0.0099	0.9998
2	0.2	0.98	0.9615	0.0185	1.9241
3	0.3	0.9416	0.9174	0.0242	2.6379
4	0.4	0.8884	0.8621	0.0263	3.0507
5	0.5	0.8253	0.8	0.0253	3.1625

EXAMPLE 8.8

Apply the Euler's method to the ordinary differential equation $\dfrac{dy}{dx} = x + y$, $y(0) = 1$ using increments of size $h = 0.2$. The exact solution is $y = -1 - x + 2e^x$. Determine the error and the percentage error at each step.

Solution:

$$\frac{dy}{dx} = x + y$$

when $x = 0$ and $y(0) = 1$.

Hence $\dfrac{dy}{dx} = x + y = 0 + 1 = 1$ or $y_0 = 1$.

Now, $h = 0.2$ and $y_1 = y_0 + hf(x_n, y_n)$

or $y_1 = y_0 + hf(x_0, y_0) = 1 + 0.2(1.0) = 1.2$

The process is continued as shown in Table 8.4.

Exact value at $x_n = 0.2$ is

$$y_{0.2} = -1 - 0.2 + 2e^{0.2} = 1.2428$$

Table 8.4 gives the summary of the computations.

Error = Exact value – value from Euler's method.

$$\text{Percentage error} = \frac{|\text{error}|}{\text{exact value}} \times 100$$

TABLE 8.4

n	t	y_n	Exact value	Error	Percentage error
0	0	1	1	0	0
1	0.2	1.2	1.2428	0.0428	3.4438
2	0.4	1.48	1.5836	0.1036	6.5421
3	0.6	1.856	2.0442	0.1882	9.2065
4	0.8	2.3472	2.6511	0.3039	11.4632
5	1.0	2.9766	3.4366	0.46	13.3853

EXAMPLE 8.9

Use Euler's method to solve the initial value problem $\dfrac{dy}{dt} = 1 - t + 4y, y(0) = 1$, in the interval $0 \le t \le 0.5$ with $h = 0.1$. The exact value is

$$y = \frac{-9}{16} + \frac{1}{4}t + \frac{19}{16}e^{4t}$$

Compute the error and the percentage error.

Solution:

Here, $f(t_n, y_n) = 1 - t_n + 4y_n$ and thus

$$y_{n+1} = y_n + (0.1)(1 - t_n + 4y_n)$$

For $n = 0$: $\quad y_1 = y_0 + (0.1)(1 - t_0 + 4y_0) = 1 + (0.1)(1 - 0 + 4) = 1.5$

$\quad n = 1$: $\quad y_2 = y_1 + 0.1(1 - t_1 + 4y_1) = 1.5 + (0.1)(1 - 0.1 + 6) = 2.19$

The exact value are computed from

$$y = \frac{-9}{16} + \frac{1}{4}t + \frac{19}{16}e^{4t}$$

Error = exact value – value from Euler's method

$$\text{Percentage error} = \frac{|\text{error}|}{\text{exact value}}$$

Table 8.5 summarizes the computed results.

TABLE 8.5

n	t	y_n	Exact value	Error	Percentage error
0	0	1	1	0	0
1	0.1	1.5	1.6090	0.109	6.7768
2	0.2	2.19	2.5053	0.3153	12.5864
3	0.3	3.146	3.8301	0.6841	17.8620
4	0.4	4.4744	5.7942	1.3192	22.7783
5	0.5	6.3242	8.7120	2.3878	27.4082

EXAMPLE 8.10

Use Euler's method to solve the following differential equation $\dfrac{dy}{dx} = \dfrac{1}{2}y$, $y(0) = 1$ and $0 \le x \le 1$. Use $h = 0.1$.

Solution:

Using Eq. (8.26)

$$y_1 = y_0 + h f(x_0, y_0) = 1 + 0.1 f(0, 1)$$
$$f(0, 1) = f(x_0, y_0) = \frac{1}{2}y_0 = \frac{1}{2}(1) = 1/2$$

Hence $\qquad y_1 = 1 + 0.1(1/2) = 1.05.$

For $n = 1$: $\qquad y_2 = y_1 + hf(x_1, y_1) = 1.05 + 0.1\,f(0.1, 1.05)$

where $\quad f(0.1, 1.05) = \dfrac{1}{2}(1.05) = 0.525$

Therefore, at $x_2 = 2h = 2(0.1) = 0.2\ y_2$ is

$$y_2 = 1.05 + 0.1(0.525) = 1.1025$$

The exact values of $y = e^{x/2}$ (from direct integration).

This procedure is repeated for $n = 2, \ldots, 5$ and a summary of the results obtained is given in Table 8.6.

TABLE 8.6 Euler's method versus exact solution

n	x_n	y_n	$f(x_n, y_n)$	y_{n+1} (Euler)	$y_{n+1} = e^{x/2}$ (exact)
0	0	1	0.5	1.05	1.0513
1	0.1	1.05	0.525	1.1025	1.1052
2	0.2	1.1025	0.5513	1.1576	1.1619
3	0.3	1.1576	0.5788	1.2155	1.2214
4	0.4	1.2155	0.6077	1.2763	1.2840
5	0.5	1.2763	0.6381	1.3401	1.3499

8.3.2 Modified Euler's Method

The modified Euler's method (also called Heun's method) is a single-step, explicit, numerical technique for solving a first-order ordinary differential equation. The method is a modification of Euler's explicit method. In section 8.3.1 on Euler's method, we stated the assumption in that method is that in each subinterval or step, the derivative or the slope between points (x_n, y_n) and (x_{n+1}, y_{n+1}) is constant and equal to the slope of $y(x)$ at point (x_n, y_n). This assumption causes error. In the modified Euler's method, the slope used for computing the value of y_{n+1} is modified to include the effect of that the slope changes within the subinterval. This slope is the average of the slope at the beginning of the interval and an estimate of the slope at the end of the interval.

Hence, the slope at the beginning of the interval is given by

$$\dfrac{dy}{dx}\bigg|_{x=x_n} = \text{slope at } x = x_n = f(x_n, y_n) \qquad (8.27)$$

The slope at the end of the interval is estimated by finding first an approximate value for y_{n+1}, written as y_{n+1}^m using Euler's explicit method.

That is $\quad y_{n+1}^m = y_m + f(x_n, y_n)h$ (8.28)

The estimation of the slope at the end of interval is obtained by substituting the point (x_{n+1}, y_{n+1}^m) in the equation for $\dfrac{dy}{dx}$,

or $\qquad \dfrac{dy}{dx}\bigg|_{\substack{y=y_{n+1}^m \\ x+x_{n+1}}} = f(x_{n+1}, y_{n+1}^m)$ (8.29)

y_{n+1} is then estimated using the average of the two slopes.

That is, $\quad y_{n+1} = y_n + \dfrac{f(x_n, y_n) + f(x_{n+1}, y_{n+1}^m)}{2}h$ (8.30)

The modified Euler's method is illustrated in Figure 8.3. The slope at the beginning of the interval (given by Eq. [8.27]) and the value of y_{n+1}^m as per Eq. (8.28) are shown in Figure 8.3(a). Figure 8.3(b) shows the estimated slope at the end of the interval as per Eq. (8.29). The value of y_{n+1} obtained using Eq. (8.30) is shown in Figure 8.3(c).

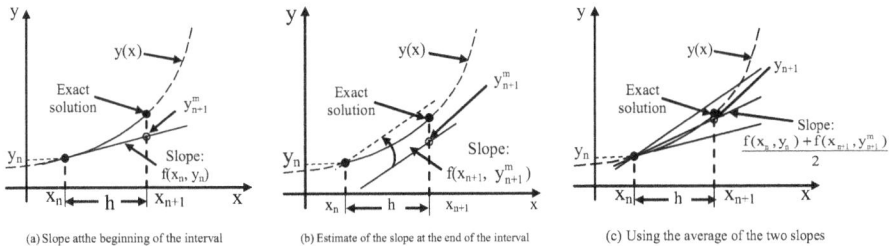

(a) Slope atthe beginning of the interval (b) Estimate of the slope at the end of the interval (c) Using the average of the two slopes

FIGURE 8.3 The modified Euler's method.

In modified Euler's method, instead of approximating (x, y) by $f(x_0, y_0)$ in Eq. (8.22), the integral in Eq. (8.23) is approximated using the trapezoidal rule.

Therefore $\qquad y_1^{(1)} = y_0 + \dfrac{h}{2}\left[f(x_0, y_0) + f(x_1, y_1^{(0)}) \right]$ (8.31)

where $y_1^{(0)} = y_0 + h\, f(x_0, y_0)$ obtained using Euler's formula.

Similarly, we obtain

$$y_1^{(2)} = y_0 + \dfrac{h}{2}\left[f(x_0, y_0) + f(x_1, y_1^{(1)}) \right]$$

$$y_1^{(3)} = y_0 + \frac{h}{2}\left[f(x_0, y_0) + f(x_1, y_1^{(2)}) \right]$$

$$y_1^{(4)} = y_0 + \frac{h}{2}\left[f(x_0, y_0) + f(x_1, y_1^{(3)}) \right] \tag{8.32}$$

and so on.

Therefore, we have

$$y_1^{(n+1)} = y_0 + \frac{h}{2}\left[f(x_0, y_0) + f(x_1, y_1^{(n)}) \right], \quad n = 0, 1, 2, 3, \ldots \tag{8.33}$$

where $y_1^{(n)}$ is the nth approximation to y_1.

The iteration formula given by Eq. (8.33) can be started by selecting $y_1^{(0)}$ from the Euler's formula. The formula given by Eq. (8.33) is terminated at each step if the condition $\left| y_n^{(n)} - y_n^{(n-1)} \right| < \epsilon$, where ϵ is a very small arbitrary value selected depending on the level of accuracy to be accomplished is satisfied. If this happens for sa, $n = k$, then we consider $y_n = y_n^{(k)}$ and continue to compute the value of y at the next point by repeating the procedure described above. Equation (8.33) can also be written as

$$y_{n+1} = y_n + \frac{1}{2}(K_1 + K_2) + (O)h^3 \tag{8.33a}$$

where $\quad K_1 = h\, f(x_n, y_n) \tag{8.33b}$

$$K_2 = h + (x_{n+1}, y_n + K_1) \tag{8.33c}$$

EXAMPLE 8.11

Use the modified Euler's method to solve the differential equation $\dfrac{dy}{dx} = x + y^2$ with $y(0) = 1$. Take the step size $h = 0.1$.

Solution:

From Eq. (8.31), we have

$$y_1^{(1)} = y_0 + \frac{h}{2}\left[f(x_0, y_0) + f(x_1, y_1^{(0)}) \right]$$

where $\quad y_1^{(0)} = y_0 + h\, f(x_0, y_0)$

Therefore $\quad y_1^{(1)} = 1 + \frac{h}{2}\left[(0 + 1^2) + (0.1 + (1 + 0.1(0 + 1^2)^2)^2) \right]$

$$= 1 + 0.05[1 + (0.1 + 1.1^2)] = 1.1155$$

is the improved Euler's estimate.

Similarly
$$y_1^{(2)} = y_0 + \frac{h}{2}\left[f(x_0, y_0) + f(x_1, y_1^{(1)})\right]$$

where
$$y_1^{(1)} = 1.1155$$

$$y_1^{(2)} = y_1^{(1)} + \frac{h}{2}\left[f(x_1, y_1^{(1)}) + f(x_2, y_1^{(1)} + h f(x_1, y_1^{(1)}))\right]$$

$$= 1.1155 + \frac{0.1}{2}[(0.1 + 1.1155^2)$$

$$+(0.2 + (1.1155 + 0.1(0.1 + 1.1155^2)))] = 1.2499$$

is the Euler's method estimate starting from $(x_1, y_1^{(1)})$. Now, starting from $[x_1, y_0 + h f(x_0, y_0)]$, we have

$$y_1^{(2)} = 1.1155 + 0.05[(0.1 + 1.1155^2) + (0.2 + 1.2499^2)] = 1.2708$$

is the improved Euler's estimate.

EXAMPLE 8.12

Use the modified Euler's method to obtain an approximate solution of $\frac{dy}{dt} = -2ty^2, y(0) = 1$, in the interval $0 \le t \le 0.5$ using $h = 0.1$. Compute the error and the percentage error. Given the exact solution is given by $y = \frac{1}{(1+t^2)}$.

Solution:

For $n = 0$:
$$y_1^{(1)} = y_0 - 2h\, t_0\, y_0^2 = 1 - 2(0.1)\,(0)\,(1)^2 = 1$$

Now
$$y_1^{(1)} = y_0 + \frac{h}{2}\left[-2t_0 y_0^2 - 2t_1 y_1^{(1)2}\right]$$

$$= 1 - (0.1)[(0)\,(1)^2 + (0.1)\,(1)^2] = 0.99$$

Table 8.7 shows the remaining calculations. Table 8.7 also shows the values obtained from the Euler's method, the modified Euler's method, the exact values, and the percentage error for the modified Euler's method.

TABLE 8.7

n	t_n	Euler y_n	Modified Euler y_n	Exact value	Error	Percentage Error
0	0	1	1	1	0	0
1	0.1	1	0.9900	0.9901	0.0001	0.0101
2	0.2	0.9800	0.9614	0.9615	0.0001	0.0104
3	0.3	0.9416	0.9173	0.9174	0.0001	0.0109
4	0.4	0.8884	0.8620	0.8621	0.0001	0.0116
5	0.5	0.8253	0.8001	0.8000	0.0001	0.0125

In Table 8.7,

Error = Exact value – value from the modified Euler's method

$$\text{Percentage error} = \frac{|\text{error}|}{\text{exact value}}.$$

EXAMPLE 8.13

Use the modified Euler's method to find the approximate value of $y(1.5)$ for the solution of the initial value problem $\dfrac{dy}{dx} = 2xy$, $y(1) = 1$. Take $h = 0.1$. The exact solution is given by $y = e^{x^2-1}$. Determine the relative error and the percentage error.

Solution:

With $x_0 = 1$, $y_0 = 1$, $f(x_n, y_n) = 2x_n y_n$, $n = 0$ and $h = 0.1$, we first compute $y_1^{(0)} = y_0 + h f(x_0, y_0)$ from Eq. (8.31).

$$y_1^{(0)} = y_0 + (0.1) \, 2(x_0, y_0) = 1 + (0.1) \, 2(1)(1) = 1.2$$

We use this value in Eq. (8.33) along with

$$x_1 = 1 + h = 1 + 0.1 = 1.1$$

$$y_1^1 = y_0 + \left(\frac{0.1}{2}\right) 2x_0 y_0 + 2x_1 y_1 = 1 + \left(\frac{0.1}{2}\right) 2(1)(1) + 2(1.1)(1.2) = 1.232$$

Table 8.8 gives the values computed for the modified Euler's method, exact value, relative error, and the percentage error. Exact value is calculated from $y = e^{x^2-1}$.

Error = exact value – value from the modified Euler's method

$$\text{Percentage relative error} = \frac{|\text{error}|}{\text{exact value}}.$$

TABLE 8.8

n	x_n	y_n	Exact value	Absolute error	Percentage Relative error
0	1	1	1	0	0
1	1.1	1.2320	1.2337	0.0017	0.14
2	1.2	1.5479	1.5527	0.0048	0.31
3	1.3	1.9832	1.9937	0.0106	0.53
4	1.4	1.5908	2.6117	0.0209	0.80
5	1.5	3.4509	3.4904	0.0394	1.13

EXAMPLE 8.14

Repeat Example 8.10 using the modified Euler's method.

Solution:

From Eqs. (8.33a) to (8.33c), we have

$$K_1 = h\,f(x_0, y_0) = h\left(\frac{1}{2}y_0\right) = 0.1\left(\frac{1}{2}\right) = 0.05$$

and $\quad K_2 = h\,f(x_1, y_0 + K_1) = h\left[\dfrac{y_0 + K_1}{2}\right] = 0.1\left[\dfrac{1+0.05}{2}\right] = 0.0525$

The functional approximate at $x_1 = 0.1$ $(n = 1)$ is given by

$$y_1 = y_0 + \frac{1}{2}(K_1 + K_2) = 1 + \frac{1}{2}(0.05 + 0.0525) = 1.05125 \approx 1.0513$$

Hence, at $x_2 = 0.2$, we have

$$K_1 = 0.1\left[\frac{0.05125}{2}\right] = 0.0526$$

$$K_2 = 0.1\left[\frac{1.0513 + 0.0526}{2}\right] = 0.0552$$

$$y_2 = 1.0513 + \frac{1}{2}(0.0526 + 0.0552) = 1.1051$$

This procedure is repeated for $n = 2, 3, 4$, and 5 to give the functional approximations shown in Table 8.9.

TABLE 8.9

n	x_n	y_n	K_1	K_2	y_{n+1} (modified Euler)	y_{n+1} (exact)
0	0	1	0.05	0.0525	1.0513	1.0513
1	0.1	1.0513	0.0526	0.0552	1.1051	1.1052
2	0.2	1.1051	0.0526	0.0581	1.1618	1.1619
3	0.3	1.1618	0.0581	0.0699	1.2213	1.2214
4	0.4	1.2213	0.0611	0.0641	1.2839	1.2840
5	0.5	1.2839	0.0642	0.0674	1.3513	1.3499

Table 8.9 clearly shows that the modified Euler's method gives better accuracy for the same h interval when compared with the basic Euler's method.

8.3.3 Runge-Kutta Methods

Runge-Kutta methods are a family of single-step, explicit, numerical techniques for solving a first-order ordinary differential equation. Various types of Runge-Kutta methods are classified according to their order. The order identifies the number of points within the subinterval that are utilized for finding the value of the slope in Eq. (8.7). For instance, second-order Runge-Kutta methods use the slope at two points, third-order methods use three-points, and so on. The classical Runge-Kutta method is of order four and uses four points. Runge-Kutta methods give a more accurate solutions compared to the simpler Euler's explicit method. The accuracy increases with increasing order of the Runge-Kutta method.

8.3.3.1 Runge-Kutta Method of Order Two

In the Runge-Kutta method of order two, we consider up to the second derivative term in the Taylor series expansion and then substitute the derivative terms with the appropriate function values in the interval.

Consider the Taylor series expansion of the function about y_n.

$$y_{n+1} = y_n + hy'(x_n, y_n) + \frac{h^2}{2} y''(x_n, y_n)$$

$$y_{n+1} = y_n + hg(x_n, y_n) + \frac{h^2}{2} g'(x_n, y_n)$$

$$y_{n+1} = y_n + h\left[g(x_n, y_n) + \frac{h}{2} g'(x_n, y_n) \right] \qquad (8.34)$$

Now, substituting

$$g'(x_n, y_n) = \frac{\partial g}{\partial x} + \frac{\partial g}{\partial y} g(x_n, y_n)$$

where $\qquad \dfrac{dy}{dx} = g(x_n, y_n)$

From the differential equation, we obtain

$$y_{n+1} = y_n + h\left[g(x_n, y_n) + \frac{h}{2}\frac{\partial g}{\partial x} + \frac{h}{2}\frac{\partial g}{\partial y} g(x_n, y_n) \right] \tag{8.35}$$

It should be noted here that the factor inside the square brackets consisting of the derivatives may be substituted with a function of the type $ag(x + \alpha, y + \beta)$ in a Taylor series expansion, such that from Eq. (8.34), we have

$$y_{n+1} = y_n + h[ag(x_n + \alpha, y_n + \beta)] \tag{8.36}$$

Now, expanding the function $g(x_n + \alpha, y_n + \beta)$ in Eq. (8.36) in a Taylor series expansion with two variables about (x_n, y_n) and considering only the first derivative terms, we obtain

$$y_{n+1} = y_n + ha\left[g(x_n, y_n) + \alpha\frac{\partial g}{\partial x} + \beta\frac{\partial g}{\partial y} \right] \tag{8.37}$$

Now, equating the coefficients of the respective terms on the right-hand side of Eqs. (8.35) and (8.37), we obtain

$$a = 1$$
$$\alpha = h/2 \tag{8.38}$$

and $\qquad \beta = h/2\, g(x_n, y_n)$

Therefore, Eq. (8.36) becomes

$$y_{n+1} = y_n + hg\left[x_n + \frac{h}{2}, y_n + \frac{h}{2} g(x_n, y_n) \right] \tag{8.39}$$

Equation (8.39) can also be rewritten as

$$y_{n+1} = y_n + hK_2 \tag{8.40}$$

where $\qquad K_2 = hg\left[x_n + \dfrac{h}{2}, y_n + \dfrac{K_1}{2} \right] \tag{8.41}$

in which $K_1 = hg(x_n, y_n)$ $\qquad\qquad\qquad\qquad\qquad$ (8.42)

The Runge-Kutta method of order two is also known as the *midpoint method* because the derivative is replaced by functions evaluated at the midpoint $x_n + h/2$.

The midpoint method is illustrated schematically in Figure 8.4. The determination of the midpoint with Euler's explicit method using $y_m = y_n + f(x_n, y_n)h/2$ is shown in Figure 8.4(a). Figure 8.4(b) shows the estimated slope that is computed with the equation

$$\frac{dy}{dx}\bigg|_{x=x_m} = f(x_m, y_m).$$

Figure 8.4(c) shows the value of y_{n+1} obtained using

$$y_{n+1} = y_n + f(x_m, y_m)h.$$

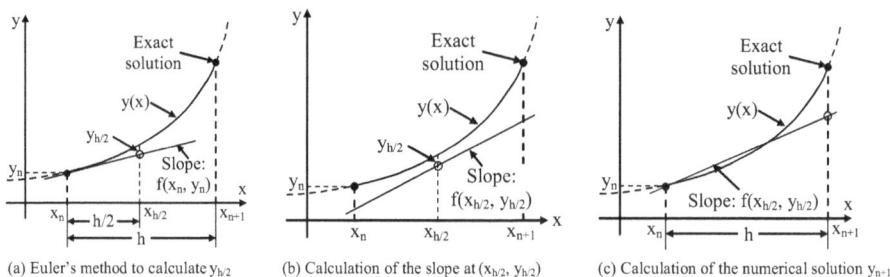

(a) Euler's method to calculate $y_{h/2}$ (b) Calculation of the slope at $(x_{h/2}, y_{h/2})$ (c) Calculation of the numerical solution y_{n+1}

FIGURE 8.4 The midpoint method.

The local truncation error in the Runge-Kutta method of order two is $O(h^3)$, and the global truncation error is $O(h^2)$. Note that this is smaller by a factor of h than the truncation errors in Euler's explicit method. In other words, for the same accuracy, a larger step size can be used. However, in each step, the function $f(x, y)$ in the Runge-Kutta method of order two is computed twice.

EXAMPLE 8.15

Use the second-order Runge-Kutta method with $h = 0.1$, find y_1 and y_2 for

$$\frac{dy}{dx} = -xy^2, y(2) = 1.$$

Solution:

For $f(x, y) = -xy^2$, the modified Euler's method, Eq. (8.40) is

$$y_{n+1} = y_n - 0.1(x_n + 0.05)[y_n + 0.05f_n]^2,$$

where $\quad f_n = -x_n y_n^2.$

$\quad n = 0:$

Here $\quad x_0 = 2$ and $y_0 = 1$, hence $f_0 = -2(1)^2 = -2$

$$y_1 = 1 - 01(2 + 0.05)[1 + 0.05(-2)]^2 = 0.83395$$

$\quad n = 1:$

Now $\quad x_1 = 2.1$ and $y_1 = 0.83395$; hence $f_1 = -x_1 y_1^2 = -1.46049$

Hence, $\quad y_2 = 0.83395 - 0.1(2.1 + 0.05)[0.83395 + 0.05(-1.46049)]^2 = 0.70946.$

Relative error when $n = 0$ is

$$E_1(0.1) = 0.8299 - 0.83395 - 0.00405$$

and $\quad E_2(0.1) = 0.7042 - 0.70946 - 0.00526.$

Comparing these values $(y_1$ and $y_2)$ with the exact values obtained in Table 8.1, we see that the second-order Runge-Kutta method do indeed give accuracy comparable to the second-order Taylor's series method without requiring partial derivatives.

EXAMPLE 8.16

Use the Runge-Kutta method of order two to integrate $\dfrac{dy}{dx} = \sin y$ with $y(0) = 1$ from $x = 0$ to 0.5 in steps of $h = 0.1$. Keep four decimal places in the calculations.

Solution:

Here $\quad g(x, y) = \sin y$

Hence, the integration formulae in Eqs. (8.41) to (8.42) are

$$K_1 = hg(x, y) = 0.1 \sin y$$

$$K_2 = hf\left(x + \frac{h}{2}, y + \frac{1}{2}K_1\right) = 0.1\sin\left(y + \frac{K_1}{2}\right)$$

$$y(x + h) = y(x) + K_2$$

Given that $y(0) = 1$, we can carry out the integration as follows:

$$K_1 = 0.1 \sin(1) = 0.0841$$

$$K_2 = 0.1 \sin\left(1 + \frac{0.0841}{2}\right) = 0.0863$$

$$y(0.1) = 1 + 0.0863 = 1.0863$$

$$K_1 = 0.1 \sin(1.0863) = 0.0885$$

$$K_2 = 0.1 \sin\left(1.0863 + \frac{0.0885}{2}\right) = 0.0905$$

$$y(0.2) = 1.0863 + 0.0905 = 1.1768$$

and so on.

The computations are summarized in Table 8.10 to four decimal places.

TABLE 8.10

x	y	K_1	K_2
0	1	0.0841	0.0863
0.1	1.0863	0.0885	0.0905
0.2	1.1768	0.0925	0.0940
0.3	1.2708	0.0955	0.0968
0.4	1.3676	0.0979	0.0988
0.5	1.4664		

8.3.3.2 Runge-Kutta Method of Order Four

In the classical Runge-Kutta method of order four, the derivatives are evaluated at four points, once at each end and twice at the interval midpoint as given below:

$$y(x_{n+1}) = y(x_n) + \frac{h}{6}(K_1 + 2K_2 + 2K_3 + K_4) \tag{8.43}$$

where $\quad K_1 = g[x_n, y_n(x_n)]$

$$K_2 = g\left[x_n + \frac{h}{2}, \ y(x_n) + \frac{1}{2}K_1 h\right]$$

$$K_3 = g\left[x_n + \frac{h}{2}, y(x_n) + \frac{1}{2}K_2 h\right]$$

and $\quad K_4 = g[x_n + h, y(x_n) + K_3 h] \tag{8.44}$

The classical Runge-Kutta method of order four is illustrated schematically in Figure 8.5, and each phase shows the determination of the slopes in Eq. (8.4). Figure 8.5(a) shows the slope K_1 and how it is used to compute slope K_2. Figure 8.5(b) shows how slope K_2 is used to find the slope K_3. Figure 8.5(c) shows how slope K_3 is used to find the slope K_4. Figure 8.5(d) shows the application of Eq. (8.43) where the slope used for evaluating y_{n+1} is a weighted average of the slopes K_1, K_2, K_3, and K_4.

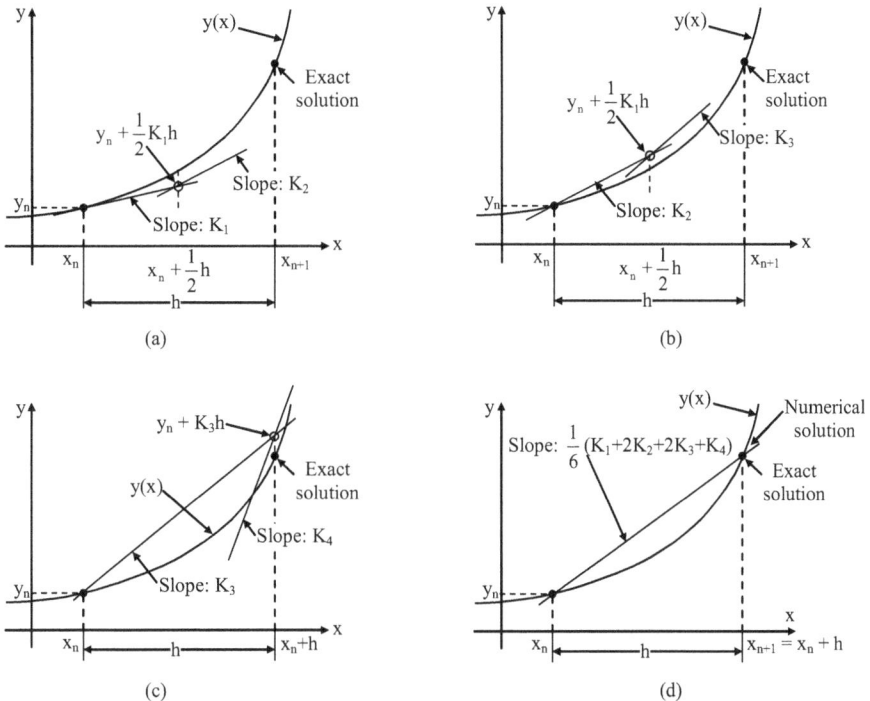

FIGURE 8.5 The classical fourth-order Runge-Kutta method.

The local truncation error in the classical Runge-Kutta method of order four is $O(h^5)$, and the global truncation error is $O(h^4)$. This method gives the most accurate solution compared to the other methods. Equation (8.44) is the most accurate formula available without extending outside the interval $[x_n, x_{n+1}]$.

Equations (8.43) and (8.44) can also be written as

$$y_{n+1} = y_n + \frac{1}{6}[K_1 + 2K_2 + 2K_3 + K_4 \qquad (8.44a)$$

where $\quad K_1 = hf(x_n, y_{n_0})$

$$K_2 = hf\left(x_n + \frac{h}{2}, \ y_n + \frac{h}{2}\right)$$

$$K_3 = hf\left(x_n + \frac{h}{2}, \ y_n + \frac{K_2}{2}\right)$$

and $\quad K_4 = hf(x_n + h, y_n + K_3) \qquad (8.44b)$

EXAMPLE 8.17

Use the Runge-Kutta method of order four with $h = 0.1$ to obtain an approximation to $y(1.5)$ for the solution of $\dfrac{dy}{dx} = 2xy$, $y(1) = 1$. The exact solution is given by $y = e^{x^2-1}$. Determine the relative error and the percentage relative error.

Solution:

For $n = 0$, from Eq. (8.44), we have

$$K_1 = g(x_0, y_0) = 2x_0 y_0 = 2$$

$$K_2 = g\left[x_0 + \frac{1}{2}(0.1), y_0 + \frac{1}{2}(0.1)(2)\right] = 2\left[x_0 + \frac{1}{2}(0.1)\right]\left[y_0 + \frac{1}{2}(0.2)\right]$$

$$= 2.31$$

$$K_3 = g\left[x_0 + \frac{1}{2}(0.1), y_0 + \frac{1}{2}(0.1)2.31\right] = 2\left[x_0 + \frac{1}{2}(0.1)\right]\left[y_0 + \frac{1}{2}(0.231)\right]$$

$$= 2.3426$$

$$K_4 = g[x_0 + 0.1, y_0 + 0.1(2.3426) = 2(x_0 + 0.1)(y_0 + 0.2343) = 2.7154$$

Hence, $\quad y_1 = y_0 + \dfrac{h}{6}[K_1 + 2K_2 + 2K_3 + K_4$

$$= 1 + \frac{0.1}{6}[2 + 2(2.31) + 2(2.3426) + 2.7154]$$

$$= 1.2337$$

Table 8.11 summarizes the computations. In Table 8.11, exact value is computed from $y = e^{x^2-1}$. The absolute error = exact value minus the value from the Runge-Kutta method. Percentage relative error = |error|/exact value.

TABLE 8.11

n	x_n	y_n	Exact value	Absolute error	Percentage relative error
0	1	1	1	0	0
1	1.1	1.2337	1.2337	0	0
2	1.2	1.5527	1.5527	0	0
3	1.3	1.9937	1.9937	0	0
4	1.4	2.6116	2.6117	0.0001	0
5	1.5	3.4902	3.4904	0.0001	0

EXAMPLE 8.18

Use the Runge-Kutta method of order four with $h = 0.1$ on (2, 3) for the initial problem $\dfrac{dy}{dx} = -xy^2$, $y(2) = 1$. Compute the results obtained with the exact solution $y(x) = \dfrac{2}{x^2 - 2}$.

Solution:

Starting with $t_0 = 2$, $y_0 = 1$, Eq. (8.44) gives

$$K_1 = f(2.0,\ 1) = -(2)(1)^2 = -2$$

$$K_2 = f(2.05,\ 1 + 0.05(-2)) = -(2.05)(0.9)^2 = -1.6605$$

$$K_3 = f(2.05,\ 1 + 0.05(-1.6605)) = -(2.05)(0.916975)^2 = -1.72373$$

$$K_4 = f(2.1,\ 1 + 0.1(-1.72373)) = -(2.1)(0.82763)^2 = -1.43843$$

$$y_1 = y_0 - \frac{0.1}{6}\{2 + 2(1.6605 + 1.72373) + 1.43843\} = 0.829885$$

Table 8.12 shows both the Runge-Kutta method of order 4 values and the exact values of $y(2.0)$, $y(2.1)$, ..., $y(3.0)$ rounded to six decimal places. The exact values in Table 8.12, y_n were computed from $y(x) = \dfrac{2}{x^2 - 2}$.

TABLE 8.12

x_n	y_n	$y(x_n)$
2.0	1.000000	1.000000
2.1	0.829885	0.829876
2.2	0.704237	0.704225
2.3	0.607914	0.607903
2.4	0.531924	0.531915
2.5	0.470596	0.470588
2.6	0.420175	0.420168
2.7	0.378078	0.378072
2.8	0.342471	0.342466
2.9	0.312017	0.312012
3.0	0.285718	0.285714

The reasons for the popularity of the Runge-Kutta method of order 4 are evident from Table 8.12. Clearly the method is more accurate. However, four slope values must be computed at each step. This is a short coming of the method.

EXAMPLE 8.19

Using the Runge-Kutta method of order four and with $h = 0.2$ to obtain an approximate solution of $\dfrac{dy}{dt} = -2ty^2$, $y(0) = 1$, in the initial $0 \le t \le 1$ with $h = 0.2$. The exact value of y is given by $y = \dfrac{1}{1+t^2}$. Compute the relative error and the percentage relative error.

Solution:

Here
$$K_1 = -2t_n\, y_n^2$$
$$K_2 = -2(t_n + 0.1), (y_n + 0.1\,K_1)^2$$
$$K_3 = -2(t_n + 0.1), (y_n + 0.1K_2)^2$$
$$K_4 = -2(t_{n+1})(y_n + 0.2K_3)^2$$

For $n = 0$:

$$K_1 = 0,\ K_2 = -0.2,\ K_3 = -0.192 \text{ and } K_4 = -0.37.$$

Therefore, $$y_1 = 1 - \frac{0.2}{6}[2(0.2) + 2(0.192) + 0.37] = 0.9615.$$

Table 8.13 gives the summary of the calculations. In Table 8.13, the exact values are calculated using $y = \dfrac{1}{1+t^2}$. The absolute error = exact value minus the value from the Runge-Kutta method. Percentage relative error = |error|/ exact value.

TABLE 8.13

n	x_n	y_n	Exact value	Absolute error	Percentage relative error
0	0	1.0	1.0	0	0
1	0.2	0.9615	0.9615	0	0
2	0.4	0.8621	0.8621	0	0
3	0.6	0.7353	0.7353	0	0
4	0.8	0.6098	0.6098	0	0
5	1.0	0.5	0.5	0	0

EXAMPLE 8.20

Find an approximate solution to the initial value problem $\dfrac{dy}{dt} = 1 - t + 4y$, $y(0) = 1$, in the initial $0 \le t \le 1$ using Runge-Kutta method of order four with $h = 0.1$. Compute the exact value given by $y = \dfrac{-9}{16} + \dfrac{1}{4}t + \dfrac{19}{16}e^{4t}$. Compute the absolute error and the percentage relative error.

Solution:

For $n = 0$, from Eq. (8.44), we have

$$K_1 = f(x_0, y_0) = 5$$
$$K_2 = f(0 + 0.05, 1 + 0.25) = 5.95$$
$$K_3 = f(0 + 0.05, 1 + 0.2975) = 6.14$$
$$K_4 = f(0.1, 1 + 0.614) = 7.356$$

Hence $\quad y_1 = 1 + \dfrac{0.1}{6}[5 + 2(5.95) + 2(6.14) + 7.356] = 1.6089$

Table 8.14 gives a summary of all the calculations for y_n, exact value, absolute error, and the percentage relative error.

TABLE 8.14

n	t_n	Runge-Kutta y_n	Exact value	Absolute error	Percentage relative error
0	0	1	1		
1	0.1	1.6089	1.6090	0.0001	0.0062
2	0.2	2.5050	2.5053	0.0002	0.0119
3	0.3	3.8294	3.8301	0.0007	0.07
4	0.4	5.7928	5.7942	0.0014	0.14
5	0.5	8.7093	8.7120	0.0027	0.27

The superiority of the Runge-Kutta method of order four is clearly demonstrated in this Table 8.14 in comparison with the Euler's method (Table 8.5).

EXAMPLE 8.21

Use the Runge-Kutta method of order four and with $h = 0.1$ to find an approximate solution of $\dfrac{dy}{dx} = x^2 + y$ at $x = 0.1$, 0.2 and 0.4. Given that $y = -1$ when $x = 0$.

Solution:

Equation (8.44) can be written as

$$K_1 = g(x_0, y_0)h = [0^2 - 1]0.1 = -0.1$$

$$K_2 = g\left[x_0 + \frac{h}{2}, y_0 + \frac{1}{2}K_1\right]h = [(0.05)^2 - 1.05]0.1 = -0.1047$$

$$K_3 = g\left[x_0 + \frac{h}{2}, y_0 + \frac{1}{2}K_2\right]h = [(0.05)^2 - 1.0524]0.1 = -0.1050$$

$$K_4 = g[x_0 + h, y_0 + K_3]h = [(0.1)^2 - 1.105]0.1 = -0.1095$$

Let $\quad \Delta y_1 = \dfrac{1}{6}[K_1 + 2K_2 + 2K_3 + K_4]$

$$= \frac{1}{6}[-0.1 + 2(-0.1047) + 2(-0.1050) + (-0.1095)$$

$$= -0.1048$$

Hence, $\quad y_1 = y_0 + \Delta y_1 = -1.1048$

For the second increment, we have

$$K_1 = -0.1095, \, K_2 = -0.1137, \, K_3 = -0.1139 \text{ and } K_4 = -0.1179$$

$$\Delta y_2 = -0.1138$$

Therefore $\quad y_2 = y_1 + \Delta y_2 = -1.2186$

For the third increment, we have

$$K_1 = -0.1179, \, K_2 = -0.1215, \, K_3 = -0.1217 \text{ and } K_4 = -0.1250$$

and $\quad \Delta y_3 = -0.1215$

Hence, $\quad y_3 = y_2 + \Delta y_2 = -1.3401.$

EXAMPLE 8.22

Repeat Example 8.10 using the Runge-Kutta method of order four. Use $h = 1$.

Solution:

Here $\quad f(x, y) = y/2$

From Eq. (8.44b), we have

$$K_1 = hf(x_0, y_0) = 1 f(0, 1) = 1\left(\frac{1}{2}\right) = \frac{1}{2}$$

$$K_2 = hf\left(x_0 + \frac{h}{2}, \, y_0 + \frac{K_1}{2}\right) = 1 \, f\left(\frac{1}{2}, \frac{5}{4}\right) = \frac{\frac{5}{4}}{2} = \frac{5}{8}$$

$$K_3 = hf\left(x_0 + \frac{h}{2}, \, y_0 + \frac{K_2}{2}\right) = 1 f\left(\frac{1}{2}, \frac{21}{16}\right) = \frac{\frac{21}{16}}{2} = \frac{21}{32}$$

$$K_4 = hf(x_0 + h, \, y_0 + K_3) = 1 \ f\left(1, \frac{53}{32}\right) = \frac{\frac{53}{32}}{2} = \frac{53}{64}$$

From Eq. (8.44a), we have

$$y = y(1) = y_0 + \frac{1}{6} \left[K_1 + 2K_2 + 2K_3 + K_4\right]$$

$$y(1) = 1 + \frac{1}{6}\left(\frac{1}{2}\right) + 2\left(\frac{5}{8}\right) + 2\left(\frac{21}{32}\right) + \frac{53}{64} = 1.6484$$

The exact value

$$y(1) = e^{x/2} = e^{1/2} = 1.6487.$$

8.3.4 Predictor-Corrector Methods

Predictor-corrector methods refer to a family of schemes for solving ordinary differential equations using two formulae: *predictor* and *corrector* formulae. In predictor-corrector methods, four prior values are required to find the value of y at x_n. Predictor-corrector methods have the advantage of giving an estimate of error from successive approximations to y_n. The predictor is an explicit formula and is used first to determine an estimate of the solution y_{n+1}. The value y_{n+1} is calculated from the known solution at the previous point (x_n, y_n) using single-step method or several previous points (multistep methods). If x_n and x_{n+1} are two consecutive mesh points such that $x_{n+1} = x_n + h$, then in Euler's method we have

$$y_{n+1} = y_n + h f(x_0 + nh, y_n), \qquad n = 0, 1, 2, 3, \ldots \qquad (8.45)$$

Once an estimate of y_{n+1} is found, the corrector is applied. The corrector uses the estimated value of y_{n+1} on the right-hand side of an otherwise implicit formula for computing a new, more accurate, value for y_{n+1} on the left-hand side.

The modified Euler's method gives us

$$y_{n+1} = y_n + \frac{h}{2}\left[f\left(x_n, y_n\right) + f\left(x_{n+1}, y_{n+1}\right)\right] \qquad (8.46)$$

The value of y_{n+1} is first estimated by Eq. (8.45) and then utilized in the right-hand side of Eq. (8.46) resulting in a better approximation of y_{n+1}. The value of y_{n+1} thus obtained is again substituted in Eq. (8.46) to find a still better approximation of y_{n+1}. This procedure is repeated until two consecutive iterated values of y_{n+1} are very close. Here, the corrector Equation (8.46) which is an implicit equation is being used in an *explicit* manner since no solution of a nonlinear equation is required.

In addition, the application of corrector can be repeated several times such that the new value of y_{n+1} is substituted back on the right-hand side of the corrector formula to obtain a more refined value for y_{n+1}. The technique of refining an initially crude estimate of y_{n+1} by means of a more accurate formula is known as *predictor-corrector* method. Equation (8.45) is called the *predictor* and Eq. (8.46) is called the *corrector* of y_{n+1}. In what follows, we describe two such predictor-corrector methods:

1. Adams-Moulton method

2. Milne's predictor-corrector method

8.3.4.1 Adams-Moulton Predictor-Corrector Method

The Adams-Moulton method is an implicit multistep method for solving first-order ordinary differential equations. There are several versions of Adams-Moulton formulas available for computing the value of y_{n+1} by using the previously obtained solution at two or more points. These formulas are classified based on their order, that is, based on the number of points used in the formula and the order of the global truncation error. For instance, in the second-order formula, two points (x_n, y_n) and (x_{n+1}, y_{n+1}) are used. In the third-order formula, three points (x_n, y_n), (x_{n-1}, y_{n-1}) and (x_{n-2}, y_{n-2}) are used and so on.

Consider the differential equation

$$\frac{dy}{dx} = f(x,y), \quad y(x_0) = y_0 \tag{8.47}$$

Integrating Eq. (8.47), we obtain

$$y = y_0 + \int_{x_0}^{x} f(x,y)dx \tag{8.48}$$

or $\quad y_1 = y_0 + \int_{x_0}^{x} f(x,y)dx, \quad x_0 \le x \le x_1 \tag{8.49}$

Applying Newton's backward difference formula, we have

$$f(x,y) = f_0 + n\nabla f_0 + \frac{n(n+1)}{2}\nabla^2 f_0 + \frac{n(n+1)(n+2)}{6}\nabla^3 f_0 + \dots \tag{8.50}$$

where $\quad n = \dfrac{x - x_0}{h}$ and $f_0 = f\left(x_0, y_0\right)$

Now, substituting $f(x, y)$ from Eq. (8.50) into the right-hand side of Eq. (8.49), we obtain

$$y_1 = y_0 \int_{x_0}^{x_1}\left[f_1 + n\nabla f_0 + \frac{n(n+1)}{2}\nabla^2 f_0 + \dots\right]dx$$

$$y_1 = y_0 + h\int_{0}^{1}\left[f_0 + n\nabla f_0 + \frac{n(n+1)}{2}\nabla^2 f_0 + \dots\right]dx$$

or $\quad y = y_0 + h\left[1 + \frac{1}{2}\nabla + \frac{5}{12}\nabla^2 + \frac{3}{8}\nabla^3 + \frac{251}{720}\nabla^4 + \dots\right]f_0 \tag{8.51}$

We note here that the right-hand side of Eq. (8.51) depends on $y_0, y_{-1}, y_{-2}, \ldots$ all of which are known.

Hence, we can write Eq. (8.51) as

$$y_1^p = y_0 + h\left[1 + \frac{1}{2}\nabla + \frac{5}{12}\nabla^2 + \frac{3}{8}\nabla^3 + \frac{251}{720}\nabla^4 + \ldots\right]f_0 \qquad (8.52)$$

Equation (8.52) is called the *Adams-Bashforth formula* and is used as a predictor formula.

A corrector formula is derived by applying Newton's backward difference formula at f_1. Therefore,

$$f(x,y) = f_1 + n\nabla f_1 + \frac{n(n+1)}{2}\nabla^2 f_1 + \frac{n(n+1)(n+2)}{6}\nabla^3 f_1 + \ldots \qquad (8.53)$$

Now, substituting $f(x, y)$ from Eq. (8.53) into the right-hand side of Eq. (8.49), we obtain

$$y_1 = y_0 + \int_{x_0}^{x_1}\left[f_1 + n\nabla f_1 + \frac{n(n+1)}{2}\nabla^2 f_1 + \ldots\right]dx$$

$$= y_0 + h\int_{-1}^{0}\left[f_1 + n\nabla f_1 + \frac{n(n+1)}{2}\nabla^2 f_1 + \ldots\right]dx$$

or $\qquad y = y_0 + h\left[1 - \frac{1}{2}\nabla - \frac{1}{12}\nabla^2 - \frac{1}{24}\nabla^3 - \frac{19}{720}\nabla^4 + \ldots\right]f_1 \qquad (8.54)$

Equation (8.54) shows that the right-hand side depends on $y_1, y_0, y_{-1}, y_{-2}, \ldots$, where y_1^p is used for y_1. Hence, the new value of y_1 is given by

$$y_1^c = y_0 + h\left[1 - \frac{1}{2}\frac{-5}{12}\nabla^2 - \frac{3}{8}\nabla^3 - \frac{251}{720}\nabla^4\right]f_1^p \qquad (8.55)$$

$$f_1^p = f_1\left(x_1, y_1^p\right)$$

the formula, Eq. (8.55) is called the *Adams-Moulton* corrector formula. Now expressing the remaining difference operators in their functional values and neglecting the fourth and higher order differences, Eqs. (8.51) and (8.55) become respectively,

$$y_1^p = y_0 + \frac{h}{24}\left[55f_0 - 59f_{-1} + 37f_{-2} - 9f_{-3}\right] \qquad (8.56)$$

and $\qquad y_1^c = y_0 + \frac{h}{24}\left[9f_1^p + 19f_0 - 5f_{-1} - 9f_{-2}\right] \qquad (8.57)$

Equation (8.57), known as the corrector formula is repeatedly applied by computing an improved value of f_1 at each stage, to obtain a better value of y_1 unless it becomes stable and remains unchanged and then we proceed to calculate y_2.

The approximate errors in Eqs. (8.56) and (8.57) are $\dfrac{251}{720}h^5 f_0^{(4)}$ and $\dfrac{-19}{720}h^5 f_0^{(4)}$ respectively.

It should be noted here that in order to apply the Adams-Moulton method, we require four starting values of y, which can be obtained by using Picard's method of successive approximation or the Taylor series method or Euler's method or Runge-Kutta methods.

Summarizing, the Adams-Bashforth and Adam-Moulton formulae are given by

$$y_{n+1}^p = y_n + \frac{h}{24}\left[55f_n - 59f_{n-1} + 37f_{n-2} - 9f_{n-3}\right] \tag{8.58}$$

and $\qquad y_{n+1}^c = y_n + \dfrac{h}{24}\left[9f_{n+1} + 19f_n - 5f_{n-1} + f_{n-2}\right] \tag{8.59}$

respectively. The local error estimates for Eqs. (8.58) and (8.59) are

$$\frac{251}{720}h^5 y^v\left(\xi_1\right) \text{ and } \frac{-19}{720}h^5 y^v\left(\xi_2\right) \tag{8.60}$$

Let y_{n+1}^0 represents the value of y_{n+1} found using Eq. (8.58) and y_{n+1}^1 the solution obtained with one application of Eqs. (8.58) and (8.59). If $y(x_{n+1})$ represents the exact value of y at x_{n+1} and the values of f are assumed to be exact at all points including x_n, then from Eq. (8.60), we obtain the order estimates

$$y\left(x_{n+1}\right) - y_{n+1}^0 = \frac{251}{720}h^5 y^v\left(\xi_1\right) \tag{8.61}$$

$$y\left(x_{n+1}\right) - y_{n+1}^1 = \frac{-19}{720}h^5 y^v\left(\xi_2\right) \tag{8.62}$$

which leads to the estimate of y^v, based on the assumption that the over the interval of interest $y^v(x)$ is approximately constant, as

$$h^5 y^v = \frac{720}{270}\left[y_{n+1}^1 - y_{n+1}^0\right].$$

Hence, from Eq. (8.62), we obtain

$$y\left(x_{n+1}\right) - y_{n+1}^1 = \frac{-19}{720}\left[y_{n+1}^1 - y_{n+1}^0\right] \approx \frac{-1}{14}\left[y_{n+1}^1 - y_{n+1}^0\right] = D_{n+1} \tag{8.63}$$

Hence, the error of the corrected value is approximately $-1/14$ of the difference between the corrected and the predicted values.

EXAMPLE 8.23

Use the Adams-Moulton method on $(2, 3)$ with $h = 0.1$ for the initial value problem $\dfrac{dy}{dx} = -xy^2$, $y(2) = 1$. Exact solution is $y(x) = \dfrac{2}{x^2 - 2}$.

Solution:

We will try to obtain about four significant digits. To start the method, we use the following exact values to seven significant digits.

$$x_0 = 2.0 : y_0 = y(2.0) = 1.0; \qquad f_0 = -x_0 y_0^2 = -2.0$$
$$x_1 = 2.1 : y_1 = y(2.1) = 1.8298755; \quad f_1 = -x_1 y_1^2 = -1.446256$$
$$x_2 = 2.2 : y_2 = y(2.2) = 0.7042254; \quad f_2 = -x_2 y_2^2 = -1.091053$$
$$x_3 = 2.3 : y_3 = y(2.3) = 0.6079027; \quad f_3 = x_3 y_3^2 = -0.8499552$$

$$n = 3 \qquad y_4^p = y_3 + \frac{h}{24}\left[55 f_3 - 59 f_2 + 37 f_1 - 9 f_0\right] = 0.5333741$$

$$y_4^c = y_3 + \frac{h}{24}\left[9\left(-x_4\left(y_4^p\right)^2 + 19 f_3 - 5 f_2 + f_1\right)\right] = 0.5317149$$

The local truncation error estimate is from Eq. (8.62),

$$y\left(x_{n+1}\right) - y_{n+1}^1 = \frac{-19}{720}\left[y_4^c - y_4^p\right]$$

$$= \frac{-19}{720}[0.5317149 - 0.5333741] = 0.0001144$$

Since the local truncation error estimate indicates possible inaccuracy in the 4th decimal place (4th significant digit) of y_4^c, we take y_4^c as an improved y_4^p to get an improved y_4^c as follows:

$$y_4^c = y_3 + \frac{h}{24}\left[9\left[-x_4(0.5117149)^2\right] + 19 f_3 - 5 f_2 + f_1\right] = 0.5318739$$

The local truncation error estimate of this y_4^c is

$$\frac{-19}{720}[0.5318739 - 0.5317149] = -0.0000112$$

indicating that y_4^c should be accurate to about five significant digits.

$$n = 4: \qquad f_4 = f\left(x_4, y_4\right) = -(2.4)(0.5318739)^2 = -0.6789358$$

$$y_5^c = y_4 + \frac{h}{24}\left[55f_4 - 59f_3 + 37f_2 - 9f_1\right] = 0.4712642$$

$$y_5^c = y_4 + \frac{h}{24}\left[-9\left(-x_5 y_5^p\right)^2 + 19f_4 - 5f_3 + f_2\right] = 0.4704654$$

The local truncation error estimate is $\dfrac{-19}{720}\left[y_5^c - y_5^p\right] = 0.0000562$. As before, this estimate indicates possible inaccuracy in the 4th significant digit of y_5^C. Hence, we get an improved y_5 as

$$y_5^c = y_4 + \frac{h}{24}\left[-9\left(-x_5(0.4704654)^2\right) + 19f_4 - 5f_3 + f_2\right] = 0.4705358$$

The local truncation error estimate for this y_5^C is

$$\frac{-19}{270}[0.4705358 - 0.4704654] = -0.0000050$$

indicating this y_5^C should be accurate to about five significant digits.

Table 8.15 summarizes the computations and comparison with the exact solution $y(x_n)$ and the relative error $E_n(h)$.

TABLE 8.15 Adams-Moulton method value for $\dfrac{dy}{dx} = -xy^2, y(2) = 1$ with $h = 0.1$.

x_n	Exact $y(x_n)$	Using Adams-Moulton method	
		y_n	$E_n(h)$
$x_0 = 2.0$	1.000000	Exact	—
2.1	0.829876	Exact	—
2.2	0.704225	Exact	—
2.3	0.607903	Exact	—
2.4	0.531915	0.531874	0.000041
2.5	0.470588	0.470536	0.000052
2.6	0.420168	0.420114	0.000054
2.7	0.378072	0.378020	0.000052
2.8	0.342466	0.342419	0.000047
2.9	0.312012	0.311971	0.000041
$x_F = 3.0$	0.285714	0.285674	0.000040

EXAMPLE 8.24

Approximate the y value at $x = 0.4$ of the following differential equation $\dfrac{dy}{dx} = 0.5y$, $y(0) = 1.0$ using the Adams-Moulton method.

Solution:

The predicted value at $x = 0.4$ is given by Eq. (8.58)

$$y_{n+1}^p = y_n + \frac{h}{24}[55f_n - 59f_{n-1} + 37f_{n-2} - 9f_{n-3}]$$

or $\qquad y_4^p = y_3 + \dfrac{0.1}{24}[55f_0 - 59f_2 + 37f_1 - 9f_0] \qquad\qquad$ (E.1)

where f_0, f_1 and f_2 values are obtained from Table 8.6. Substituting the values of f_0, f_1 and f_2 from Table 8.6, Eq. (E.1) becomes

$$y_4^p = 1.1576 + \frac{0.1}{24}[55(0.5) - 59(0.5513) + 37(0.525) - 9(0.5)] = 1.1988$$

The corrected value is obtained by first evaluating $f(x_4, y_4)$, then substituting into Eq. (8.59). That is,

$$f_4 = f(x_4, y_4, p) = \frac{1}{2}(1.2213) = 0.6106$$

and from Eq. (8.59)

$$y_{n+1}^C = y_n + \frac{h}{24}[9f_{n+1} + 19f_n - 5f_{n-1} + f_{n-2}]$$

$$y_4^C = y_3 + \frac{0.1}{24}[9f_4 + 19f_3 - 5f_2 + f_1]$$

$$= 1.1576 + \frac{0.1}{24}[9(0.6129) + 19(0.5788) - 5(0.5513) + 0.5250] = 1.2171$$

The corrected value (1.2171) is clearly more accurate than the predicted value (1.1988) when compared with the exact value of $y_4 = 1.2214$.

8.3.4.2 Milne's Predictor-Corrector Method

Consider the differential equation

$$\frac{dy}{dx} = f(x,y) \qquad y(0) = 0 \qquad\qquad (8.64)$$

Integrating Eq. (8.64), we obtain

$$y = y_0 + \int_{x_0}^{x} f(x)\, dx$$

or $$y = y_0 + \int_{x_0}^{x_4} f(x, y)\, dx \quad \text{in the range } x_0 \le x \le x_4 \tag{8.65}$$

Applying Newton's forward difference formula, we get

$$f(x, y) = f_0 + n\Delta f_0 + \frac{n(n+1)}{2}\Delta^2 f_0 + \frac{f(n+1)(n+2)}{6}\Delta^3 f_0 + \dots \tag{8.66}$$

Substituting Eq. (8.66) into the right-hand side of Eq. (8.65), we get

$$y_4 = y_0 + \int_{x_0}^{x_4}\left[f_0 + n\Delta f_n + \frac{n(n-1)}{2}\Delta^2 f_0 + \dots \right] dx$$

$$= y_0 + h\int_{x_0}^{x_4}\left[f_0 + n\Delta f_n + \frac{n(n-1)}{2}\Delta^2 f_0 + \dots \right] dn \tag{8.67}$$

or $$y = y_1 + h\left[4f_0 + 8\Delta f_0 + \frac{20}{3}\Delta^2 f_0 + \frac{8}{3}\Delta^3 f_0 + \dots \right]$$

Neglecting the fourth and higher order differences and expressing the differences Δf_0, $\Delta^2 f_0$, and $\Delta^3 f_0$ in terms of the functional values, we get

$$y = y_0 + \frac{4}{3}h[2f_1 - f_2 + 2f_3] \tag{8.68}$$

Equation (8.68) can be used to predict the value of y_4 when those of y_0, y_1, y_2 and y_3 are known. Once we obtain y_4, we can then find a first approximation to

$$f_4 = f(x_0 + 4h,\, y_4)$$

A better value of y_4 can then be obtained by applying Simpson's rule as

$$y_4 = y_2 + \frac{h}{2}[f_2 + 4f_3 + f_4] \tag{8.69}$$

Equation (8.64) called a *corrector*. An improved value of f_4 is calculated and again the corrector is applied to obtain a still better value of y_4. The procedure is repeated until y_4 remains unchanged. After obtaining y_4 and f_4 to a desired degree of accuracy, then $y_5 = (x_0 + 5h)$ is obtained from the predicted as

$$y_5 = y_1 + \frac{4}{3}h[2f_2 - f_3 + 2f_4]$$

and $\quad f_5 = f[x_0 + 5h, y_5]$

is computed.

A better approximation to the value of y_5 is then obtained from the corrector as

$$y_5 = y_3 + \frac{h}{3}[f_3 + 4f_4 + f_5]$$

This step is repeated until y_5 becomes stable and then we proceed to compute y_6 as before. This procedure is known as *Milne's predictor-corrector method*. The accuracy of the method is improved if we must improve the starting values and then sub-divide the intervals.

Summarizing, the predictor and corrector formulae are given by

$$y_{n+1}^p = y_{n-3} + \frac{4h}{3}[2f_n - f_{n-1} + 2f_{n-2}] \tag{8.70}$$

and $\quad y_{n+1}^C = y_{n-1} + \frac{h}{3}[f_{n+1}^p + 4f_n + f_{n-1}] \tag{8.71}$

The corresponding error estimates for Eqs. (8.70) and (8.71) are given by

$$e^p = \frac{28}{29}h^5 y^v(\xi_1) \tag{8.72}$$

$$e_m^p = -\frac{1}{90}h^5 y^v(\xi_2) \tag{8.73}$$

The local error estimate can be shown to be

$$D_{n+1} = \frac{-1}{29}\left[y_{n+1}^1 - y_{n+1}^0\right] \tag{8.74}$$

It should be noted here that Eq. (8.71) can be subjected to numerical instability in some cases.

EXAMPLE 8.25

Approximate the y value at $x = 0.4$ of the differential equation $dy = \frac{1}{2}y$, $y(0) = 1.0$ using the Milne predictor-corrector method.

Solution:

The predicted y value at $x = 4$ and $n = 3$ is given by Eq. (8.70). Hence,

$$y_{n+1}^p = y_{n-3} + \frac{4h}{3}[2f_n - f_{n-1} + 2f_{n-2}]$$

or $\qquad y_4^p = y_0 + \frac{4(0.1)}{3}[2f(x_1,y_1) - f(x_2,y_2) + 2f(x_3,y_3)]$

Here, we use the past values given in Example 8.10.

$$y_4^p = 1 + \frac{0.4}{3}[2(0.5250) - 0.5513 + 2(0.5788)] = 1.2259$$

The derivative at $x = 0.4$ can be approximated by using the predicted value to obtain

$$\left.\frac{dy}{dx}\right|_{x_4} = f(x_4, y_4) = \frac{1}{2}(y_4^p) = 0.6129$$

Hence, the corrected y_4 is obtained using Eq. (8.71)

$$y_{n+1}^C = y_{n-1} + \frac{h}{3}[f_{n+1}^p + 4f_n + f_{n-1}]$$

or $\qquad y_4^C = y_2 + \frac{0.1}{3}[f(x_2,y_2) + 4f(x_3,y_3) + f(x_4,y_4)]$

$$= 1.1025 + \frac{0.1}{3}[0.5513 + 4(0.5788) + 0.6129] = 1.2185$$

The predicted value 1.2259 is noted to be closer to the exact value of 1.2214 than the corrected value.

8.4 SUMMARY

Differential equations arise in scientific and engineering applications when a dependent variable y varies with the independent variable either time t or position x. In this chapter, the numerical techniques commonly used for solving ordinary differential equations are presented. There are two categories of methods to solve ordinary differential equations: *one-step methods* and *multistep methods*. In one-step methods, the value of the increment function is based on information at a single point "i." The class of methods called multistep methods use information from several previous points as the basis for extrapolating to a new value.

Among the one-step methods or single-step methods, Picard's method of successive approximation method and Taylor's series method were presented. Among the step-by-step methods or the marching methods are Euler's method, modified Euler's method, Runge-Kutta methods of order two and four, Adam-Moulton predictor-corrector method, and Milne's predictor-corrector method. These methods have been illustrated with example exercises and solutions.

EXERCISES

8.1. Use Picard's method of successive approximation to solve the equation $\dfrac{dy}{dx} = 1 + xy$, $y(0) = 1$ at $x = 0.1$.

8.2. Solve $\dfrac{dy}{dx} = x + y$ with the initial condition $x_0 = 0$, $y_0 = 1$ using Picard's method of successive approximation.

8.3. Use Picard's method of successive approximation to find $y(0.2)$ correct to five decimal places by solving $\dfrac{dy}{dx} = x - y$, with $y(0) = 1$.

8.4. Use Picard's method of successive approximation to tabulate the values of $y(0.1)$, $y(0.2)$, ..., $y(1.0)$ and form the solution of $\dfrac{dy}{dx} = x(1 + x^3 y)$, $y(0) = 3$.

8.5. Use Picard's method of successive approximation to find $y(0.1)$ from the equation $\dfrac{dy}{dx} = \dfrac{y - x}{y + x}$, $y(0) = 1$.

8.6. Use Picard's method of successive approximation to find $y(0.2)$ by solving the equation $\dfrac{dy}{dx} = x + y^2$ with $y(0) = 0$.

8.7. Using the Taylor's series method for $y(x)$, and find $y(0.1)$ correct to four decimal places from the initial value problem $y' = xy + 1$, $y(0) = 1$.

8.8. Find the values of $y(1.1)$ and $y(1.2)$ correct to three decimal places given that $\dfrac{dy}{dx} = xy^{1/3}$, $y(1) = x(1) = 1$ using the first three terms of the Taylor's series expansions.

8.9. Find the value of y at $x = 0.1$ and $x = 0.2$ using the Taylor's series method from $\dfrac{dy}{dx} = x^2 y - 1$, $y(0) = 1$ accurate to five decimal places.

8.10. Given that $y' + 4y = x^2$, $y(0) = 1$ Determine $y(0.2)$ with the fourth-order Taylor's series method using a single integration step.

8.11. Using the Taylor's series method for $y(x)$ given that $y' = y^2 - x$, $y(0) = 1$, find $y(0.1)$ and $y(0.3)$ correct to four decimal places.

8.12. Use the Taylor's series method to solve the differential equation

$\dfrac{dy}{dx} = \dfrac{1}{x^2 + y}$, $y(4) = 4$ to find $y(4)$ and $y(4.2)$.

8.13. Use Euler's method to find $y(1)$ from the differential equation

$\dfrac{dy}{dx} = x + y$, $y(0) = 1$.

8.14. Use Euler's method to solve $\dfrac{dy}{dx} = -1.2y + 7e^{-0.3x}$ from $x = 0$ to $x = 2$

with the initial condition $y = 3$ at $x = 0$. Take $h = 0.5$.

8.15. Solve using Euler's method to solve $\dfrac{dy}{dx} = x + y^2$, $y(1) = 0$ at $x = 1.3$ with

$h = 0.5$ and at $x = 1.175$ with $h = 0.025$.

8.16. Solve the following differential equation using Euler's method for $x = 0$

to 0.4. $\dfrac{dy}{dx} = 3yt^2 + 2yt = 1$ with $y(0) = 1$. Take step size $h = 0.1$.

8.17. Use Euler's method to approximate the solution of the following initial

value problem. $\dfrac{dy}{dt} = te^{3t} - 2y$, $0 \le t \le 1$, $y(0) = 0$, with $h = 0.5$.

8.18. Solve $\dfrac{dy}{dx} = x^2(1 + y)$ with $y(1) = 1$ to find $y(1.1)$ by using Euler's

method and taking $h = 0.025$.

8.19. Use the modified Euler's method to find an approximate value of y

when $x = 0.3$. Given that $\dfrac{dy}{dx} = x + y$, $y(0) = 1$.

8.20. Repeat Exercise 8.14 using the modified Euler's method.

8.21. Use the modified Euler's method to find the value of y at $x = 0.1$, given

that $y(0) = 1$ and $y' = x^2 + y$.

8.22. Using the modified Euler's method to find the value of $y(2)$ in steps of

0.1, given that $\dfrac{dy}{dx} = 2 + \sqrt{xy}$, $y(1) = 1$.

8.23. Solve $\dfrac{dy}{dx} = \sqrt{(xy+1)}$ with $y(0) = 1$ for finding $y(0.075)$ by using the modified Euler's method taking $h = 0.025$.

8.24. Use the modified Euler's method to approximate the solution to the following initial value problem.

$$\frac{dy}{dx} = 1 + (t - y)^2, \ 2 \leq t \leq 3, \ y(2) = 1, \text{ with } h = 0.5.$$

8.25. Find $y(0.1)$, $y(0.2)$, $y(0.3)$, and $y(0.4)$ correct to four decimal places given that $\dfrac{dy}{dx} = y - x$, $y(0) = 2$. Take $h = 0.1$ and use the second-order Runge-Kutta method.

8.26. Use the second-order Runge-Kutta method to solve the equation $\dfrac{dy}{dx} = \sin y$, $y(0) = 1$ from $x = 0$ to 0.5 in steps of $h = 0.1$. Keep four decimal places in the computations.

8.27. Use the second-order Runge-Kutta method to solve the equation $\dfrac{dy}{dt} = t^2 - y + 1; \ 0 \leq t \leq 0.5$ with $y(0) = 1$ and $h = 0.1$. Keep five decimal places in the computations.

8.28. Using the Runge-Kutta method of order 2, find y for $x = 0.1$, given that $y = 1$ when $x = 0$ and $\dfrac{dy}{dx} = x + y$. Use $h = 0.1$ and keep five decimal places in the computations.

8.29. Use the second-order Runge-Kutta method to solve the equation $\dfrac{dy}{dx} = y - x$, $y(0) = 2$ and find $y(0.1)$ correct to four decimal places with $h = 0.1$ and keep four decimal places.

8.30. Solve $\dfrac{dy}{dx} = \dfrac{(1 + xy)}{(x + y)}$, $y(1) = 1.2$ by the Runge-Kutta method of order 2. Take h = 0.1 for $y(1.2)$.

8.31. Use the classical Runge-Kutta method of fourth order to find the numerical solution at $x = 0.8$ for $\dfrac{dy}{dx} = \sqrt{x+y}$, $y = (0.4) = 0.41$. Assume a step length of $h = 0.2$.

8.32. Use the Runge-Kutta fourth-order method to find the value of y when $x = 1$ given that $\dfrac{dy}{dx} = \dfrac{y - x}{y + x}$, $y(0) = 1$.

8.33. Use the Runge-Kutta fourth-order method to solve the equation
$\dfrac{dy}{dx} = \dfrac{y^2 - x^2}{y^2 + x^2}$ with $y(0) = 1$ at $x = 0.2, 0.4$. Take $h = 0.2$.

8.34. Use the classical fourth-order Runge-Kutta method to solve
$\dfrac{dy}{dx} = -1.2y + 7e^{-0.3x}$ from $x = 0$ to $x = 1.5$ with the initial condition $y = 3$ at $x = 0$. Take $h = 0.5$.

8.35. Use the classical fourth-order Runge-Kutta method to integrate
$f(x, y) = -2x^3 + 12x^2 - 20x + 8.5$ using a step size of $h = 0.5$ and an initial condition of $y = 1$ at $x = 0$. Compute $y(0.5)$.

8.36. Use the Runge-Kutta fourth-order method to find $y(0.2)$, $y(0.4)$, and $y(0.6)$ given that $\dfrac{dy}{dx} = 1 + y^2$, $y(0) = 0$ and take $h = 0.2$.

8.37. Use the Adams-Moulton method to solve the differential equation
$\dfrac{dy}{dx} = x^2 - y + 1$, $0 \leq x \leq 1$ with $y(0) = 1$. Use $h = 0.1$ and find the solution of $y(0.2)$ accurate to six digits.

8.38. Use the Adams-Moulton method to solve the differential equation
$\dfrac{dy}{dx} = x - y^2$, $y(0) = 1$ to find $y(0.4)$. Given that $y(0.1) = 0.9117$, $y(0.2) = 0.8494$ and $y(0.3) = 0.8061$.

8.39. Use the Adams-Moulton method to find $y(0.8)$ given that $\dfrac{dy}{dx} = 1 + y^2$, $y(0) = 0$.

8.40. Use the Adams-Moulton method to solve the differential equation
$\dfrac{dy}{dx} = x^2 - y + 1$, $0 \leq x \leq 1$ with $y(0) = 1$. Find $y(0.4)$ given that $y(0.1) = 1.0003$, $y(0.2) = 1.00243$ and $y(0.3) = 1.00825$.

8.41. Use the Adams-Moulton method to find $y(1.4)$ given that $\dfrac{dy}{dx} = x^2(1 + y)$ and that $y(1) = 1$, $y(1.1) = 1.233$, $y(1.2) = 1.543$ and $y(1.3) = 1.979$.

8.42. Use the Adams-Moulton method to approximate the solution to the following initial value problem. Use the exact starting values

$$\frac{dy}{dx} = 1 + (t - y)^2, \; 2 \le t \le 3, \; y(2) = 1, \text{ with } h = 0.2.$$

Actual solution is $y(t) = t + \dfrac{1}{1-t}$.

8.43. Use Milne's predictor-corrector method to find $y(0.8)$ taking $h = 0.2$. Given that $\dfrac{dy}{dx} = y + x^2$ with $y(0) = 1$.

8.44. Use Milne's predictor-corrector method to solve $\dfrac{dy}{dx} = x + y$, with the initial condition $y(0) = 1$, from $x = 0.2$ to $x = 0.3$.

8.45. Use Milne's method to compute the solution at $x = 0.4$ given that $\dfrac{dy}{dx} = xy + y^2$, $y(0) = 1$. Take $h = 0.1$ and obtain the starting values for Milne's method using the Runge-Kutta method of order 4.

8.46. Use Milne's method to solve the differential equation for $x = 0.4$ and $x = 0.5$ given that $\dfrac{dy}{dx} = x^2 - y + 1$, $0 \le x \le 1$ and $y(0) = 1$, Given $y(0.1) = 1$, $y(0.2) = 1.0024$ and $y(0.3) = 1.0083$.

8.47. Use Milne's method to find $y(0.8)$ and $y(1.0)$ given that $\dfrac{dy}{dx} = 1 + y^2$, $y(0) = 0$ and $y(0.2) = 0.2027$, $y(0.4) = 0.4228$ and $y(0.6) = 0.6841$. Take $h = 0.2$.

8.48. Solve $\dfrac{dy}{dx} = -y$ with $y(0) = 1$ by using Milne's predictor-corrector method for $x = 0.5$ to 0.7 with $h = 0.1$.

*B*IBLIOGRAPHY

There are several outstanding text and reference books on numerical methods that merit consultation for those readers who wish to pursue these topics further. The following list is but a representative sample of many excellent books.

Numerical Methods

Abramowitz, M. and Stegun, I., *Handbook of Mathematical Functions*, Dover Press, New York, 1964.

Ahlberg, J.H., Nilson, E.N. and Walsh, J.L., *The Theory of Splines and Their Applications*, Academic Press, New York, 1967.

Akai, T.J., *Applied Numerical Methods for Engineers*, Wiley, New York, 1993.

Al-Khafaji, A.W. and Tooley, J.R., *Numerical Methods in Engineering Practice*, Holt, Rinehart and Winston, New York, 1986.

Allen, M. and Iaacson, E., *Numerical Analysis for Applied Science*, Wiley, New York, 1998.

Ames, W.F., *Numerical Methods for Partial Differential Equations*, 3rd ed., Academic Press, New York, 1992.

Ascher,U., Mattheij, R. and Russell, R., *Numerical Solution of Boundary Value Problems for Ordinary Differential Equations*, Prentice Hall, Englewood Cliffs, NJ, 1988.

Atkinson, K.E., *An Introduction to Numerical Analysis*, 2nd ed., Wiley, New York, 1993.

Atkinson, K.E. and Han, W., *Elementary Numerical Analysis*, 3rd ed. Wiley, New York, 2004.

Atkinson, L.V. and Harley, P.J., *Introduction to Numerical Methods with PASCAL*, Addison Wesley, Reading, MA, 1984.

Atkinson, L.V., Harley, P.J. and Hudson, J.D., *Numerical Methods with FORTRAN 77*, Addison Wesley, Reading, MA, 1989.

Axelsson, K., *Iterative Solution Methods*, Cambridge University Press, Cambridge, UK, 1994.

Ayyub, B.M. and McCuen, R.H., *Numerical Methods for Engineers*, Prentice Hall, Upper Saddle River, NJ, 1996.

Baker, A.J., *Finite Element Computational Fluid Mechanics*, McGraw-Hill, New York, 1983.

Balagurusamy, E., *Numerical Methods*, Tata McGraw-Hill, New Delhi, India, 2002.

Bathe, K.J. and Wilson, E.L., *Numerical Methods in Finite Element Analysis*, Prentice Hall, Englewood Cliffs, NJ, 1976.

Bhat, R.B. and Chakraverty, S., *Numerical Analysis in Engineering*, Narosa Publishing House, New Delhi, India, 2004.

Bhat, R.B. and Gouw, G.J., *Numerical Methods in Engineering*, Simon and Schuster, Needham Heights, MA, 1996.

Bjorck, A., *Numerical Methods for Least Squares Problems*, Society for Industrial and Applied Mathematics (SIAM), Philadelphia, PA, 1996.

Booth, A.D., *Numerical Methods*, Academic Press, New York, 1958.

Brice, C., Luther, H.A and Wilkes, J. O., *Applied Numerical Methods*, New York, 1969.

Buchanan, J.L. and Turner, P.R., *Numerical Methods and Analysis*, McGraw-Hill, New York, 1992.

Burden, R.L. and Faires, J.D., *Numerical Analysis*, 6th ed., Brooks/Cole, Pacific Grove, CA 1997.

Butcher, J.C., *The Numerical Analysis of Ordinary Differential Equations: Runge-Kutta and General Liner Methods*, Wiley, New York, 1987.

Carnahan, B., Luther, A. and Wilkes, J.O., *Applied Numerical Methods*, Wiley, New York, 1969.

Chapra, S.C. and Canale, R.P., *Introduction to Computing for Engineers*, 2nd ed., McGraw-Hill, New York, 1994.

Chapra, S.C. and Canale, R.P., *Numerical Methods for Engineers with Personal Computers*, McGraw-Hill, New York, 1985.

Chapra, S.C., *Applied Numerical Methods with MATLAB for Engineers and Scientists*, McGraw-Hill, New York, 2005.

Chapra, S.C., *Numerical Methods for Engineers with Software and Programming Applications*, 4th ed., McGraw-Hill, New York, 2002.

Cheney, W. and Kincaid, D., *Numerical Mathematics and Computing*, 2nd ed., Brooks/Cole, Monterey, CA, 1994.

Chui, C., *An Introduction to Wavelets*, Academic press, Burlington, MA, 1992.

Collatz, L., *Numerical Treatment of Differential Equations*, 3rd ed., Springer-Verlag, Berlin, 1966.

Consatantinides, A., *Applied Numerical Methods with Personal Computers*, McGraw-Hill, New York, 1987.

Conte, S.D. and DeBoor, C.W., *Elementary Numerical Analysis: An Algorithm Approach*, 2nd ed., McGraw-Hill, New York, 1972.

Dahlquist, G. and Bjorck, A., *Numerical Methods*, Prentice Hall, Englewood Cliffs, NJ, 1974.

Davis, P. and Rabinowitz, P., *Methods of Numerical Integration,* Academic Press, 2nd ed, New York, 1998.

Demmel, J.W., *Applied Numerical Linear Algebra*, Society for Industrial and Applied Mathematics (SIAM), Philadelphia, PA, 1997.

Dennis, J.E. and Schnabel, R.B., *Numerical Methods for Unconstrained Optimisation and Non-linearEquations,* Society for Industrial and Applied Mathematics (SIAM), Philadelphia, PA, 1996.

Dukkipati, R.V., *Numerical Methods through Solved Problems*, New Age International Publishers (P) Ltd., New Delhi, India, 2009.

Dukkipati, R.V., *Applied Numerical Methods with MATLAB*, New Age International Publishers (P) Ltd., New Delhi, India, 2010.

Dukkipati, R.V., *Applied Numerical Methods with MATLAB through Solved Problems*, New Age International Publishers (P) Ltd., New Delhi, India, 2010.

Epperson, J.F., *An Introduction to Numerical Methods and Analysis*, Wiley, New York, NY, 2001.

Fadeev, D.K. and Fadeeva, V.N., *Computational Methods of Linear Algebra*, Freeman, San Francisco, 1963.

Fadeeva, V.N., (Trans. Curtis D. Benster)., *Computational Methods of Linear Algebra*, Dover, New York, 1959.

Fatunla, S.O., *Numerical Methods for Initial Value Problems in Ordinary Differential Equations*, Academic Press, San Diego, 1988.

Ferziger, J.H., *Numerical Methods for Engineering Applications*, 2nd ed., Wiley, New York, 1998.

Forbear, C.E., *Introduction to Numerical Analysis*, Addison-Wesley, Reading, MA, 1969.

Forsythe, G.E. and Wasow, W.R., *Finite Difference Methods for Partial Differential Equations*, Wiley, New York, 1960.

Forsythe, G.E., Malcolm, M.A. and Moler, C.B., *Computer Methods for Mathematical Computation*, Prentice Hall, Englewood Cliffs, NJ, 1977.

Fox, L., *Numerical Solution of Ordinary and Partial Differential Equations*, Pergamon Press, London, 1962.

Froberg, C.E., *Introduction to Numerical Analysis*, Addison-Wesley, Reading, MA, 1965.

Gautschi, W., *Numerical Analysis: An Introduction*, Birkhauser, Boston, MA, 1997.

Gear, C.W., *Numerical Initial Value Problems in Ordinary Differential Equations*, Prentice Hall, Englewood Cliffs, NJ, 1971.

Gerald, C.F. and Wheatley, P.O., *Applied Numerical Analysis*, 5th ed., Addison-Wesley, Reading, MA, 1994.

Gladwell, J. and Wait, R., *A Survey of Numerical Methods of Partial Differential Equations*, Oxford University Press, New York, 1979.

Goldberg, D.E., *Genetic Algorithms in Search, Optimisation and Machine Learning*, Addison-Wesley, Reading, MA, 1989.

Golub, G.H. and Van Loan, C.F., *Matrix Computations*, 3rd ed., Johns Hopkins University Press, Baltimore, MD, 1996.

Greenbaum, A., *Iterative Methods for Solving Linear Systems*, Society for Industrial and Applied Mathematics (SIAM), Philadelphia, PA, 1997.

Griffiths, D.V. and Smith, I.M., *Numerical Methods for Engineers*, Oxford University Press, New York, 1991.

Guest, P.G., *Numerical Methods of Curve Fitting*, Cambridge University Press, New York, 1961.

Gupta, S.K., *Numerical Methods for Engineers*, New Age International Publishers (P) Ltd., New Delhi, India, 1995.

Hager, W.W., *Applied Numerical Algebra*, Prentice Hall, Upper Saddle River, NJ, 1998.

Hamming, R.W., *Numerical Methods for Scientists and Engineers*, 2nd ed., McGraw-Hill, New York, 1973.

Henrici, P.H., *Elements of Numerical Analysis*, Wiley, New York, 1964.

Higham, N.J., *Accuracy and Stability of Numerical Algorithms*, Society for Industrial and Applied Mathematics (SIAM), Philadelphia, PA, 1996.

Hildebrand, F.B., *Introduction to Numerical Analysis*, 2nd ed., McGraw-Hill, New York, 1974.

Hoffman, J., *Numerical Methods for Engineers and Scientists*, McGraw-Hill, New York, 1992.

Hornbeck, R.W., *Numerical Methods*, Quantum, New York, 1975.

Householder, A.S., *Principles of Numerical Analysis*, McGraw-Hill, New York, 1953.

Householder, A.S., *The Theory of Matrices in Numerical Analysis*, Blaisdell, New York, 1964.

Iserles, A., *A First Course in the Numerical Analysis of Differential Equations*, Cambridge University Press, New York, 1996.

Issaccson, E., and Keller, H.B. and Bishop, H., *Analysis of Numerical Methods*, Wiley, New York, 1966.

Jacobs, D. (Ed.)., *The State of the Art in Numerical Analysis*, Academic Press, London, 1977.

Jacques, I. and Colin, J., *Numerical Analysis*, Chapman and Hall, New York, 1987.

Jain, M.K., Iyengar, S.R.K. and Jain, R.K., *Numerical Methods for Scientific and Engineering Computations,*New Age International Publishers (P) Limited, New Delhi, India, 1984.

Jain, M.K., *Numerical Analysis for Scientists and Engineers*, S.B.W. Publishers, New Delhi, India, 1971.

Jain, M.K., *Numerical Analysis for Scientists and Engineers*, S.B.W. Publishers, New Delhi, India, 1971.

James, M.L., Smith, G.M. and Wolford, J.C., *Applied Numerical Methods for Digital Computations with FORTRAN and CSMP*, 3rd ed., Harper & Row, New York, 1985.

Johnson, L.W. and **Riess, R.D.,** *Numerical Analysis*, 2nd ed., Addison-Wesley, Reading, MA, 1982.

Johnston, R.L., *Numerical Methods: A Software Approach*, Wiley, New York, 1982.

Kahaneer, D., Moher, C. and Nash, S., *Numerical Methods and Software*, Prentice Hall, Englewood Cliffs, NJ, 1989.

Keller, H.B., *Numerical Methods for Two-Point Boundary Value Problems*, Wiley, New York, 1968.

Kelley, C.T., *Iterative Methods of Optimisation*, Society for Industrial and Applied Mathematics (SIAM), Philadelphia, PA, 1999.

Kharab, A. and Guenther, R.B., *An Introduction to Numerical Methods–A MATLAB Approach*, CRC Press, Boca Raton, FL, 2001.

Kincaid, D. and Cheney, W., *Numerical Analysis: Mathematics of Scientific Computing*, Brooks/Cole, Pacific Grove, CA, 1996.

Kress, R., *Numerical Analysis*, Springer-Verlag, New York, 1998.

Krishnamurthy, E.V. and Sen, S.K., *Numerical Algorithms*, East West Publishers, New Delhi, India, 1986.

Krommer, A.R. and Ueberhuber, C.W., *Computational Integration*, Society for Industrial and Applied Mathematics (SIAM), Philadelphia, PA, 1998.

Lambert, J.D., *Numerical Methods for Ordinary Differential Equations— The Initial Value Problems*, Wiley, New York, NY, 1991.

Lapidus, L. and Pinder, G.F., *Numerical Solution of Ordinary Differential Equations in Science and Engineering*, Wiley, New York, 1981.

Lapidus, L. and Seinfield, J.H., *Numerical Solution of Partial Differential Equations*, Academic Press, New York, 1971.

Lastman, G.J. and Sinha, N.K., *Microcomputer Based Numerical Methods for Science and Engineering,* Saunders College Publishing, New York, 1989.

Levy, H. and Baggott, E.A., *Numerical Solutions of Differential Equations*, Dover, New York, 1950.

Maron, M.J., *Numerical Analysis: A Practical Approach*, Macmillan, New York, 1982.

Mathews, J.H., *Numerical Methods for Mathematics, Science and Engineering*, 2nd ed, Prentice Hall of India, New Delhi, India, 1994.

Milne, W.E., *Numerical Solution of Differential Equations*, Wiley, New York, 1953.

Mitchell, A.R. and Griffiths, D.F., *The Finite Difference Method in Partial Differential Equations*, Wiley, New York, 1980.

Moin, P., *Fundamentals of Engineering Numerical Analysis*, Cambridge University Press, New York, 2001.

Morton, K.W. and Mayers, D.F., *Numerical Solution of Partial Differential Equations: An Introduction*, Cambridge University Press, Cambridge, UK, 1994.

Myron, A. and Issacson, E.L., *Numerical Analysis for Applied Science*, Wiley, Hoboken, NJ, 1998.

Na, T.Y., *Computational Methods in Engineering Boundary Value Problems*, Academic Press, New York, 1979.

Nakamura, S., *Computational Methods in Engineering and Science*, Wiley, New York, 1977.

Nielson, K.L., *Methods in Numerical Analysis*, Macmillan Company, New York, 1964.

Noble, B., *Numerical Methods,* Vol. 2, Oliver and Boyd, Edinburgh, Scotland 1964.

Nocedal, J. and Wright, S.J., *Numerical Optimisation*, Springer-Verlag, New York, 1999.

Ortega, J.M., *Numerical Analysis—A Second Course*, Academic Press, New York, 1972.

Powell, M., *Approximation Theory and Methods*, Cambridge University Press, Cambridge, UK, 1981.

Press, W.H., Teukolsky, S.A., Vetterling, W.T. and Flannery, B.P., *Numerical Recipes: The Art of Scientific Computing*, 2nd ed., Cambridge University Press, New York, 1992.

Quarteroni, A., Sacco, R. and Saleri, F., *Numerical Mathematics*, Springer-Verlag, New York, 2000.

Ralston, A. and Rabinowitz, P., *A First Course in Numerical Analysis*, 2nd ed., McGraw-Hill, New York, 1978.

Ralston, A. and Wilf, H.S., (Eds.), *Mathematical Methods for Digital Computers*, Vols. 1 and 2, Wiley, New York, 1967.

Rao, K.S., *Numerical Methods for Scientists and Engineers*, Prentice Hall, New Delhi, India, 2001.

Rao, S.S., *Applied Numerical Methods for Engineers and Scientists*, Prentice Hall, Upper Saddle River, NJ, 2002.

Ratschek, H. and Rokne, J., *Computer Methods for the Range of Functions*, Ellis Horwood, Chichester England, 1984.

Rice, J.R., *Numerical Methods, Software and Analysis*, McGraw-Hill, New York, 1983.

Sastry, S.S., *Introductory Methods of Numerical Analysis*, Prentice Hall of India, New Delhi, India, 2001.

Scarborough, J.B., *Numerical Mathematical Analysis*, 6th ed., Johns Hopkins University Press, Baltimore, MD, 1966.

Scheid, F., *Schaum's Outline of Theory and Problems in Numerical Analysis*, 2nd ed., Schaum's Outline Series, McGraw-Hill, New York, 1988.

Schiesser, W.E., *Computational Mathematics in Engineering and Applied Science*, CRC Press, Boca Raton, FL, 1994.

Shampine, L.F., *Numerical Solution of Ordinary Differential Equations*, Chapman and Hall, New York, 1994.

Sharma, J.N., *Numerical Methods for Engineers and Scientists*, Narosa Publishing House, New Delhi, India, 2004.

Smith, G.D., *Numerical Solution of Partial Differential Equations: Finite Difference Methods*, 3rd ed., Oxford University Press, Oxford, 1985.

Smith, W.A., *Elementary Numerical Analysis*, Prentice Hall, Englewood Cliffs, NJ, 1986.

Snyder, M.A., *Chebyshev Methods in Numerical Approximation*, Prentice Hall, Englewood Cliffs, NJ, 1966.

Somasundaram, R.M. and Chandrasekaran, R.M., *Numerical Methods with C++ programming*, Prentice Hall of India, New Delhi, India, 2005.

Stanton, R.G., Numerical Methods for Science and Engineering, Prentice Hall of India, New Delhi, India, 1967.

Stark, P.A., *Introduction to Numerical Methods*, Macmillan, New York, 1970.

Stewart, G.W., *Matrix Algorithms*, Vol. 1, *Basic Decompositions*, Society for Industrial and Applied Mathematics (SIAM), Philadelphia, PA, 1998.

Stoer, J. and Bulirsch, R., *Introduction to Numerical Analysis*, Springer-Verlag, New York, 1980.

Stroud, A., and Secrets, D., *Gaussian Quadrature Formulas*, Prentice Hall, Englewood Cliffs, NJ, 1966.

Stroud, A.H., *Numerical Quadrature and Solution of Ordinary Differential Equations*, Springer-Verlag, New York, 1974.

Taylor, J.R., *An Introduction to Error Analysis*, University Science Books, Mill Valley, CA, 1982.

Todd, Y., *Survey of Numerical Analysis*, McGraw-Hill, New York, 1962.

Traub, J.F., *Iterative Methods for the Solution of Equations*, Prentice Hall, Englewood Cliffs, NJ, 1964.

Trefethen, L.N. and Bau, D., *Numerical Linear Algebra*, Society for Industrial and Applied Mathematics (SIAM), Philadelphia, PA, 1997.

Tyrtyshnikov, E.E., *A Brief Introduction to Numerical Analysis*, Birkhauser, Boston, 1997.

Ueberhuber, C.W., *Numerical Computation 1: Methods, Software and Analysis*, Springer-Verlag, New York, 1997.

Ueberhuber, C.W., *Numerical Computation 2: Methods, Software and Analysis*, Springer-Verlag, New York, 1997.

Vemuri, V. and Karplus, W.J., *Digital Computer Treatment of Partial Differential Equations*, Prentice Hall, Englewood Cliffs, NJ, 1981.

Vichnevetsky, R., *Computer Methods for Partial Differential Equations,* Vol. 1: *Elliptic Equations and the Finite Element Method,* Prentice Hall, Englewood Cliffs, NJ, 1981.

Vichnevetsky, R., *Computer Methods for Partial Differential Equations,* Vol. 2: *Initial Value Problems,* Prentice Hall, Englewood Cliffs, NJ, 1982.

Wait, R., *The Numerical Solution of Algebraic Equations,* Wiley, New York, 1979.

Wendroff, B., *Theoretical Numerical Analysis,* Academic Press, New York, 1966.

Wilkinson, J.H., *Rounding Errors in Algebraic Processes,* Dover, New York, 1994.

Yokowitz, S. and Szidarovsky, F., *An Introduction to Numerical Computation,* Macmillan, New York, 1986.

Yong, D.M. and Gregory, R.T., *A Survey of Numerical Mathematics,* Vol. 1 and 2, Addison-Wesley, Reading, MA, 1972.

Young, D., *Iterative Solution for Large Linear Systems,* Academic Press, New York, 1971.

Young, D.M. and Gregory, R.T., *A Survey of Numerical Mathematics,* Addison-Wesley, Reading, MA, 1972.

Young, D.M., *Iterative Solution of Large Linear Systems,* Academic Press, New York, 1971.

PARTIAL FRACTION EXPANSIONS

In obtaining the solution of many engineering problems, we encounter rational algebraic fractions that are ratio of two polynomials in s, such as

$$F(s) = \frac{P(s)}{Q(s)} = \frac{b_0 s^m + b_1 s^{m-1} + \dots + b_m}{a_0 s^n + a_1 s^{n-1} + \dots + a_n} \tag{A.1}$$

In practical systems, the order of polynomial in numerator is equal to or less than that of denominator. In terms of the orders m and n, rational algebraic fractions are subdivided as follows:

(i) Improper fraction if $m \geq n$

(ii) Proper fraction if $m < n$

An improper fraction can be separated into a sum of a polynomial in s and a proper fraction, i.e.,

$$F(s) = \frac{P(s)}{Q(s)} = d(s) + \frac{P(s)}{Q(s)} \tag{A.2}$$

Improper   Proper

This can be achieved by performing a long division. To obtain the partial fraction expansion of a proper factorize, first of all we factorize the polynomial $Q(s)$ into n first-order factors. The roots may be real, complex, distinct or repeated. Several cases are discussed below:

Case-I   Partial Fraction Expansion when Q(s) has Distinct Roots

In this case, Eq. (A.1) may be written as

$$F(s) = \frac{P(s)}{Q(s)} = \frac{P(s)}{(s+p_1)(s+p_2)\cdots(s+p_k)\cdots(s+p_n)} \tag{A.3}$$

which when expanded, gives

$$F(s) = \frac{A_1}{s+p_1} + \frac{A_2}{s+p_2} + \cdots + \frac{A_k}{s+p_k} + \cdots + \frac{A_n}{s+p_n} \tag{A.4}$$

where A_k ($k = 1, 2, \ldots, n$) are constants.

To evaluate the coefficients A_k, multiply $F(s)$ in Equation (A.3) by $(s+p_k)$ and let $s = -p_k$. This gives

$$A_k = (s+p_k)\frac{P(s)}{Q(s)}\bigg|_{S=-p_k} = \frac{P(s)}{\frac{d}{ds}Q(s)}\bigg|_{S=-p_k} \tag{A.5}$$

$$= \frac{P(-p_k)}{(p_1-p_k)(p_2-p_k)\cdots(p_{k-1}-p_k)\cdots(p_n-p_k)} \tag{A.6}$$

EXAMPLE A.1

Find the partial fraction expansion of the function

$$F(s) = \frac{3}{(s+1)(s+2)}$$

Solution:

The roots of the denominator are distinct.

Hence, $$F(s) = \frac{3}{(s+1)(s+2)} = \frac{A}{(s+1)} + \frac{B}{(s+2)}$$

or $$\frac{3}{(s+2)} = A + \frac{B(s+1)}{(s+2)}$$

Let $\qquad s = -1$, then $A = 3$

Let $\qquad s = -2$, then

$$\frac{3}{(s+1)} = \frac{A(s+2)}{(s+1)} + B$$

or $$\frac{3}{(-2+1)} = B$$

or $$B = -3$$

Hence, $\qquad F(s) = \frac{3}{s+1} - \frac{3}{s+2}$

EXAMPLE A.2

Find the partial fraction expansion of the function

$$F(s) = \frac{5}{(s+1)(s+2)^2}$$

Solution:

The roots of $(s+2)^2$ in the denominator are repeated. We can write

$$F(s) = \frac{5}{(s+1)(s+2)^2} = \frac{A}{(s+1)} + \frac{B}{(s+2)^2} + \frac{C}{(s+2)} \qquad (A.7)$$

Multiplying Eq. (A.7) by $(s+1)$ on both sides and letting $s = -1$

$$\frac{5}{(s+2)^2} = A + \frac{B(s+1)}{(s+2)^2} + \frac{C(s+1)}{(s+2)}$$

$$5 = A$$

Multiplying Eq. (A.7) by $(s+2)^2$ and letting $s = -2$

$$\frac{5}{(s+1)} = \frac{A(s+2)^2}{(s+1)} + B + C\ (s+2) \qquad (A.8)$$

or $\qquad\qquad B = -5$

To find C, we differentiate Eq.(A.8) with respect to s and letting $s = -2$

$$\frac{-5}{(s+1)^2} = \frac{(s+2)s}{(s+1)^2}\ A + C$$

or $\qquad\qquad C = -5$

Hence, $\qquad F(s) = \dfrac{5}{(s+1)(s+2)^2} = \dfrac{5}{(s+1)} + \dfrac{-5}{(s+2)^2} + \dfrac{-5}{(s+2)}$

EXAMPLE A.3

Find the partial fraction expansion of the function

$$F(s) = \frac{4\left(s^2 + 5s + 2\right)}{s(s+1)(s+3)}$$

Solution:

$$\frac{4\left(s^2 + 5s + 2\right)}{s(s+1)(s+3)} = \frac{A}{s} + \frac{B}{s+1} + \frac{C}{s+2}$$

$$4\left(s^2 + 5s + 2\right) \equiv A(s+1)(s+2) + Bs(s+2) + Cs(s+1)$$

Substituting $s = 0$, -1 and -2 in the above equation we get the values of A, B, and C as 4, 8, and -8 respectively. Hence,

$$F(s) = \frac{4\left(s^2 + 5s + 2\right)}{s(s+1)(s+3)} = \frac{4}{s} + \frac{8}{s+1} + \frac{-8}{s+2}$$

Case-II   Partial Fraction Expansion When Q(s) has Complex Conjugate Roots

Suppose that there is a pair of complex conjugate roots in $Q(s)$, given by

$$s = -a - j\omega \text{ and } s = -a + j\omega$$

Then $F(s)$ may be written as

$$F(s) = \frac{P(s)}{Q(s)} = \frac{P(s)}{(s+a+j\omega)(s+a-j\omega)(s+p_3)(s+p_4)...(s+p_n)} \qquad (A.9)$$

which when expanded gives

$$F(s) = \frac{A_1}{(s+a+j\omega)} + \frac{A_2}{(s+a-j\omega)} + \frac{A_3}{s+p_3} + \frac{A_4}{s+p_4} + \cdots\cdots + \frac{A_n}{s+p_n} \qquad (A.10)$$

where A_1 and A_2 are the coefficients at $s = -(a+j\omega)$ and $s = -(a-j\omega)$ respectively. As per Eq. (A.5), the coefficient A^1 is given by

$$A_1 = \frac{P(s)}{Q(s)}(s+a+j\omega)\Big|_{s=-(a+j\omega)} \qquad (A.11)$$

B

Basic Engineering Mathematics

B.1 ALGEBRA

B.1.1 Basic Laws

Cumulative law:

$$a + b = b + a; \quad ab = ba$$

Associative law:

$$a + (b + c) = (a + b) + c; \; a(bc) = (abc)$$

Distributive law:

$$a(b + c) = ab + ac$$

B.1.2 Sums of Numbers

The sum of the first n numbers:

$$\sum_{1}^{n}(n) = \frac{n(n+1)}{2}$$

The sum of the squares of the first n numbers:

$$\sum_{1}^{n}(n^2) = \frac{n(n+1)(2n+1)}{6}$$

The sum of the cubes of the first n numbers:

$$\sum_{1}^{n} (n^3) = \frac{n^2 (n+1)^2}{4}$$

B.1.3   Progressions

Arithmetic Progression

$$a, a + d, a + 2d, a + 3d, \ldots.$$

where

a = first term

d = common difference

n = number of terms

S = sum of n terms

ℓ = last term

$\ell = a + (n - 1)d$

$S = (n/2)(a + \ell)$

$(a + b)/2$ = arithmetic mean of a and b.

Geometric Progression

$$a, ar, ar^2, ar^3, \ldots.$$

where

a = first term

r = common ratio

n = number of terms

S = sum of n terms

ℓ = last term

$\ell = ar^{n-1}$

$$S = a\frac{r^{n1} - 1}{r - 1} = \frac{r\ell - a}{r - 1}$$

$$S = \frac{a}{1-r} \text{ for } r^2 < 1 \text{ and } n = x$$

$$\sqrt{ab} = \text{geometric mean of } a \text{ and } b.$$

B.1.4 Powers and Roots

$$a^x a^y = a^{x+y}$$

$$\frac{a^x}{y^x} = a^{x-y}$$

$$(ab)^x = a^x b^x$$

$$(a^x)^y = a^{xy}$$

$$a^0 = 1 \text{ if } a \neq 0$$

$$a^{-x} = 1/a^x$$

$$a^{x/y} = \sqrt[y]{a^x}$$

$$a^{1/y} = \sqrt[y]{a}$$

$$\sqrt[x]{ab} = \sqrt[x]{a}\sqrt[x]{b}$$

$$\sqrt[x]{a/b} = \sqrt[x]{a} / \sqrt[x]{b}$$

B.1.5 Binomial Theorem

$$(a \pm b)^n = a^n \pm na^{n-1}b + \frac{n(n-1)}{2!}a^{n-2}b^2 + \frac{n(n-1)(n-2)}{3!}a^{n-3}b^3 + \ldots$$

$$+(\pm 1)^m \frac{n(n-1)\ldots(n-m+1)}{m!}a^{n-m}b^m + \ldots$$

where

$$m! = 1.2.3. \ldots (m-1)m$$

The series is finite if n is a positive integer. If n is negative or fractional, the series is infinite and will converge for $|b| < |a|$ only.

B.1.6   Absolute Values

The numerical or absolute value of a number n is denoted by $|n|$ and represents the magnitude of the number without regard to algebraic sign. For example,

$$|-5| = |+5| = 5$$

B.1.7   Logarithms

Laws of Logarithms

$$\log_b MN = \log_b M + \log_b N$$

$$\log_b 1 = 0$$

$$\log_b \frac{M}{N} = \log_b M - \log_b N$$

$$\log_b b = 1$$

$$\log_b N^m = m \log_b N$$

$$\log_b 0 = +\infty, \, 0 < b < 1$$

$$\log_b \sqrt[r]{N^m} = m/r \log_b N$$

$$\log_b 0 = -\infty, \, 1 < b < \infty$$

Important Constants

$$\log_{10} e = 0.4342944819$$

$$\log_{10} x = 0.4343 \log_e x = 0.4343 \ln x$$

$$\ln 10 = \log_e 10 = 2.3025850930$$

$$\ln x = \log_e x = 2.3026 \log_{10} x$$

B.2   TRIGONOMETRY

$$\sin \theta = 0 \quad \Rightarrow \quad \theta = n\pi, \, n \text{ is an integer}$$

$$\cos \theta = 0 \quad \Rightarrow \quad \theta = (2n + 1)\pi/2, \, n \text{ is an integer}$$

$\sin \theta = \sin \alpha \quad \Rightarrow \quad \theta = n\pi + (-1)^{n}; \, \alpha, \, n \text{ is an integer}$

$\cos \theta = \cos \alpha \quad \Rightarrow \quad \theta = 2n\pi \pm \alpha, \, n \text{ is an integer}$

$\tan \theta = 0 \quad\quad\quad \Rightarrow \quad \theta = n\pi, \, n \text{ is an integer}$

$\tan \theta = \tan \alpha \quad \Rightarrow \quad \theta = n\pi + \alpha, \, n \text{ is an integer}$

$$\cosh x = \frac{e^{x} + e^{-x}}{2}$$

$$\sinh x = \frac{e^{x} - e^{-x}}{2}$$

$$\tanh x = \frac{\sinh x}{\cosh x} = \frac{e^{x} - e^{-x}}{e^{x} + e^{-x}}$$

$$\coth x = \frac{e^{x} + e^{-x}}{e^{x} - e^{-x}} = \frac{1}{\tanh x} = \frac{\cosh x}{\sinh x}$$

$$\operatorname{sech} x = \frac{1}{\cosh x} = \frac{2}{e^{x} + e^{-x}}$$

$$\operatorname{cosec} x = \frac{1}{\sinh x} = \frac{2}{e^{x} - e^{-x}}$$

B.2.1   Trigonometric Identities

$\sin^{2} \alpha + \cos^{2} \alpha = 1$

$$1 + \tan^{2} \alpha = \frac{1}{\cos^{2} \alpha}$$

$\tan \alpha \cot \alpha = 1$

$$1 + \cot^{2} \alpha = \frac{1}{\sin^{2} \alpha}$$

$\sin (\alpha \pm \beta) = \sin \alpha \cos \beta \pm \cos \alpha \sin \beta$

$\cos (\alpha \pm \beta) = \cos \alpha \cos \beta \mp \sin \alpha \sin \beta$

$$\tan(\alpha \pm \beta) = \frac{\tan \alpha \pm \tan \beta}{1 \mp \tan \alpha \tan \beta}$$

$$\cot(\alpha \pm \beta) = \frac{\cot \alpha \cot \beta \mp 1}{1 \pm \cot \alpha \cot \beta}$$

$$\sin\alpha + \sin\beta = 2\sin\frac{\alpha+\beta}{2}\cos\frac{\alpha-\beta}{2}$$

$$\sin\alpha - \sin\beta = 2\cos\frac{\alpha+\beta}{2}\sin\frac{\alpha-\beta}{2}$$

$$\cos\alpha + \cos\beta = 2\cos\frac{\alpha+\beta}{2}\cos\frac{\alpha-\beta}{2}$$

$$\cos\alpha - \cos\beta = -2\sin\frac{\alpha+\beta}{2}\sin\frac{\alpha-\beta}{2}$$

$$\tan\alpha \pm \tan\beta = \frac{\sin(\alpha\pm\beta)}{\cos\alpha\cos\beta}$$

$$\cot\alpha \pm \cot\beta = \frac{\sin(\beta\pm\alpha)}{\sin\alpha\sin\beta}$$

$$\sin\alpha\cos\beta = \frac{1}{2}\sin(\alpha+\beta) + \frac{1}{2}\sin(\alpha-\beta)$$

$$\cos\alpha\cos\beta = \frac{1}{2}\cos(\alpha+\beta) + \frac{1}{2}\cos(\alpha-\beta)$$

$$\sin\alpha\sin\beta = \frac{1}{2}\cos(\alpha-\beta) - \frac{1}{2}\cos(\alpha+\beta)$$

$$\tan\alpha\tan\beta = \frac{\tan\alpha+\text{tab}\beta}{\cot\alpha+\cot\beta} = -\frac{\tan\alpha-\tan\beta}{\cot\alpha-\cot\beta}$$

$$\cot\alpha\cot\beta = \frac{\cot\alpha+\cot\beta}{\tan\alpha+\tan\beta} = -\frac{\cot\alpha-\cot\beta}{\tan\alpha-\tan\beta}$$

$$\cot\alpha\tan\beta = \frac{\cot\alpha+\tan\beta}{\tan\alpha+\cot\beta} = -\frac{\cot\alpha-\tan\beta}{\tan\alpha-\cot\beta}$$

B.3 DIFFERENTIAL CALCULUS

B.3.1 List of Derivatives

(Note: u, v, and w are functions of x)

$$\frac{d}{dx}(a) = 0 \qquad a = \text{constant}$$

$$\frac{d}{dx}(x) = 1$$

$$\frac{dy}{dx} = \frac{dy}{dv}\frac{dv}{dx} \qquad y = y\,(v)$$

$$\frac{d}{dx}(av) = a\frac{du}{dx}$$

$$\frac{dy}{dx} = \frac{1}{dx\,/\,dy} \qquad \text{if}\,\frac{dx}{dy} \neq 0$$

$$\frac{d}{dx}(\pm u \pm v \pm \ldots) = \pm\frac{du}{dx}\pm\frac{dv}{dx}+\ldots$$

$$\frac{d}{dx}(u^n) = nu^{n-1}\frac{du}{dx}$$

$$\frac{d}{dx}(uv) = u\frac{dv}{dx}+v\frac{du}{dx}$$

$$\frac{d}{dx}\frac{u}{v} = \frac{v\,du\,/\,dx - u\,dv\,/\,dx}{v^2}$$

$$\frac{d}{dx}(u^v) = vu^{v-1}\frac{du}{dx}+u^v\ln u\frac{dv}{dx}$$

$$\frac{d}{dx}(a^u) = a^u\ln a\frac{du}{dx}$$

$$\frac{d}{dx}(e^u) = e^u\frac{du}{dx}$$

$$\frac{d}{dx}(\ln u) = \frac{1}{u}\frac{du}{dx}$$

$$\frac{d}{dx}(\log_a u) = \frac{\log_a}{u}\frac{du}{dx}$$

$$\frac{d}{dx}(\sin u) = \cos u\frac{du}{dx}$$

$$\frac{d}{dx}(\cos u) = -\sin u \frac{du}{dx}$$

$$\frac{d}{dx}(\tan u) = \sec^2 u \frac{du}{dx}$$

$$\frac{d}{dx}(\operatorname{cosec} u) = -\operatorname{cosec} u \cot u \frac{du}{dx}$$

$$\frac{d}{dx}(\sec u) = \sec u \tan u \frac{du}{dx}$$

$$\frac{d}{dx}(\cot u) = -\operatorname{cosec}^2 u \frac{du}{dx}$$

$$\frac{d}{dx}(versu) = \sin u \frac{du}{dx}$$

$$\frac{d}{dx}\sin^{-1} u = \frac{1}{\sqrt{1-u^2}} \frac{du}{dx}; \quad \frac{-\pi}{2} \leqq \sin^{-1} u \leqq \frac{\pi}{2}$$

$$\frac{d}{dx}\cos^{-1} u = \frac{1}{\sqrt{1-u^2}} \frac{du}{dx}; \quad 0 \leqq \cos^{-1} u \leqq \pi$$

$$\frac{d}{dx}\tan^{-1} u = \frac{1}{1+u^2} \frac{du}{dx}$$

$$\frac{d}{dx}\sinh^{-1} u = \frac{1}{\sqrt{u^2+1}} \frac{du}{dx}$$

$$\frac{d}{dx}\cosh^{-1} u = \frac{1}{\sqrt{u^2-1}} \frac{du}{dx}; \quad u > 1$$

$$\frac{d}{dx}\tanh^{-1} u = \frac{1}{1-u^2} \frac{du}{dx}$$

$$\frac{d}{dx}\operatorname{cosec}^{-1} u = \frac{-1}{u\sqrt{u^2+1}} \frac{du}{dx}$$

$$\frac{d}{dx}\operatorname{sech}^{-1} u = -\frac{1}{\sqrt{1-u^2}} \frac{du}{dx} \quad u > 0$$

$$\frac{d}{dx}\coth^{-1}u = \frac{1}{1-u^2}\frac{du}{dx}$$

$$\frac{d}{dx}\operatorname{cosec}^{-1}u = -\frac{1}{u\sqrt{u^2-1}}\frac{du}{dx}; \quad -\pi < \operatorname{cosec}^{-1}u \leqq -\frac{\pi}{2},\ 0 < \operatorname{cosec}^{-1}u \leqq \frac{\pi}{2}$$

$$\frac{d}{dx}\sec^{-1}u = \frac{1}{u\sqrt{u^2-1}}\frac{du}{dx}; \quad -\pi \leqq \sec^{-1}u < -\frac{\pi}{2},\ 0 \leqq \sec^{-1}u < \frac{\pi}{2}$$

$$\frac{d}{dx}\cot^{-1}u = \frac{-1}{1+u^2}\frac{du}{dx}$$

$$\frac{d}{dx}vers^{-1}u = \frac{1}{\sqrt{2u-u^2}}\frac{du}{dx} \quad 0 \leqq vers^{-1}u \leqq \pi$$

$$\frac{d}{dx}\sinh u = \cosh u\frac{du}{dx}$$

$$\frac{d}{dx}\cosh u = \sinh u\frac{du}{dx}$$

$$\frac{d}{dx}\tanh u = \operatorname{sech}^2 u\frac{du}{dx}$$

$$\frac{d}{dx}\operatorname{cosech} u = -\operatorname{cosech} u\coth u$$

$$\frac{d}{dx}\operatorname{sech} u = -\operatorname{sech} u\tanh u\frac{du}{dx}$$

$$\frac{d}{dx}\coth u = -\operatorname{cosech}^2 u\frac{du}{dx}$$

B.3.2   Expansion in Series

Exponential and Logarithmic Series

$$e^x = 1 + \frac{x}{1!} + \frac{x^2}{2!} + \frac{x^3}{3!} + \frac{x^4}{4!} + \dots \quad [-\infty < x < +\infty]$$

$$a^x = e^{mx} = 1 + \frac{m}{1!}x + \frac{m^2}{2!}x^2 + \frac{m^3}{3!}x^3 + \dots \quad [a > 0, -\infty < x < +\infty]$$

where $\quad m = \ln a = (2.3026 \times \log_{10} a)$

$$\ln(1+x) = x - \frac{x^2}{2} + \frac{x^3}{3} - \frac{x^4}{4} + \frac{x^5}{5} + \dots \quad [-1 < x < +1]$$

$$\ln(1-x) = -x - \frac{x^2}{2} - \frac{x^3}{3} - \frac{x^4}{4} - \frac{x^5}{5} - \dots \quad [-1 < x < +1]$$

$$\ln\left(\frac{1+x}{1-x}\right) = 2\left(x + \frac{x^3}{3} + \frac{x^5}{5} + \frac{x^7}{7} + \dots\right) \quad [-1 < x < +1]$$

$$\ln\left(\frac{x+1}{x-1}\right) = 2\left(\frac{1}{x} + \frac{1}{3x^3} + \frac{1}{5x^5} + \frac{1}{7x^7} + \dots\right) \quad [x < -1 \text{ or } +1 < x]$$

$$(1+x)^n = 1 + nx + \frac{n(n-1)}{2!}x^2 + \frac{n(n-1)(n-2)}{3!}x^3 + \dots\infty, \ |x| < 1$$

$$(1+x)^{-n} = 1 - nx + \frac{n(n+1)}{2!}x^2 - \frac{n(n+1)(n+3)}{3!}x^3 + \dots\infty, \ |x| < 1$$

$$(1-x)^{-n} = 1 \times nx + \frac{n(n+1)}{2!}x^2 + \frac{n(n+1)(n+3)}{3!}x^3 + \dots\infty, \ |x| < 1$$

$$(1+x)^{-1} = 1 - x + x^2 - x^3 + \dots, \ |x| < 1$$

$$(1+x)^{-2} = 1 - 2x + 3x^2 - 4x^3 + \dots, \ |x| < 1$$

$$(1-x)^{-1} = 1 + x + x^2 + x^3 + \dots, \ |x| < 1$$

$$(1-x)^{-2} = 1 + 2x + 3x^3 + 4x^2 + \dots, \ |x| < 1$$

$$(1-x)^{-1/2} = 1 + \frac{1}{2}x + \frac{1.3}{2.4}x^2 + \frac{1.3.5}{2.4.6}x^3 + \dots, \ |x| < 1$$

$$\frac{1}{1^2} + \frac{1}{2^2} + \frac{1}{3^2} + \dots = \frac{\pi^2}{6}$$

$$\frac{1}{1^2} + \frac{1}{3^2} + \frac{1}{5^2} + \dots = \frac{\pi^2}{8}$$

$$\frac{1}{1^4} + \frac{1}{2^4} + \frac{1}{3^4} + \dots = \frac{\pi^4}{90}$$

$$\frac{1}{1^4} + \frac{1}{3^4} + \frac{1}{5^4} + \ldots = \frac{\pi^4}{96}$$

$$\ln x = 2\left[\frac{x-1}{x+1} + \frac{1}{3}\left(\frac{x-1}{x+1}\right)^3 + \frac{1}{5}\left(\frac{x-1}{x+1}\right)^5 + \ldots\right] \quad [0 < x < \infty]$$

$$\ln(a+x) = \ln a + 2\left[\frac{x}{2a+x} + \frac{1}{3}\left(\frac{x}{2a+x}\right)^3 + \frac{1}{5}\left(\frac{x}{2a+x}\right)^5 + \ldots\right]$$

$$[0 < a < +\infty, -a < x < +\infty]$$

Series for the Trigonometric Functions

$$\sin x = x - \frac{x^3}{3!} + \frac{x^5}{5!} - \frac{x^7}{7!} + \ldots \quad [-\infty < x < +\infty]$$

$$\cos x = 1 - \frac{x^2}{2!} + \frac{x^4}{4!} - \frac{x^6}{6!} + \frac{x^8}{8!} - \ldots \quad [-\infty < x < +\infty]$$

$$\tan x = x + \frac{x^3}{3} + \frac{2x^3}{15} + \frac{17x^7}{315} + \frac{62x^9}{2835} + \ldots \quad [-\pi/2 < x < +\pi/2]$$

$$\cot x = \frac{1}{x} - \frac{x}{3} - \frac{x^3}{45} - \frac{2x^5}{945} - \frac{x^7}{4725} - \ldots \quad [-\pi < x < +\pi]$$

$$\sin^{-1} y = y + \frac{y^3}{6} + \frac{3y^5}{40} + \frac{5y^7}{112} + \ldots \quad [-1 \le y \le +1]$$

$$\tan^{-1} y = y - \frac{y^3}{3} + \frac{y^5}{5} - \frac{y^7}{7} + \ldots \quad [-1 \le y \le +1]$$

$$\cos^{-1} y = \frac{1}{2}\pi - \sin^{-1} y$$

$$\cot^{-1} y = \frac{1}{2}\pi - \tan^{-1} y$$

In these formulas, all angles are expressed in radians.

Series for the Hyperbolic Functions

$$\sinh x = x + \frac{x^3}{3!} + \frac{x^5}{5!} + \frac{x^7}{7!} + \dots \quad [-\infty < x < +\infty]$$

$$\cosh x = 1 + \frac{x^2}{2!} + \frac{x^4}{4!} + \frac{x^6}{6!} + \dots \quad [-\infty < x < +\infty]$$

$$\sinh^{-1} y = y - \frac{y^3}{6} + \frac{3y^5}{40} - \frac{5y^7}{112} + \dots \quad [-1 < y < +1]$$

$$\tanh^{-1} y = y + \frac{y^3}{3} + \frac{y^5}{5} + \frac{y^7}{7} + \dots \quad [-1 < y < +1]$$

General Formulas of Maclaurin and Taylor

If $f(x)$ and all its derivatives are continuous in the neighborhood of the point $x = 0$ (or $x = a$), then, for any value of x in this neighborhood, the function $f(x)$ may be expressed as a power series arranged according to ascending powers of x (or $x - a$), as follows:

$$f(x) = f(0) + \frac{f'(0)}{1!} x + \frac{f''(0)}{2!} x^2 + \frac{f'''(0)}{3!} x^3 + \dots + \frac{f^{(n-1)}(0)}{(n-1)!} x^{n-1} + (P_n)x^n$$

(Maclaurin)

$$f(x) = f(a) + \frac{f'(a)}{1!}(x-a) + \frac{f''(a)}{2!}(x-a)^2 + \frac{f'''(a)}{3!}(x-a)^3 + \dots$$

$$+ \frac{f^{(n-1)}(a)}{(n-1)!}(x-a)^{n-1} + (Q_n)(x-a)^n \quad \text{(Taylor)}$$

Here $(P_n) x^n$, or $(Q_n) (x - a)^n$ is called the remainder term; the values of the coefficients P_n and Q_n may be expressed as follows:

$$P_n = \frac{[f^{(n)}(sx)]}{n!} = \frac{(1-t)^{n-1} f^{(n)}(tx)}{(n-1)!}$$

$$Q_n = \frac{f^{(n)}[a+s(x-a)]}{n!} = \frac{(1-t)^{n-1} f^{(n)}[a+t(x-a)]}{(n-1)!}$$

where s and t are certain unknown numbers between 0 and 1.

B.4   INTEGRAL CALCULUS

B.4.1   List of the Most Common Integrals

$$\int a \, du = a \int du = au + C$$

$$\int (u + v) \, dx = \int u \, dx + \int v \, dx$$

$$\int u \, dv = uv - \int u \, du \quad \text{(integration by parts)}$$

$$\int dy \int f(x,y) \, dx = \int dx \int f(x,y) \, dy$$

$$\int x^n dx = \frac{x^{n-1}}{n+1} + C \quad \text{when } n \neq -1$$

$$\int \frac{dx}{x} = \ln x + C = \ln cx$$

$$\int e^x dx = e^x + C$$

$$\int \sin x \, dx = -\cos x + C$$

$$\int \cos x \, dx = \sin x + C$$

$$\int \frac{dx}{\sin^2 x} = -\cot x + C$$

$$\int \frac{dx}{\cos^2 x} = \tan x + C$$

$$\int \frac{dx}{\sqrt{1 - x^2}} = \sin^{-1} x + C = -\cos^{-1} x + c$$

$$\int \frac{dx}{1 + x^2} = \tan^{-1} x + C = -\cot^{-1} x + c$$

$$\int (a + bx)^n dx = \frac{(a + bx)^{n+1}}{(n+1)b} + C$$

$$\int \frac{dx}{a+bx} = \frac{1}{b}\ln(a+bx)+C = \frac{1}{b}\ln c(a+bx)$$

$$\int \frac{ax}{(a+bx)^2} = \frac{1}{b(a+bx)}+C$$

$$\int \frac{dx}{1-x^2} = \frac{1}{2}\ln\frac{1+x}{1-x}+C = \tanh^{-1}x+C \text{ , when } x < 1.$$

$$\int \frac{dx}{x^2-1} = \frac{1}{2}\ln\frac{x-1}{x+1}+C = -\coth^{-1}x+C \text{ , when } x < 1.$$

$$\int \frac{dx}{a+2bx+cx^2} = \frac{1}{\sqrt{ac-b^2}}\tan^{-1}\frac{b+cx}{\sqrt{ac-b^2}}+C, \quad [ac-b^2 > 0]$$

$$= \frac{1}{2\sqrt{b^2-ac}}\ln\frac{\sqrt{b^2-ac}-b-cx}{\sqrt{b^2-ac}+b+cx}+C$$

$$= -\frac{1}{\sqrt{b^2-ac}}\tanh^{-1}\frac{b+cx}{\sqrt{b^2-ac}}+C, \quad [b^2-ac > 0]$$

$$\int \sqrt{a+bx}\,dx = \frac{2}{3b}\left(\sqrt{a+bx}\right)^3+C$$

$$\int \frac{dx}{\sqrt{a+bx}} = \frac{2}{b}\sqrt{a+bx}+C$$

$$\int \frac{(m+nx)\,dx}{\sqrt{a+bx}} = \frac{2}{3b^2}(3mb-2an+nbx)\sqrt{a+bx}+C$$

$$\int \sqrt{a+2bx+cx^2}\,dx = \frac{b+cx}{2c}\sqrt{a+2bx+cx^2}$$

$$\qquad + \frac{ac-b^2}{2c}\int\frac{dx}{\sqrt{a+2bx+cx^2}}+C$$

$$\int a^x dx = \frac{a^x}{\ln a}+C$$

$$\int x^n e^{ax}\,dx = \frac{x^n e^{ax}}{a}\left[1-\frac{n}{ax}+\frac{n(n-1)}{a^2x^2}-\cdots\pm\frac{n!}{a^n x^n}\right]+C$$

$$\int \ln x\,dx = x\ln x - x + C$$

$$\int \frac{\ln x}{x^2} dx = -\frac{\ln x}{x} - \frac{1}{x} + C$$

$$\int \frac{(\ln x)^n}{x} dx = \frac{1}{n+1} (\ln x)^{n+1} + C$$

$$\int \sin^2 x \, dx = -\frac{1}{4} \sin 2x + \frac{1}{2} x + C = -\frac{1}{2} \sin x \cos x + \frac{1}{2} x + C$$

$$\int \cos^2 x \, dx = \frac{1}{4} \sin 2x + \frac{1}{2} x + C = \frac{1}{2} \sin x \cos x + \frac{1}{2} x + C$$

$$\int \sin mx \, dx = -\frac{\cos mx}{m} + C$$

$$\int \cos mx \, dx = \frac{\sin mx}{m} + C$$

$$\int \sin mx \cos nx \, dx = \frac{\cos(m+n)x}{2(m+n)} - \frac{\cos(m-n)x}{2(m-n)} + C$$

$$\int \sin mx \sin nx \, dx = \frac{\sin(m-n)x}{2(m-n)} - \frac{\sin(m+n)x}{2(m+n)} + C$$

$$\int \cos mx \cos nx \, dx = \frac{\sin(m-n)x}{2(m-n)} + \frac{\sin(m+n)x}{2(m+n)} + C$$

$$\int \tan x \, dx = -\ln \cos x + C$$

$$\int \cot x \, dx = \ln \sin x + C$$

$$\int \frac{dx}{\sin x} = \ln \tan \frac{x}{2} + C$$

$$\int \frac{dx}{\cos x} = \ln \tan \left(\frac{\pi}{4} + \frac{\pi}{2} \right) + C$$

$$\int \frac{dx}{1 + \cos x} = \tan \frac{x}{2} + C$$

$$\int \frac{dx}{1 - \cos x} = -\cot \frac{x}{2} + C$$

$$\int \sin x \cos x \, dx = \frac{1}{2}\sin^2 x + C$$

$$\int \frac{dx}{\sin x \cos x} = \ln \tan x + C$$

$$\int \frac{\cos x \, dx}{a + b \cos x} = \frac{x}{b} - \frac{a}{b}\int \frac{dx}{a + b \cos x} + C$$

$$\int \frac{\sin x \, dx}{a + b \cos x} = -\frac{1}{b}\ln(a + b \cos x) + C$$

$$\int \frac{A + B\cos x + C\sin x}{a + b\cos x + c\sin x}\, dx = A\int \frac{dy}{a + p\cos y} + (B\cos u + C\sin u)\int \frac{\cos y \, dy}{a + p\cos y}$$

$$-(B\sin u - C\cos u)\int \frac{\sin y \, dy}{a + p\cos y}, \text{ where } b - p\cos u, \, c = p\sin u \text{ and } x - u = y.$$

$$\int e^{ax}\sin bx \, dx = \frac{a\sin bx - b\cos bx}{a^2 + b^2}e^{ax} + C$$

$$\int e^{ax}\cos bx \, dx = \frac{a\cos bx + b\sin bx}{a^2 + b^2}e^{ax} + C$$

$$\int \sinh x \, dx = \cosh x + C$$

$$\int \tanh x \, dx = \ln \cosh x + C$$

$$\int \cosh x \, dx = \sinh x + C$$

$$\int \coth x \, dx = \ln \sinh x + C$$

C

CRAMER'S RULE

Cramer's Rule: Cramer's rule can be used to solve a system of simultaneous linear algebraic equations. Consider a general system of n linear equations in n unknowns:

$$a_{11}x_1 + a_{12}x_2 + \ldots + a_{1n}x_n = b_1$$

$$a_{21}x_1 + a_{22}x_2 + \ldots + a_{2n}x_n = b_2$$

$$\vdots \qquad \vdots \qquad \vdots$$

$$a_{n1}x_1 + a_{n2}x_2 + \ldots + a_{nn}x_n = b_n \qquad \text{(C.1)}$$

and define the determinant of such a system. We begin by defining the determinant of a 3×3 system:

$$C = \begin{vmatrix} a_{11} & a_{12} & a_{13} \\ a_{21} & a_{22} & a_{23} \\ a_{31} & a_{32} & a_{33} \end{vmatrix} = a_{11}\begin{vmatrix} a_{22} & a_{23} \\ a_{32} & a_{33} \end{vmatrix} - a_{12}\begin{vmatrix} a_{21} & a_{23} \\ a_{31} & a_{33} \end{vmatrix} + a_{13}\begin{vmatrix} a_{21} & a_{22} \\ a_{31} & a_{32} \end{vmatrix} \qquad \text{(C.2)}$$

The general definition of the determinant of the $n \times n$ system of Eq. (C.1) is simply an extension of the procedure (C.2).

$$C = \begin{vmatrix} a_{11} & a_{12} & \cdots & a_{1n} \\ a_{21} & a_{22} & \cdots & a_{2n} \\ \vdots & \vdots & \cdots & \vdots \\ a_{n1} & a_{n2} & \cdots & a_{nn} \end{vmatrix} = a_{11}A_{11} - a_{12}A_{12} + \ldots + (-1)^{n+1}a_{1n}A_{1n} \qquad \text{(C.3)}$$

where A_{1j} is the $(n-1) \times (n-1)$ determinant obtained by crossing out the first row and jth column of the original $n \times n$ determinant. Hence an $n \times 1$ determinant can be obtained by calculating $n\,(n-1)\,(n-1)$ determinants.

We shall conclude this appendix by introducing a method for obtaining solutions of the system (C.1). We define the determinants

$$C_1 = \begin{vmatrix} b_1 & a_{12} & \cdots & a_{1n} \\ b_2 & a_{22} & \cdots & a_{2n} \\ \vdots & \vdots & \cdots & \vdots \\ b_n & a_{n2} & \cdots & a_{nn} \end{vmatrix},$$

$$C_2 = \begin{vmatrix} a_{11} & b_1 & a_{13} & \cdots & a_{1n} \\ a_{21} & b_2 & a_{23} & \cdots & a_{2n} \\ \vdots & \vdots & \vdots & \cdots & \vdots \\ a_{n1} & b_n & a_{n3} & \cdots & a_{nn} \end{vmatrix}, \cdots,$$

$$C_k = \begin{vmatrix} a_{11} & a_{12} & \cdots & a_{1,k-1} & b_1 & a_{1,k+1} & \cdots & a_{1n} \\ a_{21} & a_{22} & \cdots & a_{2,k-1} & b_2 & a_{2,k+1} & \cdots & a_{2n} \\ \vdots & \vdots & \cdots & \vdots & \vdots & \vdots & \cdots & \vdots \\ a_{n1} & a_{n2} & \cdots & a_{n,k-1} & b_k & a_{n,k+1} & \cdots & a_{nn} \end{vmatrix}, \cdots,$$

$$C_n = \begin{vmatrix} a_{11} & a_{12} & \cdots & a_{1,n-1} & b_1 \\ a_{21} & a_{22} & \cdots & a_{2,n-1} & b_2 \\ \vdots & \vdots & \cdots & \vdots & \vdots \\ a_{n1} & a_{n2} & \cdots & a_{n,n-1} & b_n \end{vmatrix}, \tag{C.4}$$

obtained by replacing the kth column of C by the column

$$\begin{bmatrix} b_1 \\ b_2 \\ \vdots \\ b_n \end{bmatrix}$$

Then we have the following theorem, known as *Cramer's rule*.

Cramer's Rule: Let C and C_k, k = 1, 2, ..., n, be given as in (C.3). If $C \neq 0$, then the unique solution to the system (C.1) is given by the values

$$x_1 = \frac{C_1}{C}, \, x_2 = \frac{C_2}{C}, \, ..., \, x_n = \frac{C_n}{C} \tag{C.5}$$

EXAMPLE C.1

Obtain the solution of the following simultaneous linear equations by Cramer's rule.

(a) $\begin{bmatrix} 1 & 3 \\ 4 & -1 \end{bmatrix} \begin{bmatrix} x_1 \\ x_2 \end{bmatrix} = \begin{bmatrix} 5 \\ 12 \end{bmatrix}$

(b) $\begin{bmatrix} 1 & -3 & 2 \\ 3 & 4 & 1 \\ -4 & 2 & -9 \end{bmatrix} \begin{bmatrix} x_1 \\ x_2 \\ x_3 \end{bmatrix} = \begin{bmatrix} 8 \\ 5 \\ 2 \end{bmatrix}$

Solution:

(a) $x_i = \dfrac{|C_i|}{|C|}$

$$x_1 = \frac{\begin{vmatrix} 5 & 3 \\ 12 & -1 \end{vmatrix}}{\begin{vmatrix} 1 & 3 \\ 4 & -1 \end{vmatrix}} = \frac{-5-36}{-13} = 3.15$$

$$x_2 = \frac{\begin{vmatrix} 1 & 5 \\ 4 & 12 \end{vmatrix}}{\begin{vmatrix} 1 & 3 \\ 4 & -1 \end{vmatrix}} = \frac{12-20}{-13} = 0.62$$

b) $x_i = \dfrac{|C_i|}{|C|}$

$$x_1 = \frac{\begin{vmatrix} 8 & -3 & 2 \\ 5 & 4 & 1 \\ 2 & 2 & -9 \end{vmatrix}}{\begin{vmatrix} 1 & -3 & 2 \\ 3 & 4 & 1 \\ -4 & 2 & -9 \end{vmatrix}} = \frac{-441}{-63} = 7$$

$$x_2 = \frac{\begin{vmatrix} 1 & 8 & 2 \\ 3 & 5 & 1 \\ -4 & 2 & -9 \end{vmatrix}}{\begin{vmatrix} 1 & -3 & 2 \\ 3 & 4 & 1 \\ -4 & 2 & -9 \end{vmatrix}} = \frac{189}{-63} = -3$$

$$x_3 = \frac{\begin{vmatrix} 1 & -3 & 8 \\ 3 & 4 & 5 \\ -4 & 2 & 2 \end{vmatrix}}{\begin{vmatrix} 1 & -3 & 2 \\ 3 & 4 & 1 \\ -4 & 2 & -9 \end{vmatrix}} = \frac{252}{-63} = -4$$

EXAMPLE C.2

Consider the system

$$2x_1 + 4x_2 - x_3 = -5$$

$$-4x_1 + 3x_2 + 5x_3 = 14$$

$$6x_1 + 3x_2 - 2x_3 = 5$$

We have

$$C = \begin{vmatrix} 2 & 4 & -1 \\ -4 & 3 & 5 \\ 6 & -3 & -2 \end{vmatrix} = 112 \qquad C_1 = \begin{vmatrix} -5 & 4 & -1 \\ 14 & 3 & 5 \\ 5 & -3 & -2 \end{vmatrix} = 224$$

$$C_2 = \begin{vmatrix} 2 & -5 & -1 \\ -4 & 14 & 5 \\ 6 & 5 & -2 \end{vmatrix} = -112 \qquad C_3 = \begin{vmatrix} 2 & 4 & -5 \\ -4 & 3 & 14 \\ 6 & -3 & 5 \end{vmatrix} = 560$$

Therefore $\quad x_1 = \dfrac{C_1}{C} = 2 \qquad x_2 = \dfrac{C_2}{C} = -1 \qquad x_3 = \dfrac{C_3}{C} = 5$

EXERCISES

Solve the following system of linear equations using Cramer's rule:

C1. (a) $x + 2y + z = 0$

$3x + y - z = 0$

$x - y + 4z = 3$

(b) $2x + y - 3z = 11$

$4x - 2y + 3z = 8$

$-2x + 2y - z = -6$

(c) $x + 3y - z = -10$

$-x + 4y + 2z = -4$

$2x - 2y + 5z = 35$

(d) $y - 3z = -5$

$2x + 3y - z = 7$

$4x + 5y - 2z = 10$

(e) $\begin{bmatrix} 1 & 1 & 1 & -1 \\ 1 & -1 & -1 & 2 \\ 4 & 4 & 1 & 1 \\ 2 & 1 & 2 & -2 \end{bmatrix} \begin{Bmatrix} x_1 \\ x_2 \\ x_3 \\ x_4 \end{Bmatrix} = \begin{Bmatrix} 2 \\ 0 \\ 11 \\ 2 \end{Bmatrix}$

(f) $\begin{bmatrix} 10 & 7 & 8 & 7 \\ 7 & 5 & 6 & 5 \\ 8 & 6 & 10 & 9 \\ 7 & 5 & 9 & 10 \end{bmatrix} \begin{Bmatrix} x_1 \\ x_2 \\ x_3 \\ x_4 \end{Bmatrix} = \begin{Bmatrix} 32 \\ 23 \\ 33 \\ 31 \end{Bmatrix}$

ANSWERS TO SELECTED EXERCISES

CHAPTER 1: NUMERICAL COMPUTATIONS

1.1 **(a)** 1.0215   **(b)** 1.405

1.3 $f(1.2) = 1.44$

1.5 50824.6135

1.7 $(75)_{10} = (1001\ 1011)_2$

1.9 $(1235)_{10}$

1.11 $E_{rxy} = E_{rx} + E_{ry}$

1.13 **(a)** Absolute error, $\xi_a = 0.666\ldots \times 10^{-6}$; relative error, $\xi_r = 1 \times 10^{-6}$

    (b) Absolute error, $\xi_a = 0.33 \times 10^{-7}$; relative error, $\xi_r = 5 \times 10^{-7}$

1.15 **(a)** Absolute error, $E_A = 0.004$, relative error $E_R = 1.1312 \times 10^{-3}$

    (b) Absolute error, $E_A = 0.006$, relative error $E_R = 1.6968 \times 10^{-3}$

1.17 Absolute error = 0.002857; Relative error = 0.0009, Percentage relative error = 9%

1.19 Percentage relative error = 9.02%; Approximate estimate of error = 33.3% With 6 terms included approximate error falls below $E_s = 0.05\%$

1.21 **(a)** $f(x) = x - \dfrac{x^3}{3!} + \dfrac{x^5}{5!} - \dfrac{x^7}{7!} + \ldots$

    (b) Relative error, $r_6 = 0.005129\%$

    (c) Upper bound on the truncation error = 0.02305%

1.23 Upper bound on the error $|R_3(x)| \leq 0.0625$ for all $x \in [0,1]$

1.25

Order, n	$F^{(n)}(x)$	$F(\pi/3)$	ε_t
0	$\cos x$	0.707106781	−41.4
1	$-\sin x$	0.521986659	−4.4
2	$-\cos x$	0.497754491	0.449
3	$\sin x$	0.499869147	2.62×10^{-2}
4	$\cos x$	0.500007551	-1.51×10^{-3}
5	$-\sin x$	0.500000304	-6.08×10^{-5}
6	$-\cos x$	0.499999988	2.4×10^{-6}

1.27 **(a)** Condition number = −1

(b) Condition number = −0.994975

(c) Condition number = −0.00314161

1.29 **(a)** to **(d)** will not converge quadratically

CHAPTER 2: LINEAR SYSTEM OF EQUATIONS

2.1 **(a)** $A^{-1} = \begin{bmatrix} -0.7 & 0.2 & 0.3 \\ -1.3 & -0.2 & 0.7 \\ 0.8 & 0.2 & -0.2 \end{bmatrix}$

(b) $A^{-1} = \begin{bmatrix} 0.1765 & 0.3529 & -0.2353 \\ 0.4118 & -0.1765 & 0.1176 \\ 0.0588 & 0.1176 & -0.4118 \end{bmatrix}$

(c) $A^{-1} = \begin{bmatrix} 0.0136 & 0.0496 & 0.0583 \\ 0.1886 & -0.2233 & -0.0124 \\ 0.0298 & 0.0174 & -0.0546 \end{bmatrix}$

(d) $A^{-1} = \begin{bmatrix} 0 & 0 & 0.1429 \\ -2 & 1 & -0.2857 \\ 1.6667 & -0.6667 & 0.1429 \end{bmatrix}$

(e) $A^{-1} = \begin{bmatrix} 2 & -1 & 0 \\ -0.3333 & 0.6667 & -0.3333 \\ -0.6667 & 0.3333 & 0.3333 \end{bmatrix}$

(f) $A^{-1} = \dfrac{1}{9} \begin{bmatrix} 0 & 3 & 3 \\ 3 & 2 & -7 \\ 3 & -1 & -1 \end{bmatrix}$

2.3 **(a)** $x = 3, y = -1, z = -2$

(b) $x = 1, y = 2, z = 3$

(c) $x = 1, y = 1, z = 1$

(d) $x_1 = 2.7869, x_2 = 4.4918, x_3 = 2.1311, x_4 = -2.5410$

(e) $x_1 = 25.3587, x_2 = -19.6143, x_3 = -28.9058, x_4 = -7.8027$

(f) $x_1 = -1, x_2 = 1, x_3 = 2, x_4 = 1$

2.5 **(a)** $x = 2, y = 1, z = 3$

(b) $x = 3, y = 2, z = 2$

(c) $x = 1, y = 0, z = 3$

(d) $x = 35/18, y = 29/18, z = 5/18$

(e) $x = 1, y = 1, z = 2$

(f) $x_1 = 2.7778, x_2 = 4.2222, x_3 = -0.5556, x_4 = 6.8889$

2.7 **(a)** $x_1 = 1, x_2 = 2, x_3 = 3$

(b) $x_1 = 1, x_2 = 1, x_3 = 1, x_4 = 2$

(c) $x_1 = 1, x_2 = 1, x_3 = 1, x_4 = 2$

(d) $x_1 = 1, x_2 = 1, x_3 = 1, x_4 = 2$

(e) $x_1 = 1, x_2 = -1, x_3 = 1, x_4 = -1$

(f) $x_1 = -1, x_2 = 1, x_3 = 1, x_4 = -1$

2.9 **(a)** $x = 4, y = -1, z = 3$

(b) $x = 1, y = 2, z = 3$

(c) $x = 5, y = 4, z = 1$

(d) $x_1 = 0.999, x_2 = 1.9999, x_3 = 2.9999, x_4 = -0.0001$

(e) $x_1 = 1.155, x_2 = -0.311, x_3 = 0.088, x_4 = -0.044$

(f) $x_1 = 1, x_2 = 0, x_3 = -1, x_4 = 4$

CHAPTER 3: SOLUTION OF ALGEBRAIC AND TRANSDENTAL EQUATIONS

3.1 4.4932

3.3 1.51092

3.5 2.7119

3.7 0.3807

3.9 7.7253

3.11 0.73909

3.13 2.7392

3.15 2.798

3.17 0.5177

3.19 1.488

3.21 0.7346

3.23 2.1163

3.25 With $x_0 = 0$, the root is 0.96434

3.27 4.4934

3.29 0.56714

3.31 0.6071

3.33 3.7892

3.35 1.172

3.37 −0.70346

3.39 2.0

3.41 3.051

3.43 −1.3247

3.45 4.217163

3.47 4.730041

3.49 1.13472

3.51 7.06858

3.53 2.85234

3.55 2.95

3.57 2.1080

3.59 0.68

3.61 0.7710

CHAPTER 4: NUMERICAL DIFFERENTIATION

4.1 **(a)** $y'(1.0) = 3.0$; $y''(1.0) = 6.0$

(b) $y'(1.0) = 1.0$; $y''(1.0) = 4$

(c) $y'(1.0) = 5.0004$; $y''(1.0) = 16$

(d) $y'(1.0) = -8.2002$; $y''(1.0) = -23.6$

(e) $y'(1.0) = -7.6009$; $y''(1.0) = -22.4$

(f) $y'(1.0) = 1.6002$; $y''(1.0) = 5.2$

(g) $y'(1.0) = 7.7021$; $y''(1.0) = 19.4040$

4.3 **(a)** $y'(6.0) = -1.68$;
$y''(6.0) = -0.3520$

(b) $y'(6.0) = -2.0537$;
$y''(6.0) = -0.7126$

4.5 $y'(4) = 12.75; y''(4) = 9.75$

4.7 **(a)** $y'(3) = 1.7263$;
$y''(3) = 4.6309$

(b) $y'(3) = 0.0958$;
$y''(3) = 0.0367$

4.9 **(a)** $y'(1.2) = 0.7917$;
$y''(1.2) = 6.9667$

(b) $y'(1.2) = 0.1193$;
$y''(1.2) = -0.7767$

(c) $y'(1.2) = 2.8964$;
$y''(1.2) = -18.3050$

4.11 $f'(3.4) = 0.12864$;
$f'(5.2) = -0.00836$;
$f'(5.6) = 3.930432$

4.13 $f'(2) = 14; f'(2.5) = 20.75$

4.15 $y'(2) = 2.4667$;
$y''(2) = 6.9667$

4.17 $y'(1.2) = 0.1193$;
$y''(1.2) = -0.7767$

CHAPTER 5: FINITE DIFFERENCES AND INTERPOLATION

5.3 **(a)**

x	y	Δy	Δ²y	Δ³y
45	20			
		40		
55	60		20	
		60		-20
65	120		0	
		60		
75	180			

(b)

x	y	Δy	Δ²y	Δ³y	Δ⁴y
40	204				
		20			
50	224		2		
		22		0	
60	246		2		0
		24		0	
70	270		2		0
		26		0	
80	296		2		
		28			
90	324				

5.5 -49

5.7 **(a)** 35 **(b)** 77 **(c)** 6

5.9 **(a)** $f(x) = 3x^{(3)} + 10x^{(2)}$
$+ 5x^{(1)} + 1$

(b) $f(x) = x^{(4)} + x^{(3)} - 8x^{(2)}$
$- x^{(1)} + 4$

5.11 **(a)** 11

(b) 37

(c) -33

(d) -25

(e) 38

5.13 112.32097

5.15 0.66913

5.17 48

5.19 $y = f(x) = x^2 + 3x + 1$

5.21 8.875

5.23 $f(0.12) = 0.1197$;
$f(0.26) = 0.2571$

5.25 0.0000392

5.27 $y = x^3 - 8x^2 - 20$

5.29 $y = x^3 - x^2 + 3x + 2$

5.31 200

5.33 196

5.35 3.54672

5.37 2.5

5.39 38

5.41 32495.1328

5.43 36.05469

5.45 24.46875

5.47 36.625

5.49 478.375

5.51 $y_{1.6} = 0.36663$

5.53 1.11864

5.55 31846.8528

5.57 1.54397

5.59 46614.25781

5.61 –5.83

5.63 1.6778

5.65 5.528961

5.67 $f(1.5) = 0.7679$

5.69 $f_{2,3}(0) = 0.7212$

5.71 $f''(3.4) = -0.02856$

$f'(5.2) = -0.0927$

$f(5.6) = 3.9065$

5.73 $y(2.6) = 0.18714$

CHAPTER 6: CURVE FITTING, REGRESSION AND CORRELATION

6.1 $a = -22.0007, b = 3.7484$

6.3 $a = 0.4449, b = 0.2747$

6.5 $a = 22.917, b = 3.3883$

6.7 $a = 80.7777, b = 1.138$

6.9 $a = 1.1574, b = 0.2566$

6.11 (a) $a = -22.0007$ and
$b = 3.7484$

(b) $s_e = 30.8722$

(c) SSE = 7624.7219

(d) SST = 23973.6

(e) SSR = 16348.8781

(f) $r^2 = 0.682$

(g) $r = 0.8258$

(h) when $x = 42$, $y = 135.4307$

6.13 (a) $a = 0.4449$ and $b = 0.2747$

(b) $s_e = 1.0550$

(c) SSE = 5.5648

(d) SST = 66

(e) SSR = 60.4352

(f) $r^2 = 0.9157$

(g) $r = 0.9569$

(h) when $x = 42$, $y = 11.9825$

6.15 (a) $a = 22.917$ and $b = 3.3883$

(b) $s_e = 5.6669$

(c) SSE = 256.9085

(d) SST = 4573.6

(e) SSR = 4316.6915

(f) $r^2 = 0.9438$

(g) $r = 0.9715$

(h) when $x = 15$, $y = 73.7415$

6.17 (a) $a = 80.7777$ and $b = 1.138$

(b) $s_e = 7.0176$

(c) SSE = 492.4669

(d) SST = 2500.6667

(e) SSR = 2008.1998

(f) $r^2 = 0.8031$

(g) $r = 0.8961$

(h) when $x = 45$, $y = 131.988$

6.19 (a) $a = 1.1574$ and $b = 0.2556$

(b) $s_e = 1.0438$

(c) SSE = 5.4473

(d) SST = 58.2143

(e) SSR = 52.767

(f) $r^2 = 0.9064$

(g) $r = 0.9521$

(h) when $x = 41$, $y = 11.6778$

6.21 $b = 0.8671$ and $m = 1.3471$

6.23 $b = 0.3990$ and $m = 0.9057$

6.25 $b = 18.9926$ and $m = 0.4983$

6.27 $a = -291.9285, b = 417.3647$ and $c = -146.3105; S_t = 8.2232;$ $S_r = 0.0406; r = 0.9975; S_{y/x} = 0.1163$

6.29 $a = 2.5095, b = 1.2, c = 0.7333, S_{y/x} = 0.2024, S_t = 332.0571,$ $S_r = 0.1638, r^2 = 0.9995, r = 0.9998$

6.31 $a = 14.4217, b = 8.9904, c = -5.6087, S_t = 1051.3356, S_r = 4.3466,$

$S_{y/x} = 0.7244, r^2 = 0.9959, r = 0.9979$

6.33 $a = 2.2561, b = 3.8171, c = 5.3333, S_t = 577.84, S_r = 4.2007,$

$S_{y/x} = 0.6001, r^2 = 0.9927, r = 0.9964$

6.35 $a = 29.4487, b = -5.7449, c = 19.8449, S_t = 3463.61, S_r = 2.0574,$

$S_{y/x} = 0.2939, r^2 = 0.9994, r = 0.9997$

6.37 $a = 2.7077$ and $b = 0.0369$

6.39 $a = 63.5770$ and $b = 0.3548$

6.41 $a = 4.0672$ and $b = 0.7823$

6.43 $a = 13.2275$ and $b = 37.5036$

6.45 $a = -0.0359$ and $b = 7.6495; r = 0.9998$

6.47 $a = 3.3385; b = 2.9169;$ Sum of squares of errors $= 7.8421$

6.49 $a = 1.2045; b = 0.2599;$ Sum of squares of errors $= 1.6846$

6.51 $a = 21.5949; b = 3.5681;$ Sum of squares of errors $= 271.3282$

6.53 $a = 5.2116; b = 2.8682;$ Sum of squares of errors $= 5.6054$

6.55 $a = -1.2576; b = 0.3166;$ Sum of squares of errors $= 1.1108$

6.57 $a = 18.6429; b = 3.4286;$ Sum of squares of errors $= 14.1888$

CHAPTER 7: NUMERICAL INTEGRATION

7.1 $I = 0.9023$

7.3 (a) $I = 1.8591,$
rel error $= 0.0819$

(b) $I = 1.7539,$
rel error $= 0.0207$

(c) $I = 1.7272,$
rel error $= 0.0052$

(d) $I = 1.7205,$
rel error $= 0.0052$

7.5 (a) $I = 2.5268,$
rel error $= 0.2$

(b) $I = 2.1122,$
rel error $= 0.0086$

7.7 $I = -1.81285$

7.9 $I = 0.9045$

7.11 $I = 1.718$

7.13 $I = 1.187$

7.15 $I = -2.56622$

7.17 $I = 1.3571$

7.19 $I = 0.7853959$

7.21 $I = 3.1832$

7.23 $I = 0.091111$

7.25 $I = 1.30859$

7.27 $I = 0.78535$

7.29 $I = 1.827858$

7.31 $I = 0.52359895$

7.33 1.8278

7.35 0.264241

7.37 18.6667

7.39 (a) 0.16061

    (b) 0.19226

    (c) 0.64269

    (d) 2.58797

7.41 $R_{3,3} = 11.5246$

CHAPTER 8: NUMERICAL SOLUTION OF ORDINARY DIFFERENTIAL EQUATIONS

8.1 $y(0.1) = 1.1053$

8.3 $y(0.2) = 0.83746$

8.5 $y(0.1) = 0.9828$

8.7 1.1053

8.9 $y(0.2) = 0.80227$

8.11 $y(0.1) = 1.1057, y(0.3) = 1.3694$

8.13 $y(1) = 3.18$

8.15 $y(1.25) = 0.34565$ with $h = 0.5$, $y(1.175) = 0.1897$ with $h = 0.025$

8.17 $y(0.5) = 0.2836165, y(1) = 3.2190993$

8.19 $y(0.3) = 1.2432$

8.21 $y(0.1) = 1.1055$

8.23 $y(0.075) = 1.076$

8.25 $y(0.1) = 2.2050, y(0.2) = 2.421, y(0.3) = 2.6492, y(0.4) = 2.8909$

8.27 $[t, y] = [0.1\ 1.00025, 0.2\ 1.00243, 0.3\ 1.00825, 0.4\ 1.01926, 0.5\ 1.03688]$

8.29 $y(0.1) = 2.2050$

8.31 $y(0.8) = 0.84899$

8.33 $y(0.2) = 1.196, y(0.4) = 1.3752$

8.35 $y(0.5) = 3.21875$

8.37 $y(0.2) = 1.00265$

8.39 $y^p(0.8) = 1.02337, y^C(0.8) = 1.0296$

8.41 $y(1.4) = 2.575$

8.43 $y(0.8) = 2.4366$

8.45 $y(0.4) = 1.8392$

8.47 $y(0.8) = 1.0294, y(1.0) = 1.5549$

APPENDIX A

(a) $\dfrac{(-1/2)}{s-2} + \dfrac{5/3}{(s-1)} + \dfrac{(-1/6)}{s-4}$

(b) $\dfrac{(3/8)}{s} + \dfrac{(1/2)}{s^2} + \dfrac{(1/2)}{s^3} + \dfrac{(-3/8)}{s-2} + \dfrac{(5/4)}{(s-2)^2}$

(c) $\dfrac{(1/8)}{s} + \dfrac{(3/4)}{s^2} - \dfrac{(1/2)}{s^3} + \dfrac{(-s/8)-3/4}{s^2+4}$

(d) $\dfrac{(-1/16)}{s} + \dfrac{(1/8)}{s^2} + \dfrac{(1/16)}{s+2} - \dfrac{(1/4)}{(s+2)^3}$

(e) $\dfrac{(1/6)}{s} + \dfrac{(1/3)}{s+1} - \dfrac{1}{2}\dfrac{(s+2)}{(s+2)^2+2} - \dfrac{2}{3}\dfrac{1}{(s+2)^2+2}$

(f) $\dfrac{6}{s^3} - \dfrac{6}{s^4} + \dfrac{1}{8} - \dfrac{2}{s+1}$

(g) $\dfrac{2}{(s+1)^3} + \dfrac{1}{s+1}$

(h) $\dfrac{10}{(s+1)^2+2^2} + \dfrac{2(s+1)}{(s+1)^2+2^2}$

(i) $\dfrac{(s+3)}{(s+1)(s+2)}$

APPENDIX C

(a) $x = 0.3334, y = -0.4444, z = 0.5555$

(b) $x = 3, y = -1, z = -2$

(c) $x = 2, y = -3, z = 5$

(d) $x = -1, y = 4, z = 3$

(e) $x_1 = 1, x_2 = 2, x_3 = -1, x_4 = 0$

(f) $x_1 = 1, x_2 = 1, x_3 = 1, x_4 = 1$

INDEX

www.ingramcontent.com/pod-product-compliance
Lightning Source LLC
Chambersburg PA
CBHW080129220326
41598CB00032B/5007